*P. Lynn Kennedy, PhD*
*Won W. Koo, PhD*
*Editors*

# Agricultural Trade Policies in the New Millennium

*Pre-publication*
*REVIEWS,*
*COMMENTARIES,*
*EVALUATIONS . . .*

Food Products Press®
An Imprint of The Haworth Press, Inc.
New York • London • Oxford

# Agricultural Trade Policies
# in the New Millennium

## FOOD PRODUCTS PRESS
Agricultural Commodity Economics,
Distribution, and Marketing

Andrew Desmond O'Rourke, PhD
Senior Editor

New, Recent, and Forthcoming Titles:

*Marketing Livestock and Meat* by William H. Lesser

*Understanding the Japanese Food and Agrimarket: A Multifaceted Opportunity* edited by A. Desmond O'Rourke

*The World Apple Market* by A. Desmond O'Rourke

*Marketing Beef in Japan* by William A. Kerr, Kurt K. Klein, Jill E. Hobbs, and Masaru Kagatsume

*Effects of Grain Marketing Systems on Grain Production: A Comparative Study of China and India* by Zhang-Yue Zhou

*Competition in Agriculture: The United States in the World Market* edited by Dale Colyer, P. Lynn Kennedy, William A. Amponsah, Stanley M. Fletcher, and Curtis M. Jolly

*Agricultural Trade Policies in the New Millennium* edited by P. Lynn Kennedy and Won W. Koo

# Agricultural Trade Policies in the New Millennium

P. Lynn Kennedy, PhD
Won W. Koo, PhD
Editors

Food Products Press®
An Imprint of The Haworth Press, Inc.
New York • London • Oxford

Published by

Food Products Press®, an imprint of The Haworth Press, Inc., 10 Alice Street, Binghamton, NY 13904-1580.

Tables 17.1 and 17.2 from Miranda, J., Torres, R., and Ruiz, M., "The International Use of Antidumping: 1987-1997," *Journal of World Trade 32*(5), 1998, pp. 65-66, reprinted with kind permission of Kluwer Law International.

Cover design by Anastasia Litwak.

**Library of Congress Cataloging-in-Publication Data**

Agricultural trade policies in the new millennium / P. Lynn Kennedy, Won W. Koo, editors.
    p. cm.
   Includes bibliographical references and index.
   ISBN 1-56022-932-2 (alk. paper)—ISBN 1-56022-933-0 (pbk. : alk. paper)
   1. Produce trade—Government policy. 2. Agriculture and state. 3. Uruguay Round (1987-1994). 4. World Trade Organization. I. Kennedy, P. Lynn II. Koo, Won W.

HD9000.6 .A46 2002
382'41—dc21

2001040742

# CONTENTS

**About the Editors**        **xii**

**Contributors**        **xiii**

**Preface**        **xvii**

**Chapter 1. Issues and Obstacles for Agricultural Trade Policy in the New Millennium**     **1**
*P. Lynn Kennedy*
*Won W. Koo*

Issues and Perspectives for Future Trade Agreements    2
Scope of This Volume    5

**PART I: KEY ISSUES INFLUENCING AGRICULTURAL TRADE NEGOTIATIONS**

**Chapter 2. Progress in Agricultural Trade Negotiations**    **9**
*Andrew Schmitz*

Introduction    9
Policy Harmonization and Side Agreements    9
State Trading Enterprises (STEs) and the WTO    13
Keeping the Borders Open    14
The Green Box    15
R&D and Trade    23
Multinationals and Vertical Market Structures    26
Gainers and Losers    27
Future Directions    29
Conclusion    31

**Chapter 3. Issues in the WTO Agricultural Negotiations: An Overview**    **35**
*Tim Josling*

Introduction    35
Building on the Uruguay Round    36
The Agenda for the Agricultural Negotiations    37
Market Access    38

TRQ Expansion 40
Curbing Export Subsidies 41
Domestic Support 45
Health, Safety, and Environmental Issues 47
The Timing of the Negotiations 50

**Chapter 4. Obstacles to Progress in Multilateral
Agricultural Trade Negotiations: The European
Viewpoint** **55**
*David Blandford*

EU Agricultural Policy Prior to the Uruguay Round
Negotiations 56
EU Policy and the Uruguay Round Agreement 57
EU Perspectives on Future Agricultural Policy 61
Implications for the Trade Negotiations 64
Conclusion 66

**Chapter 5. One Perspective on U.S. Policy and Agricultural
Progress in World Trade Organization Negotiations** **71**
*Daniel A. Sumner*

General Obstacles to Progress in Current WTO Negotiations
in Agriculture 72
The Negotiating Position of the United States and Progress
in the Current WTO Round 76
Trade and Agricultural Policies of the United States
As Obstacles to Progress in the Current WTO Round 80
Concluding Remarks 85

**Chapter 6. Agricultural Negotiations in the Context
of a Broader Round: A Developing Country Perspective** **89**
*Thomas W. Hertel*
*Bernard M. Hoekman*
*Will Martin*

Agricultural Liberalization 90
Industrial Tariffs 93
Services 97
Industrial, Investment, and Export Development Policies 100
Achieving Balance: Rule Making and Implementation 102
Labor, Environmental, and Related Standards Issues 104
Developing Country Participation 106
Concluding Remarks 107

**Chapter 7. Agricultural Trade Liberalization and the Environment: Issues and Policies**    **113**
*Wesley Nimon*
*Utpal Vasavada*

Introduction    113
The Impact of Trade Liberalization on the Environment    114
The Positions of Fourteen Major Environmental NGOs    119
Countries' Agricultural Negotiating Positions
   with Environmental Implications    124
Relationship Between Fourteen Environmental NGOs
   and Countries' WTO Proposals    130
The Multifunctionality of Agriculture
   and Production-Linked Supports    131
Conclusion    133

**PART II: COMMODITY TRADE ISSUES**

**Chapter 8. Trade Liberalization in Rice**    **141**
*Eric J. Wailes*

Progress in Rice Trade Liberalization    144
Issues for Rice in the Current WTO Round    149

**Chapter 9. Major Issues in the U.S. Sugar Industry Under 2000 WTO Negotiations and NAFTA**    **155**
*Won W. Koo*
*P. Lynn Kennedy*

Introduction    155
Overview of the World Sugar Industry    156
The U.S. Sugar Program and Policies    159
Sugar and Sweetener Trade under NAFTA    162
Foreign Sugar Policies and Practices    166
Major Issues in the U.S. Sugar Industry    167
Concluding Remarks    174

**Chapter 10. Major Issues for the U.S. Wheat Industry: The Implications of China's Entry into the WTO**    **177**
*Won W. Koo*

Introduction    177
World Wheat Industry    178

Trade Agreements and Policies   182
Major Issues in the 2000 Round of WTO Negotiations   186
The Impacts of China's WTO Entry on the World
   and U.S. Wheat Industries   188
Results   191
Conclusion   195

**Chapter 11. New WTO Agricultural Trade Negotiations:
Issues for the U.S. Coarse Grain Market**   **197**
   *Linwood Hoffman*
   *Erik Dohlman*

Introduction   197
Global Coarse Grain Trade   198
Accomplishments of the Uruguay Round and Issues
   for New Agricultural Negotiations   199
Continuing Issues   201
Other Issues   207
Conclusion   211

**Chapter 12. The World Trade Organization and Southern
Agriculture: The Cotton Perspective**   **215**
   *Darren Hudson*

The Current Cotton Situation   216
The WTO and Cotton   218
Conclusion   222

**PART III: MULTILATERAL TRADE NEGOTIATIONS:
ISSUES AND CONCERNS**

**Chapter 13. The Impacts of Export Subsidy Reduction
Commitments in the Agreement on Agriculture
and International Trade: A General Assessment**   **227**
   *Lilian Ruiz*
   *Harry de Gorter*

Introduction   227
The Economics of Volume versus Value Commitments
   on Export Subsidies   229
Final Considerations and Conclusion   242

**Chapter 14. Agricultural Trade Liberalization Beyond Uruguay: U.S. Options and Interests**    **245**
  *John Gilbert*
  *Thomas Wahl*

Introduction                                                        245
Methodology                                                         246
Unfinished Business                                                 250
China and the WTO                                                   252
Regional Alternatives                                               254
Concluding Comments                                                 256


**Chapter 15. Regional versus Multilateral Trade Arrangements: Which Way Should the Western Hemisphere Go on Trade?**    **259**
  *Karen M. Huff*
  *James Rude*

Introduction                                                        259
The Free Trade Area of the Americas                                 260
The Prospects for a New Multilateral Trade Agreement                261
The Modeling Framework                                              261
Results                                                             264
Conclusion and Policy Recommendations                               268


**Chapter 16. Regionalism and Trade Creation: The Case of NAFTA**    **273**
  *Dragan Miljkovic*
  *Rodney Paul*

Introduction                                                        273
Determination of the Break Period: Method                           275
What Happened to Trade Creation?                                    278
Regionalism, Regionalization, and NAFTA:
  Concluding Remarks                                                282


**Chapter 17. Increased Use of Antidumping Weakens Global Trade Liberalization**    **287**
  *Anita Regmi*

Introduction                                                        287
What Is Dumping?                                                    288

Use of Antidumping by Countries                                289
Examination of the Current AD Law                             295
Agriculture and Antidumping                                   296
Potential Impacts on Future Agricultural Trade                299

**Chapter 18. Implementation of the SPS Agreement            305**
*Suzanne D. Thornsbury*
*Tara M. Minton*

Institutional Changes                                        306
Standard-Setting Procedures                                  310
Issues for the Next Round                                    313
Conclusion                                                   316

**Chapter 19. Modeling Impacts of the Macroeconomic
and Political Environment on Long-Term Prospects
for Agricultural World Markets                               321**
*Martin von Lampe*

Introduction                                                 321
Brief Overview of Other Agricultural Market Models           322
Outline of the WATSIM Modeling System                        324
Model Results: Prospects and Sensitivity of Agricultural
    World Markets                                            331
Further Developments                                         337
Summary and Conclusion                                       337

**Chapter 20. Price Volatility: A Bitter Pill of Trade
Liberalization in Agriculture?                               341**
*Robert D. Weaver*
*William C. Natcher*

Introduction                                                 341
Approach                                                     342
Empirical Results: The History of Price Volatility
    and Evidence of Change                                   348
Conclusion                                                   352

**Chapter 21. Challenges and Prospects for Agricultural Trade Policy in the New Millennium**          **357**
*P. Lynn Kennedy*
*Won W. Koo*

Challenges for Agricultural Trade Policy          357
Prospects for Agricultural Trade Policy          359

**Index**          **361**

# ABOUT THE EDITORS

**P. Lynn Kennedy, PhD,** is Alexander Regents Professor of Agricultural Economics and Agribusiness at Louisiana State University in Baton Rouge, where his teaching efforts focus on international trade and agribusiness. Dr. Kennedy has been a presenter at over 20 international, national, and regional agricultural economics meetings and was a Rotary Foundation Graduate Scholar at the University of Oxford. His current research interests involve analyzing how trade and economic welfare impacts domestic and multilateral policies on trade-in commodities that are of importance to the United States.

**Dr. Won W. Koo** is Professor of Agricultural Economics and Director of the Center for Agricultural Policy and Trade Studies at North Dakota State University in Fargo. His teaching and research areas include international trade, agricultural marketing, and demand analysis. Dr. Koo received the Quality in Research Discovery Award from the American Agricultural Economics Association in 1981 and the outstanding published research award from the Western Agricultural Economics Association in 1983. He received the Eugene Dahl Excellence in Research Award from the College of Agriculture, North Dakota State University in 2000.

# CONTRIBUTORS

**David Blandford, PhD,** is Professor and Head of the Department of Agricultural Economics and Rural Sociology at The Pennsylvania State University.

**Harry de Gorter, PhD,** is Associate Professor in the Department of Applied Economics and Management at Cornell University, Ithaca, New York.

**Erik Dohlman, PhD,** is an agricultural economist in the Market and Trade Economics Division, Economic Research Service, of the U.S. Department of Agriculture.

**John Gilbert, PhD,** is Assistant Professor in the Department of Economics at Utah State University.

**Thomas W. Hertel, PhD,** is Professor in the Department of Agricultural Economics at Purdue University.

**Bernard M. Hoekman, PhD,** is a policy and research manager for International Trade at The World Bank.

**Linwood Hoffman, PhD,** is an agricultural economist in the Market and Trade Economics Division, Economic Research Service, of the U.S. Department of Agriculture.

**Darren Hudson, PhD,** is Associate Professor in the Department of Agricultural Economics, Mississippi State University.

**Karen M. Huff, PhD,** is Assistant Professor in the Department of Agricultural Economics and Business, University of Guelph.

**Tim Josling, PhD,** is Professor in the Institute for International Studies, Stanford University.

**Will Martin, PhD,** is a lead economist in the Development Research Group, Trade, at The World Bank.

**Dragan Miljkovic, PhD,** is Assistant Professor in the Department of Agriculture, Southwest Missouri State University.

**Tara M. Minton** is a master of agribusiness, Coordinator of Economic Analysis at the Indian River Research and Education Center, University of Florida.

**William C. Natcher** is a research assistant in the Department of Agricultural Economics, Pennsylvania State University.

**Wesley Nimon, PhD,** is an economist with the Economic Research Service, United States Department of Agriculture.

**Rodney Paul, PhD,** is Assistant Professor in the Department of Finance, Saint Bonaventure University.

**Anita Regmi, PhD,** is an economist with the Economic Research Service, United States Department of Agriculture.

**James Rude, PhD,** is a research scientist with the Department of Agricultural Economics at the University of Saskatchewan.

**Lilian Ruiz, MS,** is an economist with the Coffee Division, Louis Dreyfus Corporation.

**Andrew Schmitz, PhD,** is Eminent Scholar and the Ben Hill Griffin Jr. Endowed Chair in the Department of Food and Resource Economics, University of Florida.

**Daniel A. Sumner, PhD,** is Frank H. Buck Jr. Professor, Department of Agricultural and Resource Economics, University of California, Davis, and Director of the University of California Agricultural Issues Center.

**Suzanne D. Thornsbury, PhD,** is Assistant Professor at the Indian River Research and Education Center, University of Florida.

**Utpal Vasavada, PhD,** is an economist with the Economic Research Service, United States Department of Agriculture.

**Martin von Lampe, PhD,** is an analyst with the Market and Trade Division of the Directorate for Food, Agriculture and Fisheries, Organisation for Economic Co-operation and Development (OECD).

**Thomas Wahl, PhD,** is Director of the IMPACT Center at Washington State University.

**Eric J. Wailes, PhD,** is Professor in the Department of Agricultural Economics and Agribusiness at the University of Arkansas.

**Robert D. Weaver, PhD,** is Professor in the Department of Agricultural Economics, Pennsylvania State University.

# Preface

The Southern Agricultural Trade Research Committee (S-287) convened a symposium, Global Agricultural Trade in the New Millennium, in New Orleans, Louisiana, on May 26-27, 2000. The purpose of the symposium was to assemble leading agricultural economists in the field from the United States, Canada, and the European Union to discuss emerging issues in agricultural trade under ongoing multinational and regional trade negotiations, including the World Trade Organization and Free Trade Area of the Americas negotiations. This book contains the full set of invited papers and a number of selected papers from the conference, all of which were the subject of critical discussion at the conference and further revisions by the authors. The first chapter was prepared by the editors to provide a summary of the chapters in the book and it raises questions and concerns for future consideration by researchers in the field. The final chapter includes major findings presented and issues raised during the conference. It is important to note that since these papers were presented at the symposium, several important events have taken place. Among these, the Doha Ministerial set a timeline for continued WTO negotiations and the United States Congress approved its 2002 Farm Legislation. While these events have slightly altered the environment in which agricultural trade occurs, the majority of issues and trading environment have, for the most part, not changed.

The editors wish to thank all authors, discussants, and other participants in the symposium for their thoughtful and creative inputs. This conference was sponsored by several research centers across the United States. These include: The Center for North American Studies, Texas A&M University and Louisiana State University; The International Agricultural Trade and Development Center, University of Florida; The International Trade Center, North Carolina A&T University; The National Center for Peanut Competitiveness, University of Georgia; The Northern Plains Trade Research Center, North Dakota State University; The Arkansas Global Rice Project, University of Arkansas; and The Farm Foundation, Oak Brook, Illinois. Particular debts of gratitude are owed to the sponsors of the conference for their contributions. Walter Armbruster and Steve Halbrook of the Farm Foundation deserve special recognition for their financial support of the conference. Special recognition also goes to C. Parr Rosson III, Flynn Adcock, and the staff at Texas A&M University and Louisiana State University for making the conference a success. Steve Chriss is to be commended for his work in formatting the manuscript. Most important, special recognition is deserved for the S-287 Regional Research Project "Impacts of Trade Agreements and Economic Policies on Southern Agriculture" for their dedication to this important area of research.

# Chapter 1

# Issues and Obstacles for Agricultural Trade Policy in the New Millennium

P. Lynn Kennedy
Won W. Koo

According to former U.S. Trade Representative Barshefsky, agriculture is a field where the current World Trade Organization (WTO) negotiations offer immense potential for direct, concrete benefits via further reduction of tariffs, export subsidies, and domestic supports linked to production. Although providing policy analysis for the upcoming round of WTO negotiations is a significant task in itself, several other international policy issues will be considered within the next few years. The Free Trade of the Americas (FTAA) negotiations began in 2000, and a new U.S. farm bill may be completed by late 2002. The timing of these policy decisions will have major implications for the agricultural sector. As a result, policy analysts must be prepared to provide proactive support to the debate rather than playing a reactionary role.

If these goals are to be accomplished, agricultural policy analysts must identify the relevant issues in the policymaking process and produce research products that are available to the appropriate policymakers in a timely fashion. The chapters in this volume are intended to serve as a forum to initiate the dialogue and research necessary for that process. It brings together agricultural economists concerned with the WTO and other trade agreements in order to identify and discuss the issues and implications currently associated with agricultural trade negotiations. Specific themes include (1) accomplishments of the Uruguay Round Agreement and critical issues for the upcoming the WTO negotiations; (2) impacts of regionalism, such as the Free Trade Area of the Americas and the European Union on agricultural trade; and (3) analysis of major trade issues specific to individual commodities.

The current round of multilateral trade negotiations promises to be a topic of critical importance for agriculture. The initial round of WTO negotiations, which began in late 1999 in Seattle, Washington, was mandated in

1993 by the Uruguay Round Agreement of the General Agreement on Tariffs and Trade (GATT).

Several persistent trade issues must be addressed. Among these, agricultural protection levels are high relative to those of other goods, while certain nontariff barriers remain. In addition, export subsidies, credits, and limited market access continue to distort world agricultural markets. Domestic support disciplines have little effect in most countries, and some commodities remain largely outside the reform process.

Regional trade agreements, state trading enterprises, agricultural biotechnology, and environmental concerns are key issues to be examined. In addition to these selected issues, other topics (e.g., sanitary and phytosanitary issues) combined with the diverse positions of developed, developing, and potential members of the WTO stand to influence the outcome of the current round. As the diversity of issues considered within the WTO increases, analysts will be challenged to provide comparisons across this broad set of concerns. If the agricultural economics profession is to have meaningful input in this dialogue, it must be prepared to develop new and creative methods that address these key issues and provide meaningful information to aid in the negotiating process.

## ISSUES AND PERSPECTIVES FOR FUTURE TRADE AGREEMENTS

Several key issues are critical to the success of agricultural trade negotiations. These include the relationship between regional trade agreements and the WTO, state trading enterprises, agricultural biotechnology, and the interface between trade and the environment. Regional trade agreements (RTA) play a significant role in the global trading system, with most WTO member nations involved in at least one regional trade agreement. The incentive for countries to form a free trade area include economic enhancement through expanding trade volumes among RTA member countries while maintaining protection from the rest of the world. Current or potential RTAs that may impact the next round of multilateral trade negotiations include the North American Free Trade Agreement (NAFTA), FTAA, the European Union (EU), Asia-Pacific Economic Cooperation (APEC) Region, and the Transatlantic Free Trade Area, among others.

An additional important issue for upcoming WTO negotiations involves state trading. State trading is defined as the activities of quasi-governmental agencies that influence the level and direction of imports or exports. Concerns regarding state trading include the potential distortions in world markets caused by state trading enterprise (STE) activities. While the agenda of upcoming WTO negotiations is uncertain with respect to STEs, it is clear that restrictions on STE operations are needed to promote fair trade. Examples include the following: STEs should be transparent in terms of their operation and marketing practices; STEs should be subsidy-neutral, implying that STEs should not circumvent domestic and export subsidies; and finally, STEs should be restricted in exercising market power, including the use of price discrimination and price distortions.

Biotechnology used in commercial agriculture raises new issues for the next trade round. Agricultural biotechnology has significant potential for consumers and producers by helping to guarantee a global food supply through increased agricultural production while conserving the environment. Examples of these genetically modified organisms (GMOs) include corn and soybeans that are insect resistant and herbicide tolerant. The United States leads the world in acreage planted in GMOs and in regulatory approvals. Differences among countries' GMO regulations pose potential barriers to these exports. Consequently, these differences raise the need for mutual recognition of countries' regulations, harmonization of existing regulations among countries, or negotiation of an international standard.

The increasing focus in international policy circles on the trade and environment interface is motivated by the interests of both environmental and industry groups. Environmental groups have expressed the following concerns: trade and trade agreements exert pressure to reduce national environmental standards to the lowest common denominator; differences among national standards produce pollution havens; freer trade worsens pollution by stimulating economic activity of dirty industries; and trade agreements interfere with national sovereignty over environmental protection goals and legislation. Thus, institutions such as the WTO that have historically fostered global economic growth through trade and economic integration must now address environmentalists' concerns to sustain broad support.

In addition to these specific issues, diverse country and regional perspectives will influence and complicate the ability to negotiate future trade liberalization. Of particular importance are the positions of the United States, the European Union, the Cairns Group, China, and the developing countries.

Approximately 20 percent of U.S. agricultural production is exported. Thus, a strong international market is crucial for a vibrant U.S. agricultural sector. With few exceptions, U.S. agricultural commodities benefit from foreign markets. Given that its agricultural sector cannot thrive in an isolationistic environment, the U.S. has adopted an international trade agenda that seeks open agricultural markets around the world. Primary methods to accomplish this include the elimination of export subsidies, market access expansion through tariff reductions and continued liberalization of tariff rate quotas, and the decoupling of domestic support. This agenda involves negotiations completed on an accelerated timetable, provides ongoing achievements in priority areas, and results in institutional reform and capacity building at the WTO. Agriculture is viewed as a key issue, in which decreased subsidies and trade barriers combined with the use of science can provide food security while benefiting both producers and consumers. For example, one objective for U.S. agriculture with respect to trade agreements is to ensure that regulations in areas such as biotechnology are transparent, predictable, and based on sound science.

The European Commission's priorities in developing proposals for its Common Agricultural Policy (CAP) Reform provide insight into their agenda for the current round of WTO negotiations. These priorities include making EU agriculture more environmentally sensitive and more competitive in domestic and foreign markets, as well as ensuring the livelihoods of farmers. These priorities highlight the EU's commitment to the European model of agriculture, which will be "defended with vigour" in the upcoming round of WTO negotiations. Given its need for a united stance, Agenda 2000 provides the foundation for the EU negotiating position. Similarly, the EU's negotiating position will be influenced by future EU expansion under Agenda 2000.

The primary goal of the Cairns Group of Agricultural Fair Traders is to achieve a fair and market-oriented agricultural trading system. The Group's agenda for the next round of WTO negotiations involves basic reforms that seek to place trade in agricultural goods on the same level as trade in other goods. To this end, the Cairns Group supports the elimination of all trade-distorting subsidies and the substantial improvement of market access, allowing market forces to be the main determinant of agricultural trade flows.

Numerous countries are in the process of applying for WTO membership. While obligations come with membership, certain privileges will be extended. These include the extension of previously negotiated tariff bindings and market access provisions, as well as access to the WTO dispute settlement process, and input into the WTO policymaking process.

From the perspective of the developing world, several shortfalls must be addressed in the development and implementation of multilateral trade agreements. For example, existing agreements have resulted in imbalances from the perspective of many developing countries, with developed countries viewed as negligent with respect to their previous liberalization commitments. As a result, developing countries should not be expected to make concessions to correct violations of previous agreements by developed countries. Thus, the willingness of developing countries to participate in a new round of negotiations may be less than enthusiastic.

## SCOPE OF THIS VOLUME

The chapters in this volume are organized into three sections. Specific themes include accomplishments of the Uruguay Round Agreement and critical issues for the upcoming the WTO negotiations; analysis of major trade issues specific to individual commodities; and impacts of various key issues on agricultural trade.

The first section of the book, Key Issues Influencing Agricultural Trade Negotiations, provides perspectives on the current agriculture trade situation. Schmitz provides an overview of the state of "Progress in Agricultural Trade Negotiations" while Josling discusses "Issues in the WTO Agricultural Negotiations: An Overview." In addition, key issues and perspectives are discussed. The prospect for reform by the EU and United States are reviewed by Blandford in "Obstacles to Progress in Multilateral Agricultural Trade Negotiations: The European Viewpoint" and by Sumner in "One Perspective on U.S. Policy and Agricultural Progress in World Trade Organization Negotiations." In addition, a developing country viewpoint is given by Hertel, Hoekman, and Martin in the chapter "Agricultural Negotiations in the Context of a Broader Round: A Developing Country Perspective." Nimon and Vasavada review the interface between the environment and the prospect of freer trade in "Agricultural Trade Liberalization and the Environment."

Section two, Commodity Trade Issues, provides detailed overview of key issues and perspectives regarding specific agricultural commodities. Prospects for policy reform in rice are discussed by Wailes in the chapter "Trade Liberalization in Rice." Koo and Kennedy discuss "Major Issues in the U.S. Sugar Industry Under 2000 WTO Negotiations and NAFTA." Koo presents major issues for the U.S. wheat industry with a focus on the implications of China's entry into the World Trade Organization in Chapter 10. Hoffman and Dohlman discuss issues for the U.S. coarse grain market with respect to the new agricultural trade negotiations. Finally, Hudson offers the cotton perspective on the World Trade Organization and Southern Agriculture.

The final section, Multilateral Trade Negotiations: Issues and Concerns, discusses and analyzes the impacts of several key issues on agricultural trade. Discussion of the impacts of specific actions within trade agreements are presented by Ruiz and de Gorter in "The Impacts of Export Subsidy Reduction Commitments in the Agreement on Agriculture and International Trade: A General Assessment" and by Regmi in "Increased Use of Antidumping Weakens Global Trade Liberalization." Discussions of potential direction for further trade policy reform are provided by Gilbert and Wahl in "Agricultural Trade Liberalization Beyond Uruguay: U.S. Options and Interests," and Thornsbury and Minton in "Implementation of the SPS Agreement."

The impact of regionalization and its effect on trade liberalization is discussed by Huff and Rude in "Regional versus Multilateral Trade Arrangements: Which Way Should the Western Hemisphere Go on Trade?" and by Miljkovic and Paul in "Regionalism and Trade Creation: The Case of NAFTA." This section also includes a discussion of external factors on agricultural trade liberalization by von Lampe in "Modeling Impacts of the Macroeconomic and Political Environment on Long-Term Prospects for Agricultural World Markets," and a discussion of potential impacts of trade liberalization by Weaver and Natcher in "Price Volatility: A Bitter Pill of Trade Liberalization in Agriculture?" The book concludes with an overview of the challenges and opportunities for reforming agricultural trade policies in the new millennium by Kennedy and Koo.

This book consolidates the work of agricultural policy analysts concerned with the WTO and other trade agreements for the purpose of identifying and discussing key issues and their implications for agricultural trade negotiations. In order for future rounds of agricultural trade negotiations to be fruitful, it is imperative that the relevant issues in the policymaking process be identified. Further, relevant research must be produced and made available to the appropriate policymakers in a timely fashion. The chapters in this book are intended to serve as a forum to initiate the dialogue and initial research necessary for that process.

# PART I:
# KEY ISSUES INFLUENCING AGRICULTURAL TRADE NEGOTIATIONS

# Chapter 2

# Progress in Agricultural Trade Negotiations

## Andrew Schmitz

### *INTRODUCTION*

Many special interest groups, including environmentalists and labor unions, were successful in scuttling the 1999 round of trade talks in Seattle. This chapter deals specifically with key agricultural trade roadblocks to future trade talks. Freeing up agricultural trade will require ingenious schemes to deal with those sectors in agriculture that will lose under free trade. Two major players, the United States and the European Union, are once again heavily transferring income to their farm sectors. Farm programs may or may not be consistent with free trade solutions.

### *POLICY HARMONIZATION AND SIDE AGREEMENTS*

One of the arguments used under the North American Free Trade Agreement (NAFTA) and the Canada-U.S. Trade Agreement (CUSTA) was that progress toward freer trade would require policy harmonization among participating countries. But what happened to policy harmonization? (Figure 2.1 hardly suggests policy harmonization.) Policy harmonization has not happened as the United States and Canada are moving in opposite directions in the policy arena. The United States passed the 1996 FAIR Act whereby farmers were compensated over seven years for the removal of price supports. In addition, in 1999, U.S. farmers received a government cash bailout of over 8 billion dollars. For 2000, significant payments exceeding this sum were once again made under various relief packages. On the other hand, in Canada, supply management remains in place. But for major commodities such as wheat, beef, and pork, Canadian subsidies have been almost eliminated, including the Crow Rate transportation subsidy. In spite of lobbying by farm groups in western Canada, little government support will be provided to farmers in western Canada, despite record-low farm incomes.

*9*

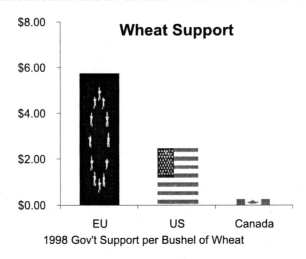

FIGURE 2.1. EU, U.S., and Canadian government support for wheat. (*Source:* Organization for Economic Cooperation and Development [1999]. Producer and Consumer Support Estimates, OECD Database, Data on CD. Paris.)

With the removal of the Crow transportation subsidy in Canada, very few subsidies remain for western Canadian farmers. The removal of the Crow was certainly a victory for taxpayers, since the government compensation package paid to producers was much lower than payments made under the Crow Rate agreement. Producers did not receive full compensation for the transfer they were receiving under the Crow Rate. The removal of the Crow was also a victory for proponents of free trade; however, many argue that even if the Crow had remained in place, it would not have brought a halt to trade talks. They argue further that the removal of the Crow was nothing more than a budgetary decision on the part of the Canadian treasury. With the Crow and other subsidies eliminated, Canada is left to compete against countries, such as the United States, who have reintroduced subsidies. Is this a policy of harmonization?

Policy decisions are clearly influenced by rent seeking. Effective rent seekers are able to convince politicians that they need financial support regardless of the impact of such support on trade and the environment. U.S. farmers were successful in lobbying government to change the 1996 farm program in their favor, particularly when, in 1999, an 8.7 billion dollar bailout was added to farmers' coffers. The U.S. government is once again heavily involved in U.S. agriculture. This involvement should not be surprising given the theory of rent seeking and public choice. Politicians' response to lobbying is in part a function of whether or not they can win votes. This is clearly demonstrated concerning policy harmonization between

Canada and the United States. For example, wheat farmers on the Canadian prairies, unlike those in the United States, have recently been unable to lobby the federal government effectively for support despite extremely low wheat prices. This was not true for the early 1990s when the level of support for Canadian prairie farmers was at least as high as support levels for American farmers. Why is support now not forthcoming? There are several reasons, including the fact that supporting prairie farmers generates very few votes.

The Canadian federal government is a Liberal government, while the Alberta provincial government is Conservative. Saskatchewan and Manitoba are governed by the New Democratic Party. There are few elected Liberals from the West in the current federal government. (Remember that when the federal government provided high support levels in the late 1980s and early 1990s, a Progressive Conservative government was in place in Ottawa and in Saskatchewan.) Unless the Liberal government feels that it can obtain additional votes in future elections through farm support, it likely will not subsidize western farmers. The Liberal government clearly supports supply management, whose primary activities are located in eastern Canada: Quebec and Ontario (Schmitz and Schmitz, 1994; Schmitz, 1995). There are also those who argue that the Ottawa civil servants who advise the federal Minister of Agriculture, including those in Agriculture and Agri-Food Canada, are not in favor of providing significant transfers to western Canada. Both arguments combined—the political and the economic—suggest why western Canada is not likely to obtain significant transfers from the federal government at anywhere close to U.S. subsidy levels.

An additional important reason helps to explain divergent farm policies among countries. In the United States, agricultural policy is a federal responsibility, which is not the case in Canada. For individual provinces to receive farm assistance, they have to provide funds that at least partly match those provided by the federal government. Consider the case for Saskatchewan, which in the late 1990s and early 2000 was one of the provinces most drastically affected by record-low commodity prices. Saskatchewan is highly dependent on agriculture; but to come up with matching funds is extremely difficult. This is not true for an agricultural state such as North Dakota, where over 70 percent of net farm income for the years 1995 through 1998 came from the federal government with little or no matching funds required from the North Dakota treasury.

The problem with policy harmonization is further illustrated with respect to NAFTA, implemented on January 1, 1994, for Mexico-U.S. sugar. The goal of NAFTA is to create a free trade zone among the countries of Canada, Mexico, and the United States. NAFTA is to be fully implemented by 2008.

NAFTA includes a sugar side agreement between Mexico and the United States. The United States agreed to change its quota system to a tariff rate quota. From 1994 to 2000, Mexico would still be allowed to export up to 25,000 metric tonnes (MT) of sugar at the tariff rate of zero. Exports in excess of 25,000 MT would be subject to the most favored nation (MFN) rate of $.18 per pound. The over-quota tariff will be reduced in equal increments over a fifteen-year period ending in 2008. At that time, both sugar and high-fructose corn syrup (HFCS) is supposed to be freely traded between Mexico and the United States. (Mexico has net sweetener exporter status if the production of sugar in Mexico exceeds consumption of sugar plus consumption of HFCS, regardless of whether or not the HFCS is imported or produced in Mexico. With the implementation of NAFTA, Mexico began to import large quantities of HFCS from the United States (Table 2.1). In spite of this, beginning in 2000, Mexico had net sweetener export status.)

Starting in October 2000, the tariff rate quota afforded to the Mexican sugar industry by the United States increased from 25,000 to 250,000 MT of sugar, if Mexico was able to maintain its status as a net surplus producer. This arrangement created a significant problem for the U.S. sugar program because of already relatively low sugar prices in the United States. In early 2000, domestic sugar prices fell drastically, even in the presence of tight import quotas. Added imports from Mexico will only amplify the problem. Ongoing discussion is taking place in Washington about how to remove excess sugar from the market in order to maintain internal U.S. price supports. In addition, there are significant lobbying efforts to have this side agreement removed.

TABLE 2.1.  The Mexican Sweetener Market, 1991-1998

| Fiscal Year | Total Supply HFCS | HFCS Imports | Domestic HFCS Production | Domestic Sugar Production | Domestic Sugar Consumption |
|---|---|---|---|---|---|
| 1991 | 8,000 | 8,000 | 0 | 3,656,000 | 4,055,700 |
| 1992 | 20,982 | 20,982 | 0 | 3,480,100 | 3,976,400 |
| 1993 | 33,181 | 33,181 | 0 | 3,987,700 | 4,000,000 |
| 1994 | 74,773 | 74,773 | 0 | 3,540,500 | 4,071,300 |
| 1995 | 58,646 | 58,646 | 0 | 4,213,500 | 4,017,700 |
| 1996 | 89,197 | 89,197 | 0 | 4,478,600 | 3,965,900 |
| 1997 | 543,531 | 206,406 | 334,125 | 4,672,200 | 3,833,300 |
| 1998 | 527,899 | 177,899 | 350,000 | 5,174,027 | 3,894,632 |

*Source:* USDA, Economic Research Service, Sugar and Sweetener/SSS-225/ May 1999b.

## STATE TRADING ENTERPRISES (STEs) AND THE WTO

The U.S. position is that STEs should be eliminated. According to Schmitz (2000), two thorny issues arise. First, the United States has to consider the role played by its own STE, the Commodity Credit Corporation (CCC). At times, the CCC's activities have been significant in international grain marketing (Schmitz, Furtan, and Baylis, 1999). In the late 1990s, because of extremely low commodity prices, the CCC again emerged as a major player. Second, despite claims to the contrary, there is little or no empirical evidence available to suggest that STEs are necessarily in violation of WTO rules. For example, a study on the Canadian Wheat Board—a major STE—argues that the CWB is not in violation of WTO rules (Schmitz, Furtan, and Baylis, 1999).

STEs have to meet certain criteria to be WTO compliant. One important criterion concerns soft price discrimination where export subsidies are involved. STEs can practice hard price discrimination (which does not involve subsidies). Generally, when the CWB price discriminates, it practices hard price discrimination, and is therefore WTO compliant. In addition, regardless of the nature of price discrimination, and whether or not the CWB is efficient in marketing, the trade-distorting effects are small indeed. STEs should be judged on the extent to which they are trade distorting, rather than on other criteria that are espoused in trade circles.

Many of the issues concerning the WTO's attempts to govern STEs are stated for both the CWB and the CCC by Schmitz et al. (1999). While the United States has been critical of STEs, now that STEs are no longer a significant state trader in grains, the U.S. seems even less friendly toward them. However, the WTO rules appear to support those STEs that practice hard-price discrimination. . . . The trade-distorting effects of the CWB are small indeed. This is the case whether the CWB earns a price premium for producers or taxes producers. In the first case, the CWB practices hard-price discrimination, which is legal under current WTO rules. In the second case, when the CWB is inefficient and lowers producer returns, its activities are still acceptable under WTO. The tax scenario makes the CWB highly favorable to its competitors. Importantly, even if the distorting effects were much larger than those reported, CWB activities would still be allowable because the WTO places no limit on the magnitude of price discrimination by STEs. This strong conclusion should cast doubt on the ability of the WTO to discipline STEs. A problem arises because the WTO does not adequately qualify its limit on quantitative restrictions. For example, since the CWB does not set Canadian grain-trade policy, to price discriminate (which the WTO permits), the CWB must use quantitative restrictions (which WTO limits). Thus, there appears to be a contradiction between the first and second requirement of the WTO. Upcoming WTO discussions must focus on more careful definitions of the criteria that limit STEs and their activities.

## *KEEPING THE BORDERS OPEN*

A gray area in international trade is the role that domestic trade laws play in the context of the WTO. Even though the WTO is set up to arbitrate and settle disputes, it appears to have little strength when it comes to disciplining a country's use of domestic trade laws. For example, in the United States, dumping and countervail laws (in addition to farmer subsidies) are in place to protect U.S. producers from foreign competition. According to Schmitz (2000), "the United States is arguing for free trade on the one hand, but practicing protectionism on the other. However, these allegations may not be true"(p. 3).

For example, consider the countervail and dumping cases leveled by U.S. beef producers against Mexico and Canada. On October 1, 1998, the Ranchers-Cattlemen Action Legal Foundation announced that it had filed three petitions in Washington, DC with the U.S. government seeking relief from unfair trade practices. It had the support of the National Farmers Union, more than twenty other state and local organizations, and nearly 8,000 individual ranchers. If it is true that the United States preaches, but does not practice, free trade, then the countervail and dumping cases against Canada would have been successful. The cases were resolved in 1999, however, *in favor* of Canada. This decision was a victory for U.S. proponents of free trade and for Canadian beef interests. Another example of the complex nature of the U.S. stance on free trade involves the injection of over 8 billion dollars in 1999 into the farm economy in the form of government disaster relief payments. This kind of bailout is not, in fact, in violation of the U.S. stance on free trade, though it may at first glance appear to be. The bailout is likely to be considered a "green box" policy by WTO rules (see next section); that is to say the bailout, under WTO rules, is considered to be production and trade neutral.

The WTO did not intervene in the cattle dispute, nor has it intervened in many of the Canada-U.S. grain disputes. This seems somewhat ironic, given the fact that many of the cases involve dumping allegations leveled by the United States. It may be true that countries that compete with the United States are dumping in international markets, but then often so is the United States. As Schlueter (1998) notes:

As "proof" of these [dumping] allegations, a grain producer from Sweetgrass, Mont., was recently quoted as saying that it cost $5.54 to produce a bushel of wheat in his area. The fact that $2.00 prices were being offered by local elevators in north central Montana, he reasoned, served as the smoking gun in proving the Canadians were guilty of illegal "dumping" practices. Excusing his errant logic for a moment, it is safer to assume that this grower was misquoted about these numbers than to accept the accuracy of his estimates. Because if these figures were anywhere close to being accurate, it would be the U.S. that stands exposed to charges of illegal dumping, as millions of tons of American wheat are currently being sold into global export markets at prices that are less than half of this fellow's suspect figures. (p. 12)

We have argued elsewhere that the $5.54 cost to produce a bushel of wheat is reasonably accurate, and that the United States dumps wheat and many other products into international markets since the sale price does not cover the cost of production (Becker, Gray, and Schmitz, 1992).

## THE GREEN BOX

Agricultural policies have been separated into multiple categories. Green box policies are not actionable for countervailing duties or other GATT challenges. Supposedly, green box criteria ensure that policies and programs in the category are production and trade neutral. However, it is my contention that so-called green box policies are a major (if not *the* major) deterrent to freeing up agricultural trade.

According to Agriculture and Agri-Food Canada (1998a,b), exempt policies include green box programs and blue box programs. For green box policies, two basic criteria apply: (1) support must be government funded, and (2) the money cannot provide price support. In addition to these criteria, a number of illustrative programs are given: research, inspection, extension and training, marketing and promotion, public stock holding for food security, domestic food aid, and decoupled income support, income insurance and safety net programs, structural adjustment assistance, regional assistance, and environmental aids. Blue box policies (which include program payments received under production limiting programs—based on fixed area and yields, a fixed number of head of livestock, or if they are made on 85 percent or less of base level of production) are considered acceptable, but only on a temporary basis. Those programs that do not fit these categories are subject to reduction commitment. For these nonexempt programs, a quantitative measure of the level of intervention is calculated using an Aggregate Measure of Support (AMS). Developed country WTO members are required to reduce their AMS to 80 percent of their 1986-1988 levels by 2000. Amber box policies are trade-distorting domestic support programs that are subject to reduction commitments (such as market price support and input subsidies). Red box programs are prohibited. There is no agreement on how to stop any domestic policies deemed to fall into the red box, so it remains empty.

During the Uruguay Round of trade negotiations, a number of authors attempted to classify domestic support policies based on the degree to which they distort trade; however, certain policies do not fit neatly into the categories (Agriculture and Agri-Food Canada, 1998a,b). For example, according to Agriculture and Agri-Food Canada, it is difficult to assign the Canadian program NISA a precise spot within the green part of the green-red spectrum. The program is not entirely neutral, since additional government contributions can be obtained through additional sales. Clearly, questions should be raised about U.S. farm programs and whether or not they fit the green box category. Many argue that they do.

In spite of CAP reform toward direct payment for EU farmers, EU compensatory payments cannot be considered green as they now stand. According to an Agriculture and Agri-Food Canada report (1998a), An Examination of Nearly Green Programs: Case Studies for Canada, The United States and the European Union,

> Although green box programs are more benign than other forms of support, it is clear that large ongoing payments, by the amount of the size and permanence, attract and keep resources in agriculture. As the green box becomes a more popular avenue for governments to provide domestic support, the size of the expenditure envelope will expand and the potential distortions will increase accordingly. Moreover, although programs may be designed to be production neutral, they are not always so in practice. Even though a program may be only marginally distorting, large program expenditures may turn a small distortion into a big impact. This raises the need for a cap on total green box spending, possibly combined with a cap on each element of the green box. (Chapter 5, unpaginated)

Do U.S. farm policies fit within the green box? For example, does the 8.7 billion dollar bailout of U.S. farmers in 1999 fit into the green box? The majority of policy analysts are silent on this issue. If this type of program significantly distorts international trade, then it does not fit the green box category. It is my opinion that it is essentially impossible to carry out decoupled farm programs. Farmers use government payments in production decisions. These payments essentially increase the price of the commodities farmers produce (our surveys show that a dollar transfer from the government trans-lated into a 75 cent increase in the price of the commodity produced). As a result, production is influenced by government transfers, and therefore trade is distorted. For example, in talking with farmers about the 1999 U.S. farm bailout, we concluded that this payment impacted both production and land-use decisions. Many farmers did not have renters for their farms until the $8.7 billion payout was announced.

The key issue of whether farm programs can be decoupled has not been seriously debated. From an economic perspective, if major farm programs cannot be decoupled (certainly leading policy experts such as Luther Tweeten [1979] argue that they cannot be), then it is fallacious to argue that nations are moving to freer trade while at the same time infusing large sums of money (for example, the U.S. 1999 bailout and EU direct support pay-ments to farmers) to help farmers cope with low farm incomes regardless of the causes. The issue of decoupling can be highlighted with reference to statements made on EU farm policy and trade, and the position of the Cairns Group concerning future trade negotiations.

According to Haniotis (2000), Agricultural Counselor with the European Commission Delegation to the United States, the position of the European Union "stresses the continuation of the present distinction of policies ac-cording to their degree of trade distortion as the essential element in deter-mining adherence to the desired move away from support linked to prices or products towards more transparent and non-distorting support policies" (p. 4). Figures 2.2 and 2.3 provide information on the evolution of EU wheat policies and EU and U.S. direct payments to farmers.

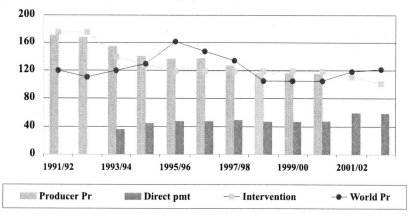

FIGURE 2.2. Evolution of EU wheat policies (in ECU/EUR per metric ton). (*Source:* Haniotis, 2000, p. 9.) Reprinted by permission.

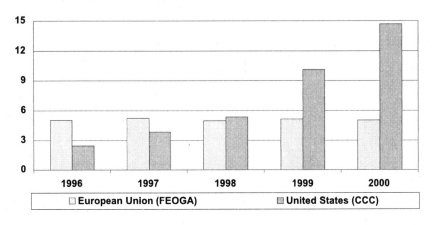

FIGURE 2.3. EU and U.S. payments per farm (fiscal years, in thousand dollars). (*Source:* Haniotis, 2000, p. 11.) Reprinted by permission.

The European model has been defined by the EU council and contains the following general set of objectives (Haniotis, 2000):

- a competitive agricultural sector, which can gradually face up to world markets without being over-subsidized;
- production methods which are sound and environmentally friendly, able to supply quality products that the public wants;
- diversity in the forms of agriculture, which maintain visual amenities and support rural communities;
- simplicity in agricultural policy, and sharing of responsibilities between the European Commission and EU member-states;
- justification of (farm) support through the provision of services that the public expects farmers to provide.

All of these objectives not only reflect generally accepted policy targets, but also fall clearly within the scope of Article 20 of the Uruguay Round Agreement on Agriculture (URAA), that sets the agenda for further agricultural policy reform. In fact, the debate about the European model of agriculture is nothing more than a reflection of the fact that within the EU, the relevant agricultural policy question is not *if*, but *how* to support agriculture (a point that, may I note, is becoming again increasingly relevant in the United States). (p. 2)

The concept of multifunctionality is pervasive in EU discussion on agricultural policy and WTO commitments. According to Haniotis (2000):

Finally under Agenda 2000 the so-called second pillar of the CAP, comprising of its environmental and rural functions, has been reinforced. This resulted in shifting part of the generic market support (the first pillar of the CAP) towards support targeted at agri-environmental and rural development measures, which were brought under a common rural development regulation. This covered the accompanying measures to the 1992 reform (agri-environment, early retirement and afforestation), aid for structural adjustment and aid to young farmers, investment aids, processing and marketing aids, diversification aids and the less favored areas (LFA) scheme to promote continued agricultural land use and low-input farming systems. Together the agri-environmental and rural measures and other direct payments such as in the LFA aim at ensuring that farmers and other rural actors continue to meet society's demand for environmental and rural services and thus contribute to safeguarding and enhancing agriculture's multi-functional role. (p. 3)

The most fundamental non-trade concern that needs to be addressed in the next WTO negotiations on agriculture is for the EU is the recognition of the multi-functional role of agriculture. As mentioned earlier, it is not just the production of food, feed and fiber, but also the preservation of the rural environment and landscape, animal welfare, and agriculture's contribution the viability of rural areas and to a balanced territorial development, that represent legitimate policy objectives. It is our intention to meet these objectives by policy measures that are tailored to meet specific goals in the least trade distorting way. (p. 6)

And further, Haniotis (2000) writes:

The EU expects agricultural negotiations to strike a balance between fundamental trade reform (by reduction of both border protection and domestic and export support) and non-trade concerns that reflect a follow-up of the 1994 URAA agreement. A future WTO agreement must lead to further agricultural trade liberalization while at the same time allowing WTO partners to maintain a policy that respects and fulfills their domestic priorities.

In order to achieve the above objective, the mandate of Article 20 of URAA provides the point of departure. Disagreements about the speed or extent of reform are natural to exist in the beginning of negotiations (although the effort to turn the agenda into an end-result itself often complicates things). But whatever accent one puts on Article 20, it is clear that it sets a long-term objective of substantial, progressive reductions in support and protection, resulting in fundamental reform, while at the same time addressing the above-mentioned wider issues based on the same mandate. (p. 4)

Other trade issues raised by Haniotis (2000) include:

### Export competition issues

This area of the URAA is often portrayed as one referring only to export subsidies, with the accent placed on the fact that almost 85 percent of all agricultural export subsidies are attributed to the EU. There is nothing new or unexpected in this fact. It is a reflection of the structure of previous EU farm policies, that has been incorporated in the EU commitments. . . . EU export subsidies came under strict rules and disciplines, have declined significantly, and are expected to decline even further as a result of the latest CAP reform (even before a new WTO agreement comes into place). The Community is willing to continue to negotiate further reduction of export subsidies, but this presupposes that *all* such support to exports is treated on a common footing.

This means that the commitment to introduce disciplines on agricultural export credits (including the provision of food aid on concessional credit) must be respected. This commitment, which relates to the major US policy tool of export support, is part of the URAA (Article 10.2) and its fulfillment is essential for a balanced agreement.

### Market access issues

The European Union is a major food exporter and the largest food importer in the world. It thus intends to share in the expansion of world trade in agricultural products. The EU will seek to obtain improvements in opportunities for its exporters, *inter alia* through greater clarity in the rules for the management of tariff rate quotas (TRQs), including imports through single desk buyers, and the removal of other unjustified non-tariff barriers.

### Non-trade concerns

A wide range of issues touches upon different WTO agreements: the Sanitary and Phytosanitary Agreement (SPS), the Agreement on Technical Barriers to Trade (TBT), the Agreement on Trade-related aspects on Intellectual Property Rights (TRIPS). From all these issues, undoubtedly the most controversial has been the area of measures related to food safety concerns and their impact on trade. Recent WTO case law has confirmed that non-discriminatory, science-based measures to achieve the level of safety determined by members are in conformity with that agreement.

It might be useful to confirm this in a more general manner in order to assure consumers that the WTO will not be used to force onto the market products about whose safety there are legitimate concerns. What the EU experience of recent years has demonstrated is that consumer perceptions on issues related to food safety, which undoubtedly have a direct impact on trade, are not viewed as such, but as health issues by the general public. Thus measures that aim at incorporating these concerns into future trade agreements should not be viewed as trade impeding. On the contrary these measures are in the long term trade enhancing.

### Other issues

Two other elements of the URAA are also considered essential by the EU to be included in a new agreement. The first relates to the need to provide legal security for the outcome of the forthcoming negotiation, just as the Uruguay Round agricultural negotiation agreement did by the inclusion of its peace clause.

The second refers to the Special Safeguard Clause, which represents a key component of agricultural liberalization agreed in the last Round by allowing abnormally low price offers or import surges to be dealt with without frequent recourse to more disruptive action under the general Safeguard Clause. A similar provision for the future should therefore be in the general interest of all members. (pp. 5-6)

The Cairns Group is a strong supporter of free trade. According to William H. Miner, the Cairns Group vision statement that will be used in any upcoming trade negotiations involves the following implicit or explicit beliefs (Miner, 1999):

- Exceptional treatment for agriculture under the trade rules should be progressively removed.
- All subsidy interventions that distort prices, production, and trade should be eliminated.
- A major expansion of market opportunities is to be achieved through tariff reductions, the removal of tariff escalation, and an increase in tariff rate quotas.
- All income and other domestic support measures must be targeted, transparent, and fully decoupled.
- Special and differential treatment for developing countries, including least-developed nations and small states, is to be an integral part of the agriculture negotiations. (p. 149)

One must wonder what these statements from EU officials and Cairns Group mean, at least with respect to direct farm-support payments (the statements on multifunctionality appear to be extremely vague with respect to having operational content under the WTO). Where is the empirical evidence to suggest that farm programs are being decoupled to make them consistent with freer trade? Ideally, for a farm program to fit the green box, it should be totally decoupled. That is, the program in question should not influence production decisions and, hence, the trade impact should be nonexistent.

We question the extent to which farm program are decoupled (Figure 2.4). The supply and demand schedules are $S$ and $D$ in the absence of farm programs. The free trade price is $P_1$ and exports total $Q_1' Q_1$. Suppose, for whatever reason, price falls to $P_2$. Without government involvement, due to the price drop, output falls to $Q_2$ and exports also fall. Total revenue declines by $(P_1 - P_2)(Q_1 - Q_2)$.

What happens if government responds to the drop in price by giving to farmers a supposedly decoupled payment to make up for the lost revenue? If the program is truly decoupled, then output responds to only price $P_2$ and remains at $Q_2$. However, if the program is not decoupled, output increases beyond $Q_2$ in response to government payments. We have shown a case in Figure 2.4 where production increases from $Q_2$ to $Q_2'$ due to government payments. Trade also increases by the amount given the fixed price $P_2$. In essence, the government payments have created a new supply curve $S'$. (The elasticity of the new supply curve associated with a government program is a function of the type of program and the size of payment. Supply becomes more inelastic as the size of the payment increases.) Note that the supply curve $S'$ is drawn with reference to price $P_2$. Alternatively, output $Q_2'$ corresponds to price $P_2'$ on the original supply curve $S$, but this is how a nondecoupled program works. Farmers perceive part and/or all of the payment as net addition to market price (i.e., $P_2' - P_2$). Farm payments thus get bid into the price of land. The corresponding Ricardian rents from the farm program are given by the cross-hatched area in Figure 2.4.

From a trade perspective, nondecoupled payments are not trade neutral. In Figure 2.4, exports are higher than what would be the case if markets were allowed to adjust. Thus, by putting most or all farm programs into the green box category, even though they may not be decoupled, should not be any indication of moving to freer trade. Large green box payments can be as trade distortionary as large explicit price supports and/or export subsidies.

Figure 2.4 can also be used to highlight a trade issue even if programs are totally decoupled. For example, what does the supply response for cereals in the European Union look like? Some argue that the supply curve is highly inelastic (i.e., $S'$ in Figure 2.4 instead of $S$). In this case, even if European support for agriculture were to be removed, output would contract very little, thus there would be no major price impacts and hence gains to major grain exporters such as Canada, Australia, and Argentina.

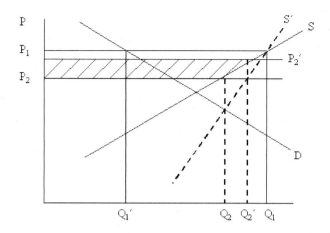

FIGURE 2.4. Decoupled payments

## R&D AND TRADE

One of the frustrations of many farmers in North America is the continued erosion of farm prices in the presence of GATT, the WTO, NAFTA, and the Uruguay Round of talks. The decline in prices, however, is due to many factors, some of which affect prices negatively.

Consider Figure 2.5 where the gains from trade liberalization are discussed in the context of the effect of technological change—one of the factors that depress prices at the farm level. The free trade price is $P_f$ prior to technological change (domestic supply and demand are given by $S$ and $D$ and total demand is $TD$). We introduce trade distortions such that the total demand is $TD'$, with the corresponding price and quantity of $P_o$ and $Q_o$. Consider the effects of both technological change and moving toward freer trade. The partial removal of trade distortions shifts total demand to $TD^*$. Technological change shifts supply to $S'$. The resulting price is $P_r$ and the corresponding output is $Q_r$. Note that prices have fallen and are below those that existed without technological change and trade distortions present (compare $P_o$ to $P_r$). However, even though prices have fallen, producers are better off than prior to technological change by an amount equal to $(P_r a'b' - P_o ab)$. There are also gains from freer trade equal to the cross-hatched area $P_r a'cP_n$. Also, note that with technological change, removing trade distortions generates larger trade gains than in the absence of technological change. Without technological change, the gains from trade from re-

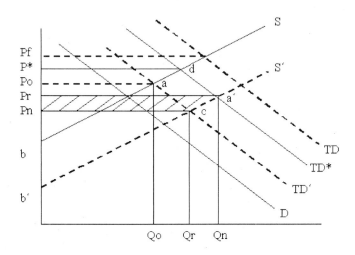

FIGURE 2.5. Technological change and trade gains

moving distortions (that is, moving to $TD^*$) are $P^*daP_o$. With technological change, the gains become the cross-hatched area.

Consider further the case of an importer (Figure 2.6). The free trade price is shown as $P_f$, and supply and demand are shown as $S$ and $D$. Imports total $Q_1Q_2$. In the presence of price supports, with technological change, supply shifts to $S'$; the county now only imports $Q_2^*Q_2$. If price supports are removed, imports increase to $Q_2Q_3$. Imports, however, are well below those prior to technological change with no distortions present.

China was once a significant importer of wheat. In recent years, wheat imports virtually disappeared, causing a significant impact on U.S. wheat prices (Figure 2.7). This was due, in part, to high internal price supports and rapid technological change. Even if price supports are lowered, the increase in Chinese wheat imports may be disappointing for U.S. producers, given the presence of rapid technological changes in agriculture. According to Schmitz and Furtan (2000):

> Grain markets are ever changing. Canada relies less heavily on the Centrally Planned Economies (CPE) as major buyers than it did in the past. Not only has the Former Soviet Union market shrunk considerably for Canadian wheat imports, but it is also very volatile. Also, China causes substantial instability in the world wheat market. For example, Chinese total wheat imports were 3.2 mmt in 1976/77, peaked in 1991/92 at 15.7 mmt, but fell to 2.0 mmt in 1997/98. (p. 108)

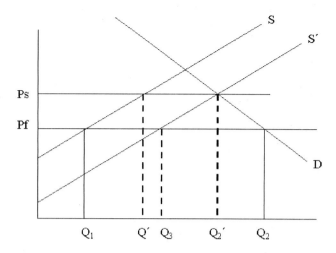

FIGURE 2.6. Importers, price supports, and technological change

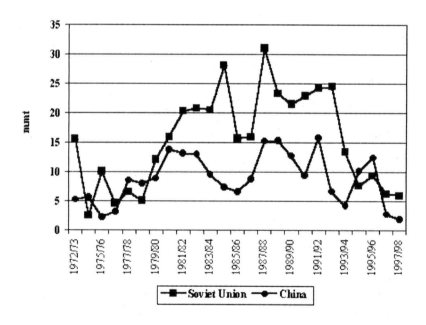

FIGURE 2.7. Wheat imports by the former Soviet Union and China, 1972/73-1997/98. (*Source:* Schmitz and Furtan, 2000.)

## MULTINATIONALS AND VERTICAL MARKET STRUCTURES

A great deal of international trade is now conducted by multinationals. Some of these firms merely trade. Others both add value to products and trade internationally. Still others are involved in the actual production of the commodity in many countries around the world and carry out trade in these products. The multinational scope of many agricultural firms has important implications for freeing up international agricultural trade. Consider, for the moment, the case discussed previously in which the United States brought charges against Canada under U.S. countervail and dumping laws (Schmitz and Furtan, 2000). Perhaps a reason why the United States ruled in favor of Canada was because of lobbying by multinationals to keep the border open, given that multinationals in the United States would be hurt if the ruling went in favor of U.S. beef producers. The majority of beef cattle fed in Canada are in Alberta. A major packing plant in Brooks, Alberta, is owned by Iowa Beef Packers Incorporated (IBP). In addition, a world-scale beef-packing plant in High River, Alberta, is owned by Cargill Incorporated under the name of Excel. Clearly, closing the borders to Canadian beef exports would have a significant negative impact on the profit picture for these multinationals. Generally, firms such as Cargill Incorporated push for free trade, since their profit picture is positively correlated with the volume of international trade they are able to conduct. This is a point often neglected in academic discussions of gainers and losers from free trade.

As another example, multinationals continue to influence U.S. sugar policy. Domestic import quotas restrict the importation of refined and raw sugar into the U.S. market. Support for this program of quotas varies by producer group (Moss and Schmitz, 1999). For example, it appears that the U.S. Sugar Corporation (a major sugar producer in Florida) supports more restrictive import quotas than does Flo-Sun (which produces in Florida and in the Dominican Republic). Flo-Sun ships sugar from the Dominican Republic into the United States under preferential quota treatment, receiving the internal U.S. sugar price for exports. Therefore, they attempt to maximize returns jointly from domestic production and from production in the Dominican Republic. The quota levels which Flo-Sun supports are above those that the U.S. Sugar Corporation supports, largely because the latter produces sugar only in the United States.

Many sectors are now becoming vertically integrated either through direct contracts between processors and growers, or by direct ownership of all stages from production to marketing. Perhaps this increasing degree of integration favors lobbying activities for free trade. On the other hand, where vertical integration is not present, key players in the marketing chain with both economic and political clout can block trade reform. For example, in the production and marketing of cotton in Turkey, there is very little vertical coordination. A major cotton textile mill is owned privately and buys the majority of its cotton on the Turkish cotton exchange. When examining cotton policy in Turkey, it is clear that certain trade instruments are in place (such as export subsidies, taxes, and import quotas on raw cotton), which maximizes returns to the private processing sector (Schmitz et al., 1999). Many of these trade instruments are highly protectionist.

The influence of vertical market structures on lobbying for or against free trade is also evident in the U.S. sugar industry. Both Flo-Sun and the U.S. Sugar Corporation can handle additional raw-sugar imports since they have the capacity to refine raw sugar, which is not true for major sugar beet producers nor for sugarcane producers in Louisiana. Therefore the opening up of trade would have a less serious impact on these firms than it would on sugar beet producers and sugarcane producers in Louisiana. The latter do not own any sugarcane refining capacity. Therefore, stronger lobbying efforts against free trade may be forthcoming from Louisiana producers relative to efforts by Flo-Sun.

### GAINERS AND LOSERS

Moving to freer trade will create both losers and gainers. As a result, losers will attempt, through rent seeking, to block free trade agreements (for example, the 1999 demonstrations in Seattle against free trade) (Schmitz, 1988a,b; Becker, Gray, and Schmitz, 1992; Gray, Becker, and Schmitz, 1995). A clear case in agriculture is Canadian supply management, in which producers were very effective in maintaining supply management under CUSTA, NAFTA, and the Uruguay Round. If Canada gave up its supply management system, Canadian producers would clearly lose, at least in the short run. Likewise, if the United States were to give up its sugar, tobacco, and peanut programs, U.S. producers would lose. One often hears at trade meetings that supply management producers in Canada are willing to give up supply management if U.S. producers are willing to give up sugar, tobacco, and peanut programs. However, one should not put a great deal of faith in such statements since, if all these programs were removed, most of the producers directly involved would lose. Therefore, why would they be interested in having their programs done away with? Supply management groups would not form a coalition with sugar, peanut, or tobacco farmers to support freer trade arrangements, since each group is in a no-win situation as far as their particular farm programs being removed. Who then in Canada, for example, would benefit from freer trade in supply management products? The gainers would have to include pork and beef producers along with grain producers, consumers, and processors. What this example suggests is that negotiators, to be successful, cannot be influenced by special interest groups, which in reality is impossible. Consider further policy harmonization between the United States and Canada in grains. If the United States removed its farm bailout programs, certainly Canadian grain farmers would benefit. Canadian grain farmers have little or nothing to offer in return, since government support for their sector is negligible. (But the United States maintains that all of its major farm programs fit the criteria set forth for the green box.)

Removing supply management in Canada and the cotton, peanut, and sugar programs in the United States should benefit consumers, provided, of course, that the complete marketing chain, which includes multinationals, remains competitive. The competitive nature of these sectors is an ongoing debate. Often, free trade policies are lobbied for by processors and handlers, who likely stand to gain the most from freeing up agricultural trade. Someone has to handle the product after it leaves the farm gate. The larger the volume of trade, the larger the gains to domestic processing firms and multinational companies.

In negotiating for freer trade in agriculture, it is not clear what individual countries are willing to give up. Consider Canada for the moment: if supply management is to remain in place, then Canada has very few chips to bargain with, since all of the remaining sectors are not subsidized to any significant extent. To remove supply management would require compensation to producers, which would approximate the transfers producers currently receive from consumers. Many argue that there would be significant gains from trade if this system were to be eliminated. Given the political climate in Canada, and the frustrating experience of western Canadian producers with respect to transfers from the federal government, compensation appears unlikely.

There is a great deal to be gained by non-EU countries if the European Union gives up subsidies, but the problem remains: unless compensation is actually paid to losers in the move to freer trade, then likely very little will happen in any upcoming trade negotiations as far as the EU moving to freer trade in agricultural products. Consider whether the United States will give up sugar, peanut, tobacco, or dairy programs. Likely not, unless compensation schemes are devised. Also, would the United States discontinue its use of bailout programs? The answer is no. It is interesting political rhetoric to argue that the world has a great deal to gain from freer trade in agriculture. But a significant question remains: Who really wants free trade anyway? My reading of the empirical literature suggests that many consumers, processors, and agribusinesses generally would gain from freer trade, but many producer groups would lose, especially if many of the current farm programs were correctly identified as not fitting green box criteria. Producers around the world still have political clout in blocking trade reform.

## FUTURE DIRECTIONS

Following the Third Ministerial Conference, held in Seattle in December 1999, the future of multilateral trade negotiations was uncertain. Despite the outcome of the Third Ministerial Conference, progress was made in November 2001 with the Fourth Ministerial Conference, held in Doha, Qatar, which sets January 2005 as the date for completing the majority of the negotiations. Progress is to be reviewed at the Fifth Ministerial Conference, to be held in Mexico in 2003. According to Coleman and Meilke (2000), the Ministerial meetings failed for two main reasons: (1) policy differences, and (2) inadequacies of WTO procedures (p. 34). The WTO could not agree on how to negotiate export subsidies and how to deal with nonagricultural topics, such as the lack of market access, concessions for textiles, and the concerns over the United States favoring WTO enforcement of labor standards. Concerning procedural inadequacy, Director General Mike Moore himself noted that the WTO's decision-making procedures were outdated and were unsuitable for such a large group of country participants.

Countries at the Ministerial meeting fell into three broad categories: the reform group, those favoring the status quo, and the developing countries. The reformers, including the United States and the Cairns group, supported substantial cuts in trade-distorting policies. In contrast, the EU and Japan, which were included in the status quo group, favored a go-slow approach (Coleman and Meilke, 2000, p. 34). Developing countries, which made up the vast majority of WTO members, sought special treatment under both new and old agreements.

Issues over which there is little agreement have been stated succinctly by other authors. According to Marchant (1999), these issues include:

> . . . further reductions in export subsidies, moving towards their elimination; expansion of markets access through tariff reductions and liberalization of tariff rate quotas; further reductions in domestic support moving towards decoupling; stricter WTO disciplines on state trading enterprises (STEs) with increased transparency of STE pricing and operations activities . . . ; and tighter restriction on the use of sanitary and phytosanitary (SPS) measures . . . [also] GMO regulations pose potential barriers to these exports and raise the need for mutual recognition of countries' regulations, harmonization of existing regulations among countries, or by negotiation of an international standard. (p. 198)

According to the USDA (1999a):

> EU consumer and environmentalists' concerns have resulted in measures that could lead to trade disputes in the future. EU concerns

with food safety have been heightened by "mad cow" disease, and, with pressure from political activists and consumers, the European Commission has enacted legislation (labeling products containing genetically engineered material, for example) that has disrupted U.S. exports to the EU. Delays in approving genetically engineered crops have significantly slowed U.S. corn exports to the EU and threaten U.S. exports of soybeans and soybean products. . . . The EU Commission has indicated that animal welfare issues should be addressed in the next round of WTO negotiations. (p. 3)

The following concerns, which resulted from the Ministerial talks, are paraphrased from Coleman and Meilke (2000):

- *Export competition:* reformers and status quo countries could not agree on how to discipline the use of export subsidies. Export credits, not subject to WTO disciplines, could be used in the future to replace explicit export subsidies. The EU wanted WTO disciplines on export credits, which is strongly opposed by the United States. Japan and others wanted disciplines on export taxes, restrictions, and embargoes.
- *Market access:* countries generally disagreed on the extent and timing of tariff reductions.
- *Domestic supports:* the debate centered over whether or not farm programs can be decoupled, and the extent to which it is possible to bring greater discipline to green box and blue box programs. The Australians argue that many green box policies allowable under the WTO are not trade neutral, and that their scope should be narrowed. The EU preferred a wider definition. The Cairns Group pressed for the elimination of the blue box, while the EU and Japan insisted on its continuation. Also, there was little agreement among the major parties over whether the Peace Clause should be extended beyond the December 31, 2002, deadline.
- *Other issues:* debate centered on state trading enterprises, biotechnology, and nontrade issues (multifunctionality). The interest in STEs was high, in part because of China and Russia, who use STEs as a vehicle for domestic support programs. No agreement was reached on how to discipline STEs. No agreement was reached on GMOs. The Europeans advocated policy that allows countries to ban imports of GMO products. Many other countries disagree, and favor a WTO biotechnology working group to determine whether specific disciplines governing trade are needed. The status quo countries strongly advocate that the new agreement reflect the "multifunctionality" of agriculture; this was strongly opposed by those who fear that multifunctionality could be used to justify open-ended support for domestic producers.

In looking at these concerns, one cannot be overly optimistic regarding future progress toward freer trade. One point overlooked in the previous discussion concerns the health of the agricultural economy in many countries. Many commodity prices going into the year 2002 were at an all-time low in real terms. (This situation is not likely to change any time soon.) At the same time, countries such as the United States have counteracted these low prices with significant government subsidies. At the beginning of 2001, U.S. subsidies were at an all-time high. Because of low market prices, caused in part by China's low imports, it is extremely difficult for negotiators to argue for freer trade if there is a general movement toward removing domestic support subsidies as a condition for freer trade (remember that major changes under the 1996 Freedom to Farm Act were possible because of the buoyant agricultural economy at that time). It is easy, however, to negotiate freer trade if in fact a country can continue the use of its domestic support programs. In this case, farmers' welfare is not affected by whether trade is restricted or not. Sumner (2000) has argued that the United States, for example, should negotiate freer trade and not discipline nor necessarily focus on domestic support programs. He argues that reduced export subsidies or greater import access have substantial trade benefits, even if farmers are compensated with payments or price supports. He uses several examples, including the impact of reforming the U.S. sugar programs by replacing tariffs and quotas with deficiency-type payments. This proposal was discussed much earlier by Leu, Schmitz, and Knutson (1987). All this assumes, of course, that one can design decoupled farm programs. Otherwise, removing trade barriers such as quotas and tariffs need not necessarily help, if they are offset by trade-distorting domestic programs. As argued earlier, programs are very difficult to decouple. Many of the programs that are now in the green box are not in fact decoupled—a position also held by the Australians.

## CONCLUSION

We have outlined some of the key roadblocks to freer trade in agriculture. Policy harmonization among countries has to focus on the issue of farm programs and their impact on trade. (NAFTA and CUSTA should be enforced.) The extent to which farm programs fit the green box category is open to debate and should be the subject of an entire conference. In addition, analysis is badly needed regarding the Free Trade Area of the Americas accord and how such an accord would influence the U.S. position on free trade. Freeing up trade among the countries comprising this block could have significant impact on U.S. agriculture. The role of special interest groups in blocking the move to freer trade cannot be overemphasized. There are losers and gainers from freer trade. Producer groups that lose obviously are going to attempt to block trade unless compensatory schemes are implemented. Although there seems to be little agreement, some argue that the biggest losers from freer trade in agriculture are the majority of farm interests. Major beneficiaries include consumers and taxpayers. Given the political climate and the mood among U.S. farmers at least, special interests seem to be lobbying the U.S. government for increased protectionism. It seems clear that this mood will impact what, in my opinion, will be a slow movement in re-entering the free-trade negotiations arena. Likely, major changes in farm policy, for example, the major U.S. farm policy to be implemented in 2002/2003, will impact freer trade outcomes more than will formal world trade negotiations. It is possible that the 2002/2003 U.S. farm program may move the United States toward protectionism rather than toward freer trade.

# REFERENCES

Agriculture and Agri-Food Canada. 1998a. An Examination of Nearly Green Programs: Case Studies for Canada, the United States and the European Union. <http://www.agr.gc.ca/policy/epad/english/pubs/wp-tp/tms/99037wp/toc.htm>.

Agriculture and Agri-Food Canada. 1998b. Green Box Criteria: A Theoretical Assessment. (January). <http://www.agr.gc.ca/policy/epad/english/pubs/wp-tp/tms/99036wp/toc.htm>.

Becker, Tilman, Richard Gray, and Andrew Schmitz. 1992. *Improving Agricultural Trade Performance Under GATT*. Wissenschaftsverlag Vaul Kiel KG.

Coleman, Jonathan and Karl Meilke. 2000. "Sleeping After Seattle." *Choices*. Third Quarter: 33-37.

Gray, Richard, Tilman Becker, and Andrew Schmitz. 1995. *World Agriculture in a Post-GATT Environment*. Regina: University of Saskatchewan, University of Extension Press.

Haniotis, Tassos. 2000. "Agriculture in the WTO: A European Union Perspective." Paper delivered before the Commission on 21st Century Production Agriculture. Washington, DC (March 7) <www.eurunion.org/news/speeches/2000/000307th.htm>.

Leu, Gwo-Jiun M., Andrew Schmitz, and Ronald Knutson. 1987. "Gains and Losses of Sugar Program Policy Options." *American Journals of Agricultural Economics*. 69(3): 591-602.

Marchant, Mary A. 1999. "Trade Negotiations and Southern Agriculture." *Journal of Agricultural and Applied Economics*. 31(2): 185-200.

Miner, William M. 1999. "The Cairns Group: Negotiating Priorities and Strategies." *2000 WTO Negotiations: Issues for Agriculture in the Plains and Rockies* (pp. 141-155). Young, Linda M., James B. Johnson, and Vincent H. Smith (Eds.). Trade Research Centre. Bozeman: Montana State University.

Moss, Charles B. and Andrew Schmitz. 1999. "The Changing Agenda for Agribusiness: Sweetener Alliances in the 21st Century." Paper presented at Sweetener Markets in the 21st Century, Miami, Florida, November 14-16.

Schlueter, Jonathan F. 1998. "American Consumers Don't Need Protection from Wild Oats." *Feedstuffs*. 70(44): 12. Reprinted by permission.

Schmitz, Andrew. 1988a. "GATT and Agriculture: The Role of Special Interest Groups." *American Journal of Agricultural Economics*. 70(5): 994-1005.

Schmitz, Andrew. 1988b. "United States Agricultural Trade: Where Are the Gains?" *Western Journal of Agricultural Economics*. 13(2): 357-364.

Schmitz, Andrew. 1995. "Supply Management and GATT: The Role of Rent Seeking." *Canadian Journal of Economics*. 43: 581-586.

Schmitz, Andrew. 2000. "The Millennium Round of Multinational Trade Negotiations." Forthcoming. *Journal of Agricultural and Applied Economics*. August.

Schmitz, Andrew, Erol Cakmak, Troy Schmitz, and Richard Gray. 1999. *Policy, State Trading, and Cooperatives in Turkish Agriculture*. Report for Agricultural Economics Research Institute, Ankara, February.

Schmitz, Andrew and Hartley Furtan. 2000. *The Canadian Wheat Board: Marketing in the New Millennium*. Regina: Canadian Plains Research Centre.

Schmitz, Andrew, Hartley Furtan, and Katherine Baylis. 1999. "State Trading and the Upcoming WTO Discussions." *Choices*. Second Quarter: 30-33.

Schmitz, Andrew and Theodore Schmitz. 1994. "Supply Management: The Past and the Future." *Canadian Journal of Agricultural Economics*. 42(2): 125-148.

Sumner, Daniel A. 2000. "Domestic Support and the WTO Negotiations." *The Australian Journal of Agricultural and Resource Economics*. 44(3): 457-474.

Tweeten, Luther. 1979. *Foundations of Farm Policy*. Second edition. Lincoln: University of Nebraska Press.

USDA. 1999a. *The European Union's Common Agricultural Policy: Pressures for Change*. Situation and Outlook Series. Economic Research Service. WRS-99-2. October.

USDA. 1999b. "Sugar and Sweetener." Economic Research Service, May.

Chapter 3

# Issues in the WTO Agricultural Negotiations: An Overview

Tim Josling

## *INTRODUCTION*

A new round of agricultural negotiations was launched on March 24, 2000, in Geneva. The WTO Agriculture Committee, meeting in special session, decided on a timetable for the first phase of the talks. Starting the talks, of course, is easier than bringing them to a successful conclusion, but after the chaos in Seattle any good news was welcomed in trade circles.

The breakdown of the Seattle Ministerial in December 1999 posed an interesting dilemma for the agricultural talks. The postponement of the start of a new round of trade negotiations should not have delayed the start of the agricultural round: the Uruguay Round Agreement on Agriculture (URAA) mandated further negotiations by the end of 1999.[1] But the lack of agreement on a more general round left the built-in agenda to languish without political impetus. The time since Seattle has been taken up with the task of gearing up for negotiations in the absence of such a stimulus.

This is not to imply that the WTO members are not well prepared for the next round of agricultural talks. In many ways the ground has been thoroughly tilled and the seeds scattered. Countries began to formulate their overall approach to the round, and to define their expectations for agriculture, before the Seattle meeting. Many tabled papers in the WTO General Council in the weeks before the Ministerial outlined their positions. One country, Canada, went further down the road, and issued a more specific statement of aims and approaches.[2] In addition, over seventy papers have been prepared in the context of the "analysis and information exchange" process authorized at the Singapore WTO Ministerial and conducted informally by the Committee on Agriculture.[3] In other words, countries were ready to start the agricultural talks if they had gotten the appropriate signal from Seattle. This does not mean that the talks would have been easy even with an enthusiastic launch of a major round. However well prepared the negotiators were, the prospect of stalemate or of minimal progress on agriculture was always in the cards. But the issue now is whether any significant progress can be achieved in agriculture in the absence of a more general set of trade talks.[4]

The next round of agricultural talks will be different in many respects from the Uruguay Round or its predecessors. In some ways the task of the negotiators will be more clear-cut, in large part because of the transparency introduced by the URAA. Tariff levels are easier to negotiate than nontariff barriers, and the defined commitments on export subsidies and domestic support can be subject to further cuts without revisiting the definitions. But clear-cut tasks can also focus opposition. There are several countries that would prefer not to pursue the path toward a more open trade system for agriculture, or at least not be pushed in that direction by international pressure. Moreover, as always, negotiations will take place in the context of contemporary events. These events could overshadow and even derail the talks. The agenda already has been influenced by a number of issues that were not on the table during the Uruguay Round. There is no reason to believe that the agenda will stop shifting with the formal start of talks.

This chapter discusses both the substantive agenda for the talks and the frictions that are emerging as countries take their stands. It is too early to predict the outcome of the discussions, but some idea of the timetable can be inferred from the political calendar. As always, a number of more political factors will condition the pace of the talks, including the ability of the European Commission to secure for itself a constructive mandate for negotiation over the objections of some of the member states; the present lack of "fast track" trade negotiating authority in the United States; the accession of China to the WTO; and the newly emerging determination of developing countries to be full partners in the WTO. One should also not forget the impact of the state of commodity markets, which can have a marked effect on the progress of agricultural talks as it impinges on the perceptions and policies of individual countries. Each of these could have a significant bearing on the agricultural talks.

## BUILDING ON THE URUGUAY ROUND

The URAA marked a turning point in agricultural trade policy. Prior to the URAA, national policies were largely unchecked by trade rules. Nontariff barriers were the norm, implying a lack of transparency in trade and little incentive for the development of competitive exports. Export subsidies made it difficult for competitive exporters to develop markets. Domestic subsidies tilted the playing field in favor of less efficient producers at home and limited the scale of specialization. The URAA established new rules that radically improved the agro-food trade system. Nontariff barriers were replaced by bound tariffs. Export subsidies have been limited in both the expenditure and the quantity that can benefit from subsidies. Domestic support is now categorized according to whether it is minimally distorting (green box), linked to production controls (blue box), or output-increasing (amber box), and this last is subject to agreed limits.

It is widely accepted that the Agreement on Agriculture did little to liberalize trade in agricultural products and improve market access. Tariffs on agricultural goods are still on average about three times as high as on manufactured goods and continue to distort trade.[5] The process of "tariffication" has produced a number of tariffs bound at such high levels that it is difficult to see how they could be reduced by conventional tariff-reduction techniques. Where tariff rate quotas (TRQs) were negotiated to pry open these markets a little, the prospect of quota rents has led governments to agree to a network of bilateral deals and firms to become concerned with market shares. This has in turn exacerbated the problem of competition between state trading enterprises and the private trade. Export subsidies still exist and are in effect legitimized to the extent of their incorporation in country schedules. The domestic farm policies of the major industrial countries have been required to make only relatively minor changes to bring them into conformity with the Agreement.

The overall objective of the next round of agricultural talks will clearly be to continue the progress made at the Uruguay Round. This implies negotiations on improved market access, further constraints on export subsidies, and, if exporters get their way, some tightening of the rules for domestic support. But if this is the core agenda, a number of other issues have emerged as a result of the experience with the Uruguay Round Agreement that can be thought of as "extensions" of the URAA core agenda, such as the administration of TRQs and the issues of state trading and of export restrictions. As if this was not enough, several other items are clamoring for a place on the agenda. Some of these issues will be dealt with in parallel to the agricultural talks, though not necessarily by the same committee, and will be part of whatever package emerges. These parallel topics include the sensitive questions of health and food safety along with a number of environmental issues relating to agriculture and biotechnology. Also important to agriculture is a resolution of the issues of regional trade agreements and preferential trade arrangements, but these will not be discussed here.

## THE AGENDA FOR THE AGRICULTURAL NEGOTIATIONS

The agenda for the upcoming WTO negotiations on agricultural trade is becoming rather congested. Those interested in the continued reform of the trade rules emphasize the "core" issues such as the liberalization of market access, the reduction or elimination of export subsidies, and the containment of domestic support that is given in a way that distorts trade. The general presumption is that the Uruguay Round set the rules but did not go far toward reducing protection. The main task of the next round is therefore to make a significant step toward the opening up of agricultural markets. These traditional trade policy issues are supplemented by others that have largely appeared as a result of the URAA. These include the administration of TRQs and the activities of State Trading Enterprises (STEs), both those that control imports and those that are engaged in export activities.[6]

For those concerned with environmental and food safety issues, a rather different set of agenda items is important. These are closely related to agricultural trade though technically outside the Agreement on Agriculture. These issues include the trade conflicts over sanitary and phytosanitary (SPS) measures and the potential conflict over genetically modified foods (GM foods), as well as questions of intellectual property rights. For developing countries there are other issues as important as those in the core, including the question of whether to press for continued "special and differential treatment" in agricultural rules. For countries more concerned with food security, issues such as the need to control the use of export restrictions in times of shortage and the ability of countries to take action against imports that threaten to disrupt markets are important, as is the future of trade preferences.

## *MARKET ACCESS*

The market access negotiations will be at the heart of the next agricultural round. The talks will not be a success unless a substantial step is taken to reduce the high levels of agricultural tariffs. With varying degrees of enthusiasm, countries have endorsed the objective of improving market access. The United States has called for an "ambitious" target for expansion of market access: the EU admits that its export interests would be served by an opening of markets but cautions that the process will take time.[7] For the Cairns Group, the negotiations "must result in deep cuts to all tariffs, tariff peaks and tariff escalation."[8] Of the major players Japan is naturally the most reticent, contributing the observation that current tariff levels "reflect particular domestic situations" and that these circumstances should be given due consideration in the negotiations.[9] NGOs are generally less enamored with market access negotiations, associating such liberalization with globalization and the pressure from multinational firms for ever wider markets over which to spread fixed costs. To many of these pressure groups, market access is a part of the problem rather than the solution. Developing countries tend to stress the importance of expanding market access in the products of export interest to themselves.

### Modalities for Tariff Reduction

The techniques of negotiating tariff reductions are well established.[10] One can choose between across-the-board tariff cuts or formulas that cut tariff peaks. One can focus on individual sectors (zero-for-zero arrangements) or agree on comprehensive coverage. One can use the "request and offer" method for identifying demands for market access, multilateralizing the results. One can attempt to reduce effective protection by making sure that processed-good tariffs come down at the same rate or faster than those of raw materials. These methods of market access each have some merit but might not be adequate in themselves. This suggests that negotiators might try a "cocktail" of the various modalities.[11]

### Other Tariff Issues

In addition to the task of tariff level reduction, two other aspects of tariff rules will be discussed. One is the common phenomenon of bound rates of tariff that are considerably higher than the rates that are applied.[12] There have been suggestions that these gaps be reduced, for instance by binding the applied rates. But this causes understandable problems for the countries involved, which will argue that the applied rates have not been negotiated in the WTO and therefore to bind these rates would be unfair to those who have undertaken unilateral liberalization. On the other hand, for those countries with such gaps, reducing the bound rate toward the applied rate is a way of getting "credit" for actions already taken.

The other aspect of tariff policy is the form of the tariff. The URAA mandated a tariff-only regime, but allowed some countries to concoct complex tariffs that involve reference prices and compound rates.[13] Moreover, the Blair House agreement between the United States and the EU obliged the EU to impose a maximum duty-paid price for cereals that acts very much like the variable levies which were outlawed in the agreement. Many countries also would like to insist on the use of ad valorem tariffs rather than specific duties, which have a somewhat more protective impact when prices are low. The United States is calling for a simplification of complex tariffs; whether any country will take aim at de facto variable levies and specific tariffs is not so clear.

## *TRQ EXPANSION*

One direct way to tackle the problem of the high levels of tariffs resulting from tariffication is to expand the guaranteed market access that forms a part of the provisions of the Agreement on Agriculture. Some position papers (though not that of the United States) mention the importance of expanding TRQs in the next round. The Cairns Group paper says that "trade volumes under tariff rate quotas must be increased substantially." Other countries suggest further improvements in the TRQ system, in addition to the administration of the quotas (discussed below). Canada argues for the elimination of the within-quota tariff whenever the above quota tariff is prohibitive (presumably to ensure that the quotas are filled, rather than merely increasing quota rents at the expense of government revenue). The same paper suggests the introduction of a TRQ whenever tariffs are higher than a specified level, and increasing the product specificity of TRQs.

### *Administering TRQs*

As a number of countries recognize in their position papers, the issue of developing a more uniform system for the administration of the TRQs is one of the most urgent tasks for the new agricultural round. TRQs for agricultural imports have created a new wave of governmental interference with trade through licensing procedures and provided a playground for rent-seeking traders—who will in turn have an incentive to lobby for the continuation of the high above-quota tariffs. The question is how to prevent the TRQs from interfering any more than necessary with the competitive development of trade.

One answer to the question lies in the method of allocation. In some cases allocation is done on a government-to-government basis, usually in accordance with historical market shares. But this perpetuates distortions in trade. To allocate the TRQs to the exporting country government, as is done for instance in the case of U.S. sugar imports, implies a deliberate attempt to influence the pattern of trade in favor of the recipient countries. This has in the past been done to target development aid or reward political friendship. Such nonmarket allocation schemes may have had their purpose. They do not, however, promote the competitive trade system that is the fundamental goal of the WTO. Efficient producers can make no headway against the assured market shares of the quota holders. Even allocating TRQs by country based on historical market shares does not ensure that the sourcing of supplies for the importer bears any necessary relation to the competitiveness of the supplier. The simple solution to the efficiency problem is to allow quotas to be auctioned, as has been suggested at various times.[14] This would seem an economically sensible solution to the problem of the capture of rents and to counteract the incentives to keep the system in place. But this is also a reason why exporters in particular are likely to resist such a move.

The type of allocation mechanism that causes most problems, however, is that which gives the import rights to domestic concerns. Exporters feel that they are neither getting assured access (as the agency or firm concerned can choose not to import the product, leading to underfill of quotas) or that they are not gaining the benefit of the access (in essence not receiving any of the quota rent). In instances where the TRQs have a deliberate purpose, such as the EU arrangements with the African, Caribbean and Pacific countries (ACP) and those of the Mediterranean basin, capture of rents by the importing firms negates much of the benefit of the scheme. When competing domestic producers receive the import entitlements (as has happened in a few cases) then the market access inherent in the TRQs may be elusive.

### State Trading Importers

The issue of state trading enterprises that have special or exclusive rights in import markets can be thought of as an extension of the problem of market access. Under WTO articles, state trading importers are not supposed to grant more protection than that given by the bound tariff (Article II: 4, GATT 47). Countries could, however, go further than just ensuring that state trading importers do not give more protection than the bound tariff. It would be possible, for instance, to link the administration of the TRQs with the import operations of state traders, perhaps converting the TRQ into an obligation to import rather than an opportunity. This could reduce the suspicion that STEs might be responsible for the underfill of the quotas. At the other extreme one could mandate that all (or a share) of the TRQ be marketed through private channels, thus providing some competition for the STE and allowing price and markup comparisons to be made.

## CURBING EXPORT SUBSIDIES

If the high level of protection sets agriculture apart, the widespread use of export subsidies is perhaps the most disruptive element in the operation of world markets. The practice of subsidizing exports of agricultural products has been constrained by the Uruguay Round, but most of the subsidies are allowed to continue in a reduced form. Countries that import agricultural products have been the gainers in economic terms from the subsidies, but even among these countries the disturbance of the domestic market has often caused problems. In the next round of negotiations, it will be more difficult than ever to persuade countries who export agricultural goods with little or no subsidy to allow countries such as the EU and the United States to continue their market-distorting practices.

### Reducing Export Subsidies

A further push to rein in these subsidies is high on the agenda of the Cairns Group, apparently supported by the United States. The Cairns Group paper declares "there is no justification for maintaining export subsidies." The United States says that members "should agree to pursue an outcome that will result in an elimination of all remaining export subsidies." Canada adds that export subsidies in agriculture should be eliminated "as quickly as possible." Developing countries also are generally in favor of the elimination of export subsidies.[15] Only the EU will have great difficulty in agreeing to the dismantlement of export subsidies, though it has already come under considerable pressure to do so.[16]

The simplest way to continue the process of reducing the incidence of export subsidies would be to extend the schedule of reductions agreed in the Uruguay Round. As with the market access improvement, this could be done using the same base. This would imply constraining the expenditure on such subsidies by another 36 percent, thus removing 72 percent of the subsidy expenditure that was used in the base period. Continuing the quantity restriction would imply that 40 percent of the volume of subsidized exports would have been removed from the market over the two periods of reform. But since the remaining 60 percent would have to be subsidized with only 29 percent of the expenditure, the disruption that could be caused by such subsidies would be significantly reduced.

The continuation of the process of reduction would be constructive, but elimination of export subsidies altogether would clearly have significant advantages. But the prerequisites for dispensing with export subsidies are a renewed confidence in world markets, with firmer and more stable price levels for the major products, and reduced dependence on intervention buying in domestic policies. The former condition itself depends on the success of the agreement in increasing trade and reducing protection; removing export subsidies may be the only way to create the conditions under which they are not needed. As for domestic programs, it is possible that practice and sentiment in the EU may have moved further away from the use of market support policies to other instruments by the end of the negotiations. If these conditions were met then a new set of negotiations could, say, set a target to phase out export subsidies over a seven-year period.

## Disciplines on Export Credits

Beside the question of export subsidies, several problems remain in the area of export competition. In the Uruguay Round, export credits were declared to be a form of export subsidy, but it did not prove possible to agree on constraints. The Organization for Economic Cooperation and Development (OECD) countries have negotiated a code for non-agricultural export credits that puts limits on credit terms and the length of credit extension, but it has not been possible to include agriculture in this agreement. This leaves this topic as one to be dealt with in the next round, though some countries have indicated that they do not wish to "pay twice" for getting rid of such policies. It should be possible to agree on the allowable terms for such credit, and hence be able to calculate the magnitude of the subsidy that is involved if softer credit terms are offered. The best way to deal with the subsidy equivalent of such concessionary credit is to charge it against the export subsidy constraints in the schedules.

## State Trading Exporters

The quantification of export subsidies and their reduction has left more visible the distinction between countries where exports are privately sold and those where a parastatal controls such exports. There is widespread concern in countries where private firms do the trading that the state trading enterprises can obtain cheap credit from their governments, offer better terms to buyers, and generally compete unfairly with the private trade. To the extent that these practices could be labeled as export subsidies, the issue is one of monitoring and transparency. But some commonly used devices such as price pooling (giving the producer an average price over several destinations or time periods) are also seen as giving the producer an unfair advantage. It might therefore be a matter for negotiation as to whether any constraints need be placed upon STEs with regard to their producer pricing policies.

The question of single-desk selling agencies for agricultural products is at the heart of this issue. On this there are some clear conflicts between the exporters. The United States has indicated that it would like to "strengthen rules . . . disciplining activities of state trading enterprises." The Cairns Group (with Canada as a member) is tactfully silent on the issue, but Canada (in its "Initial Position" document) states unequivocally that it wishes to "maintain Canada's ability to choose how to market its products, including through orderly marketing systems such as supply management and the Canadian Wheat Board." It attempts to head off a confrontation with the United States by indicating that it will "not engage in sterile debates over alternative marketing philosophies," though it also indicates a willingness to "discuss any factual concerns" over "alleged trade effects of orderly marketing systems." Demonstrating that the best defense is often a good offense, Canada adds that it will "seek to ensure that any new disciplines proposed to deal with the perceived market power of [single-desk sellers] apply equally to all entities, public or private, with similar market power." What holds for the CWB must hold for Cargill as well![17]

Different marketing practices among exporters are inevitable, and not in themselves undesirable. But international guidance is needed as to which practices of parastatal export agencies are consistent with agreed conditions of competition and which distort that competition. Now that the more clear-cut kinds of export subsidy have been identified and included in the country schedules of allowable subsidies, the main task of the negotiations will be to clarify the definition as regards the actions of state trading exporters.[18] This would ensure that such actions as dual pricing and price pooling, if deemed to be hidden subsidies, could be counted against the schedule for that country.

### Export Restrictions

In the next round, importers are likely to lead a movement to constrain the ability of exporters to restrict supplies. After all, restraints on exports are no less inconsistent with an open trade system than restraints on imports. Export taxes should be included under the same qualifications as quantitative restrictions. The argument has already surfaced in connection with the Food Security Declaration appended to the Uruguay Round Agreement (the Ministerial Decision on Measures Concerning the Possible Negative Effects of the Reform Program on Least-Developed and Net Food-Importing Developing Countries). It seems inconsistent to leave in place the possibility of export taxes and quantitative restrictions that have an immediate and harmful impact on developing country food importers.

The practice of export taxes and export restraints through quantitative controls can conveniently be thought of as an extension of the issue of export competition. Within the GATT export controls are generally disallowed, though export taxes are deemed innocuous. Article XI of GATT 1947 prohibits quantitative export restrictions but makes an explicit exception for "export prohibitions or restrictions temporarily applied to prevent or relieve critical shortages of foodstuffs or other products essential to the exporting contracting party." There is a clear conflict between the ability of exporters to withhold supplies to relieve domestic shortages and the reliability of the world market as a source of supplies for importers. The inclusion of stronger disciplines on export taxes and embargoes is likely to be part of the next round of agricultural talks.

## *DOMESTIC SUPPORT*

It is one of the ironies of the Uruguay Round that, although the biggest conceptual breakthrough was the acceptance by countries that domestic policies were a legitimate concern of trade talks, the actual disciplines imposed on those policies through the reduction of the Aggregate Measure of Support (AMS) were rather weak. The key question for the next round is therefore whether to strengthen or abandon the attempt to constrain domestic policies.

The fact that the AMS constraints have not been binding for the large majority of countries does not mean that the constraints on domestic support have been ineffective. The process of reinstrumentation of domestic support programs, away from those that most impede trade, has begun. The institution of the "green box" has in itself been useful in defining this objective. The attraction to countries of adopting green box policies is both to guard against challenge from trading partners and to avoid being counted toward the AMS. Thus the AMS constraint is of value even if not particularly onerous.

### *Further Reductions in AMS*

Some WTO members put weight on the reduction in domestic support through the AMS. The United States has called for an "ambitious target" to be set for the reduction of support. The Cairns Group points out that the "overall levels of support for agriculture remain far in excess of subsidies available to other industries." But, as with the United States, their target is clearly the trade-distorting (amber box) policies. Canada, however, indicates that it will seek "an overall limit on the amount of domestic support of all types (green, blue and amber)." This could prove difficult. The EU has announced that one of its objectives is to defend the blue box (in essence the compensation payments under the MacSharry and Agenda 2000 reforms) so as to avoid challenge to these policies and their scheduled reduction.[19] It missed the chance of changing the nature of these payments to make them compatible with the green box criteria, though this could come at some stage in the negotiations.

The AMS constraints are acknowledged to be the least effective of the Uruguay Round bindings. But this does not mean either that they will not be useful in the future or that a continued reduction would not be appropriate. A continuation from the same base would be a relatively modest move, and yet even that will eventually result in 40 percent of the "coupled" domestic support having been removed or converted into less trade-distorting types of program. But it would be even more effective to catch up with the reductions in import barriers and export subsidies. Thus one could envisage an agreed reduction of (say) 52 percent in the expenditure on price-related policies.

### The Future of the Blue Box

The "blue box" containing the U.S. and the EU direct payments which were granted exemption from challenges under the Blair House Agreement was a creature of its time, necessary to get agreement to go ahead with the broader Uruguay Round package. It is, however, still a somewhat awkward bilateral deal not appreciated in other parts of the world. Such an anomaly could possibly be removed in the next round. The policies of the United States and the EU themselves are changing for internal reasons. The U.S. Fair Act goes further than ever before to make the payments to farmers decoupled from output and therefore compatible with the green box. The EU has considered a similar move as a part of the continued reform started in 1992, as a way of making the CAP consistent with enlargement, but for now the idea has been shelved.[20] The task for the New Round will be made much easier if the EU and the United States have both modified their payments such that they meet the conditions laid down in the green box. The blue box could then be emptied and locked.

### Redefining the Green Box

The green box presently contains a number of policy instruments that, while probably less trade distorting than price or income supports, still encourage an expansion of output. Sometimes they are related to such otherwise reasonable programs such as crop insurance, but incidentally increase the incentive to produce by reducing risk. Other programs may be indirectly linked with production even though the main reason for payment is not output. This might be true of certain environmental payments, which could lead to an increase in output. But exporters fear that to reopen the definition of the green box might, however, allow countries to argue that it be expanded to include food security policies and nondecoupled support schemes designed to keep farming in certain areas.

This issue of the size of the green box appears to be where much of the prenegotiation rhetoric is targeted. The argument is usually shrouded in terms of the "multifunctionality" of agriculture. The concept of multifunctionality is not in itself particularly novel, as agriculture has always played a complex role in rural societies, and rural areas have a vital place in national social and political life. But the EU has latched onto the concept as a way of both providing cover in the WTO for policies which it would like to maintain and also providing a rationale for paying farmers in ways that are not tied to commodity output. Exporters are trying to neutralize any impact that the idea might have by pointing out that multifunctionality is neither restricted to Europe (though the EU Commission tends to link it to a European farming model, by implication different from the system of farming in competitor countries) nor indeed to agriculture. Importers are trying to link it to the "nontrade concerns" that are mentioned in Article 20 or the URAA as requiring consideration in planning the reform of the trade system.

The basic question remains: What does multifunctionality mean for trade policy? On the one hand, it could merely be a recognition that a variety of programs will be maintained in most societies that target specific aspects of rural life. For the trade system to be seen to rule out such programs would seem to be as risky as seeming to go against concerns of human health and animal welfare. On the other hand, if trade-restricting policies were to become the accepted instrument for maintaining multifunctionality, then that could signal a regression to the time of expensive commodity market distortions. The green box was intended precisely to deal with such rural concerns. It would be better to confirm the criteria for the green box and encourage multifunctional policies to conform rather than opening the green box up to be a repository for an assortment of production related payments.

One change in the constraints on domestic support that will probably be discussed is to make the AMS specific to individual commodities. This was the original intention in the Uruguay Round: it was at the Blair House negotiations between the United States and the EU that the notion of aggregating the AMS over all commodities was introduced—essentially to weaken its impact. The AMS could thus be made more binding at a stroke by defining commodity-specific amounts of coupled price support expenditure that could then be reduced over time.

## HEALTH, SAFETY, AND ENVIRONMENTAL ISSUES

Conflicts arising from different SPS standards have posed problems for the GATT for many years. Under the GATT 1947, sanitary and phytosanitary measures which impinged on trade were covered by Article XX (b), which allows countries to employ trade barriers "necessary to protect human, animal or plant life or health" which would otherwise be illegal so long as "such measures are not applied in a manner which would constitute a means of arbitrary or unjustifiable discrimination between countries where the same conditions prevail, or as a disguised restriction on international trade" (Josling, Tangermann, and Warley, 1996). But Article XX had no teeth: there was no definition of the criteria by which to judge "necessity," and there was no specific procedure for settling disputes on such matters. The attempt in the Tokyo Round to improve on this situation through the Agreement on Technical Barriers to Trade (1979), known as the Standards

Code, also failed. Though a dispute settlement mechanism was introduced and countries were encouraged to adopt international standards, relatively few countries signed the Code, and a number of basic issues were still unresolved.

### The SPS Agreement

Intensive negotiations in the Uruguay Round led eventually to a new SPS Agreement that tried to repair the faults of the existing code. This agreement defined new criteria that had to be met when imposing regulations on imports more onerous than those agreed in international standards. These included scientific evidence that the measure was needed, assessment of the risks involved, and recognition of the equivalence of different ways of testing and sampling. In addition, the dispute settlement mechanism was considerably strengthened under the WTO to make it easier to obtain an outcome that could not be avoided by the losing party. The force of the SPS Agreement comes in part from the more precise conditions under which standards stricter than international norms can be justified and partly from the strengthened dispute settlement process within the WTO. In this regard, much was expected of the panel report in the beef-hormone dispute between the EU on the one hand and Canada and the United States on the other. This was widely seen as a test case for the new SPS Agreement.

The SPS Agreement was reviewed in 1999 by the SPS Committee, which found no reason to suggest modifications. The United States and the Cairns Group do not wish to tamper with a hard-won agreement that has "science" at its core. The EU, however, has let it be known that a few amendments would not be out of place. The desire to build in the reaction to consumer confidence is natural: presumably it could be argued that the beef-hormone case would be rendered moot by a well-crafted clause written into a revised SPS Agreement. The question as to whether the trade system can tolerate regulations that take into account subjective or irrational consumer demands is one of the most contentious issues in trade policy.

### Handling the GMO Issue

One particularly contentious issue that is directly relevant to the global agrifood system is the extent to which the use of genetically modified organisms (GMOs) is harmful to the environment or indeed to consumer health. Concerns with transgenic crops, such as those with herbicide resistance built into their genetic makeup, have centered on the possibility of unpredictable crosses with wild species and hence the development of herbicide-resistant weeds. Those with genetically manipulated insect resistance give rise to concerns about the development of resistant insects and about collateral damage to harmless or beneficial insects. Clearly there needs to be vigilance to avoid the undesirable side effects of otherwise useful technology. Other fears are that consumers that suffer from plant-related allergies may react to the presence of genes from those plants to which they are allergic (Nelson et al., 1999). The most commonly recommended remedy for preventing such problems is adequate labeling, but even this creates problems for public policy.

That the GMO issue will come up in the trade talks is inevitable. How it can be resolved or at least channeled in a way that does not impede other areas of the talks is less certain. The United States has suggested that "additional approaches that address market access issues for biotechnology products" be pursued. Canada has suggested a Working Party to look at all aspect of the GMO issue, presumably including labeling and import restrictions, as a way of dealing with the issue directly. The EU has positioned itself to take the view that existing agreements (such as the SPS and TBT Agreements) may need to be revised in the light of the challenge of GMOs. The exporters have been trying to coordinate their positions, through the Cairns Group and through bilateral talks. The Asia Pacific Economic Cooperation (APEC) ministers have also been discussing the development of a coordinated plan for the regulation of trade in GMOs. The apparent aim at the moment is to isolate the EU on this issue rather to engage it in an international forum.[21]

### Animal Welfare

Creeping up behind GMOs as the next controversial issue is that of animal welfare. Few topics engender such public outrage, at least in northern Europe, and pose more serious problems of international regulation. Farmers in Europe are already having to modify their farming practices to meet new animal welfare standards. They are naturally arguing that it would be "unfair" to leave them to compete with producers who do not have to meet the same standards. The issue will revolve around whether some degree of protection at the border is allowable to compensate for the extra costs, whether border controls can keep out goods produced under conditions considered unsuitable by the importing country, and whether direct assistance can be given to domestic farmers who are burdened by such regulations without such assistance being considered amber box support. In this respect, the animal welfare debate may become part of the discussion on environmental programs, where the same choices apply. But the animal welfare issue could also take on some of the aspects of the GMO controversy, if genetically modified animal products start to be marketed.

## *THE TIMING OF THE NEGOTIATIONS*

A few fixed points give an indication as to the timing of the agricultural talks. Starting the talks does not mean that they will move fast or far without further incentives or deadlines. The WTO Agriculture Committee, meeting in negotiating mode, sketched out a series of steps for the process. The aim was to receive position papers from all members by the end of 2000, in time to allow a review of these positions in March 2001. The earliest that real negotiations could start was at the meeting in June 2001. Meanwhile, the secretariat has updated a series of papers that it had prepared on the implementation of the Agreement.[22]

One of the few deadlines for the talks to end is the year 2003, when the Peace Clause expires. Thereafter, unless the Peace Clause is renewed, the general WTO rules governing subsidies and dumping will apply to agriculture. This will presumably give a useful boost to negotiations if they are not complete by that date. The promise to renew the Peace Clause may also be a useful incentive for countries such as the EU to continue reforms.

The Peace Clause is the most apparent "internal" driving force behind further agricultural trade reform. But the impact of removal of the protection of the Peace Clause depends crucially on the effectiveness of the dispute settlement process. If countries lose their confidence in this aspect of the WTO, as the demonstrators on the streets of Seattle have clearly done, then the threat of action through panels is greatly diminished. This then could be the ultimate impact of Seattle. The agenda for further talks may well be curtailed, but progress could still occur. But the loss of political legitimacy and support for the WTO as an institution, and in particular a weakening of the dispute settlement mechanism, would be a much more fundamental blow. How WTO member governments tackle the problem of mending fences and improving procedures could well be crucial for agricultural as for other trade.

Does this mean that there will be little incentive to finish the negotiations once they have started? The United States in particular is concerned that the incentives to delay be removed. Hence the 1999 proposal, supported both by the United States and the EU, for a time limit of three years.[23] This conveniently coincides with both the expiry of the Peace Clause and the end of the current U.S. Farm Bill, as well as the date that the EU has set for a review of some of its own measures.

Perhaps the main determinant of the timing and ambitiousness of the agricultural talks, however, is the decision as to whether they should be part of a large, multisector negotiation or whether they will be self-contained. Most commentators argue that a negotiation that only included agriculture would be difficult to conclude. Countries that felt that they stood to lose would have no offsetting gains in other areas. However, no agreement has been reached at the time of writing on the scope for the next Round, and so it is uncertain what "package" will be possible.

## NOTES

1. World Trade Organization members agreed "that negotiations for continuing the process [of substantial progressive reductions in support and protection] will be initiated one year before the end of the implementation period [2000]" (Article 20, URAA). Agriculture is thus an integral part of the "Built-in Agenda" which was agreed in the Uruguay Round.

2. "Canada's Initial Negotiating Position on Agriculture," Press Release from Agriculture and Agri-Food Canada, August 19, 1999.

3. A decision has apparently been made to ensure that these useful documents will be made publicly available.

4. The United States was suspicious that the EU was pushing a broad agenda to avoid the spotlight falling on agriculture. The EU, on the other hand, undoubtedly could have moved more easily on agriculture if other sectors had been included. The impact of single-sector negotiations not embedded in a round therefore is to put the focus squarely on agriculture but also to make it more difficult to get a "good" outcome.

5. The gap for industrial countries is even greater, where the ratio of agricultural to industrial tariffs is closer to ten-to-one.

6. Tariff rate quotas were introduced to assure some minimum access in the cases where conversion of nontariff import measures to tariffs (tariffication) led to very high tariff levels, and also to preserve access under bilaterals and preferential schemes. Many of the nontariff barriers were being administered by State Trading import agencies, and the TRQs were designed in many cases to open these markets.

7. The U.S. position is contained in documents WT/GC/W/107, 115, 286, 287, 288, and 290. The EU paper is WT/GC/W/273. The EU paper contains the intriguing statement that it should "pursue an active market access policy with a view to eliminating barriers to entry in certain third country markets."

8. See the Cairns Group "Vision" statement transmitted to the WTO as WT/L/263, and the subsequent communiqué from the Buenos Aires meeting of the Cairns Group, WT/L/312. The Cairns Group members are Argentina, Australia, Brazil, Canada, Chile, Colombia, Fiji, Indonesia, New Zealand, Paraguay, Philippines, South Africa, Thailand, and Uruguay.

9. The Japanese paper is document WT/GC/W/220.

10. For a discussion of tariff reduction options see Josling and Rae (1999).

11. See Josling and Rae (1999) for an elaboration of this technique.

12. The U.S. paper specifically mentions the "lowering of bound rates to eliminate the disparity between applied and bound rates."

13. The EU tariff schedule for fruits and vegetables includes tariffs conditional on whether the offer price is below or above a reference price (IATRC, 1997).

14. Tangermann explores the arguments in favor of auctioning the TRQs (Tangermann, 1997). The issue of auctioning quotas was addressed some years ago by Bergsten and colleagues in the context of U.S. import policy. It is an interesting comment on the lack of economic rationality in trade policy—and the attraction of rents to trading interests—that such a simple device as auctioning quotas has not so far caught on with politicians and policymakers (Bergsten et al., 1987).

15. Pakistan argues for the immediate elimination of "all kinds of export subsidies by the developed countries."

16. As mentioned before, the EU was apparently prepared at Seattle to accept an objective for the agricultural talks that would imply the eventual removal of export subsidies.

17. Of course one can argue about the interpretation of "similar market power," but it remains true that many Canadian farmers claim that even if they were relieved of the obligation to sell to the CWB, they would have to sell to one of a very few U.S. multinational corporations.

18. In this regard, the outcome of the dispute over Canadian dairy exports is useful. The panel report has indicated that the use of special export grades of milk that can be sold at a lower price to processors for export of dairy products constitutes a form of export subsidy under the URAA.

19. As in the case of export subsidies, the EU authorities would not be averse to reducing direct payments over time. However, they have not yet found a politically acceptable formula for doing so.

20. The Agenda 2000 reform of the CAP did not change the conditions for the direct payments to cereal farmers. Use of the land in program crops is still required. If new members are admitted under these conditions, it is difficult to see how the EU could avoid paying the direct payments to their farmers. This would constitute a large part of the budget cost of extending the CAP to new members.

21. Despite fears to the contrary, the GMO issue did not play a major role in the breakdown of the Ministerial. There was disagreement over what should be the relationship of the WTO to discussions under the Biodiversity Convention. But a form of words had been agreed to which both the EU and the United States could accept.

22. Besides the reports on compliance, the Secretariat had to prepare papers on the various aspects of Article 20, such as the effect of the Agreement on world markets.

23. The three-year time horizon was another casualty of Seattle. The draft paper on the agricultural talks, which was never formally agreed upon, would have set a target date for the end of the talks.

# REFERENCES

Bergsten, C. Fred, Kimberly Ann Elliott, Jeffrey J. Schott, and Wendy Takacs (1987). *Auction Quotas and United States Trade Policy,* Institute for International Economics Policy Analysis Series, No. 20, Washington, DC, September.

General Agreement on Tariffs and Trade (1994). The Results of the Uruguay Round of Mulitilateral Trade Negotiations: The Legal Texts, Geneva, June 1994.

International Agricultural Trade Research Consortium (1997). "Implementation of the Uruguay Round Agreement on Agriculture and Issues for the Next Round of Agricultural Negotiations." IATRC Commissioned Paper No. 12, St. Paul, MN. November.

Josling, Tim and Allan Rae (1999). "Multilateral Market Access Issues for the Next Round of Agricultural Negotiations," paper for the World Bank Conference on the Next Round of Agricultural Talks, Geneva, October 1-2.

Josling, Tim, Stefan Tangermann, and Thorald K. Warley (1996). *Agriculture in the GATT.* Macmillan Press, London.

Nelson, Gerald, Julie Babinard, David Bullock, Carrie Cunningham, Alessandro de Pinto, Lowell Hill, Timothy Josling, Elisavet I. Nitsi, Mark Rosegrant, Laurian Unnevehr (1999). "The Economics and Politics of Genetically Modified Organisms in Agriculture." University of Illinois-Urbana Champaign Experiment Station Bulletin No. 809, November.

Tangermann, Stefan (1997). "A Developed Country Perspective of the Agenda for the Next WTO Round of Agricultural Trade Negotiations." Paper presented at a seminar in the Institute of Graduate Studies, Geneva, March 3.

Chapter 4

# Obstacles to Progress in Multilateral Agricultural Trade Negotiations: The European Viewpoint

## David Blandford

It might seem paradoxical that a long-term resident and citizen of the United States should attempt to present the European viewpoint on the current World Trade Organization negotiations on agricultural trade, although I suppose that my British origin goes some way to explaining this. In any event, since I have also written about the U.S. viewpoint on the negotiations, I feel relatively comfortable with this seeming inconsistency (Blandford, 1999). Being asked to provide observations from both sides of the fence might qualify me as a truly objective observer of agricultural policy. My record has been one of even-handed skepticism about the efficiency and effectiveness of agricultural policies on both sides of the Atlantic (Blandford and Dewbre, 1994).

In this chapter, I focus on European Union (EU) policies, rather than those in other European countries. This is justified by the fact that the Union is the leading European player in agricultural trade negotiations. Most other European countries share EU attitudes toward agricultural policy and its relationship to trade policy. Many are waiting to be drawn under the umbrella of that policy as future members of the Union.

I attempt to present the EU viewpoint, but not as an advocate or apologist for the Common Agricultural Policy (CAP). As in all industrial countries, there are major inconsistencies in EU agricultural policies and much to criticize. I limit my criticisms to the relationship between EU policy and its stance on trade and to the problems that EU policies create for achieving progress in the trade negotiations under the WTO.

In order to provide an EU perspective on the obstacles to progress in the trade negotiations, it is necessary to review briefly some recent history. It is difficult to understand the current EU position on the negotiations, and the challenges that the EU faces in concluding these, without some understanding of how the relationship between its domestic agricultural policy and trade policy has evolved.

## EU AGRICULTURAL POLICY PRIOR
## TO THE URUGUAY ROUND NEGOTIATIONS

Most countries place a high priority on achieving domestic objectives in framing their agricultural policies, but since its inception the European Union, or European Economic Community as it was initially, has treated agricultural trade as a subsidiary issue.[1] Article 39 of the Treaty of Rome specified that the major objectives of Community policy were to increase agricultural productivity, ensure a fair standard of living for the agricultural community, stabilize markets, assure the availability of supplies, and ensure that these supplies reached consumers at reasonable prices. The CAP sought to promote free trade in agricultural products between the six member states and to provide a price premium (community preference) for internal products over those from third countries.[2] High internal prices were maintained through the use of variable import levies, which also served to insulate EC markets from fluctuations in international prices.[3] In the late 1960s, EC prices for wheat were roughly double those on world markets, its sugar prices were two to three times higher, and butter prices were roughly five times higher than world prices (Ritson and Harvey, 1997).

With such high prices and a relentless increase in productivity due to technical change, the production of many agricultural commodities began to outstrip domestic demand. At the time of its formation, the Community was roughly self-sufficient in sugar, pork, and dairy products. It was a net importer of such commodities as wheat, rice, poultry, beef, and veal. By the early 1990s, when the Community had grown from six to twelve members, the production of all these commodities exceeded consumption by substantial amounts.[4] In 1992-1993, for example, EC wheat production was 33 percent higher than consumption, and its butter production was 21 percent larger than consumption (Ritson and Harvey, 1997).

As a result of the changing balance between supply and demand, the Community became a significant exporter of many agricultural products. In order to be competitive in international markets, the Community used export subsidies, much to the annoyance of other exporting countries. Despite a strong U.S. dollar during the early 1980s, which made EC products more competitive internationally, expenditure on market support programs doubled between 1981 and 1986 to the equivalent of more than 20 billion dollars (Josling, Tangermann, and Warley, 1996). The Community tried various ways to rein in production and reduce expenditures. In 1969, it began to "denature" wheat (make it unfit for human consumption) and to subsidize its use for animal feed. In 1984 delivery quotas were imposed to limit milk supplies. Adjustments were made in the arrangements for purchases at support prices (intervention purchases) in an attempt to control the accumulation of surpluses. In 1988, the policy of unlimited support purchases was effectively ended through the introduction of a "maximum guaranteed quantity" for most commodities, beyond which support prices were reduced. A variety of other measures were used to discourage sales into intervention (Hasha, 1999).

The increasing disposal of agricultural surpluses on international markets through the use of export subsidies generated a counterreaction in the United States. The Export Enhancement Program, introduced in the 1985 U.S. Food Security Act, ushered in a period of competitive export subsidization between the Union and the United States. The principal beneficiaries of the subsidy war were consumers in importing countries, who obtained commodities at subsidized prices, and the trading firms who handled the exports of these commodities. The principal losers were producers in other countries who were trying to compete without the benefit of subsidies.

## EU POLICY AND THE URUGUAY ROUND AGREEMENT

When the Uruguay Round negotiations began under the General Agreement on Tariffs and Trade (GATT) in 1986, the Union was in no position to make substantial concessions on agricultural trade. To do so would have required significant changes in domestic agricultural policies. Many farmers in Europe were violently opposed to making changes in the CAP, particularly if these were driven by the need to reach an agreement in the GATT. It was not until 1992, when European policymakers finally accepted the need to address the problem of the CAP's increasing budgetary costs, that it was possible to break the deadlock on agriculture in the Uruguay Round. A major step in finding a way forward was the release of a set of proposals for changes in the CAP by the EC Commissioner for Agriculture, Ray MacSharry. His proposals called for a reduction in support prices for key commodities to be offset by compensatory payments to farmers. The Blair House Accord, concluded with the United States in November 1992, permitted the Union to accept the components of the Uruguay Round Agreement on Agriculture (AoA) that were broadly consistent with the MacSharry proposals. The key elements of the Agreement were:

- The conversion of nontariff trade barriers to tariffs with maximum (bound) rates and an agreed-upon phased reduction in those bound rates
- Increased access for agreed-upon quantities of imports at lower preferential tariff rates (market access) under a tariff rate quota (TRQ) system
- A commitment to reduce domestic support
- Volume and value limitations on export subsidies

From the EU perspective, there were three further key provisions in the Agreement:

- The inclusion of EU direct payments (and U.S. deficiency payments) in the so-called blue box category, signifying that although they were not decoupled, they were viewed to have special status by virtue of their linkage to production controls
- The creation of a "green box" category for support that was viewed to be decoupled from production and therefore minimally trade distorting
- The Due Restraint or "peace clause" provision that shelters green box policies, blue box domestic support policies, and export subsidies from challenges under the GATT (although these are still actionable, i.e., subject to countervailing duties, if they are determined to be causing injury to domestic producers)

### Impact of the Uruguay Round Agreement on EU Agricultural Markets

It is generally acknowledged that the AoA has had a limited effect on the EU's internal market for agricultural products (e.g., Josling and Tangermann, 1999). The EU, in common with many other countries, established prohibitively high levels for the *bound tariffs* that replaced other forms of protection, such as the variable levy. In some cases, the lack of correspondence between the bound tariffs and the actual tariff equivalent of earlier policy measures led to allegations of "dirty tariffication." Certainly, there were examples of bound tariffs that were set at considerably higher levels than the average variable levies previously applied (Tangermann, 1996). Furthermore, although the variable levy system was officially replaced by bound tariffs, the EU has been able to vary the applied tariff within the ceiling provided by its bound tariff in such a way as to achieve essentially the same effect as a variable levy for key commodities, such as cereals. It also appears that the new system for fruits and vegetables operates in essentially the same way as the old system of reference prices (Josling and Tangermann, 1999).

The *market access provisions* and the creation of TRQs have produced only a marginal liberalization in access to the EU market for exporters. Many of the TRQs simply formalized the results of a number of bilateral agreements that the Union had already concluded with other countries, as the result of the expansion of the membership of the Union or the settlement of earlier trade disputes. The creation of the TRQ system, or at least the lack of discipline imposed upon it, was arguably one of the most damaging outcomes of the Uruguay Round Agreement. Not only does the system stimulate rent-seeking behavior on the part of countries and firms, it can also lack transparency. The EU created eighty-seven TRQs as a result of the Round. A recent analysis of EU implementation of the TRQ system suggests that the administration of these TRQs has been relatively transparent and fair in the Union. Fill rates for the quotas have been reasonably high and have increased over time (IATRC, 2001).[5] Nevertheless, it does not appear that the market access provisions (reduction in bound tariffs and increase in access quantities) have placed much pressure on the EU's internal markets or its Common Agricultural Policy.

Similarly, the *domestic support commitments* do not appear to have had much of an impact. In 1995, 24 percent of the Union's support expenditures (as measured by the aggregate measure of support, AMS) was classified as blue box, 21 percent was green box, and the bulk of the remainder (54 percent) was amber box—presumed to have the largest potential effects on production and trade (Young, Liapis, and Schnepf, 1998). Constraints on the level of support have not been binding in the Union, and Josling and Tangermann (1999) observe that the impact has been almost imperceptible.

The situation has been different for *export subsidy commitments*. Export subsidies were a key issue for the European Community during the Uruguay Round negotiations, and the most controversial aspect of the final agreement. In 1995, the Union accounted for 84 percent ($7.6 billion) of the value of export subsidies notified to the WTO (Leetmaa and Ackerman, 1998). In 1996, the Union exceeded its value commitment for rice and wine, and its volume commitment for rice, olive oil, beef, and wine. Since that time, the Union has had problems with other commodities, most notably cheese. The Union argued that it was not in violation of the terms of the AoA by virtue of the fact that it was employing unused portions of commitments from earlier years. It viewed unused export subsidy allowances to be bankable. It has also made other adjustments, for example by tinkering with the definition of processed cheese in order to be able to apply allowable subsidies to its components. In several key commodity areas—dairy products, beef, olive oil, poultry, and fresh fruit and vegetables—the Union has had or come very close to having problems in meeting its export subsidy commitments under the AoA.

### Pressures for Change in EU Agricultural Policy

Since the conclusion of the Uruguay Round negotiations, the major pressure for change in EU agricultural policy has not been its external commitments under the GATT, but internal and regional politics. With the collapse of the former Soviet Union and the transition of countries in central and eastern Europe to democracy and the market economy, the Union has come under increasing political pressure to expand its membership. In 1998, it launched a process of enlargement involving thirteen applicant countries.

The accession process began with ten central European countries and Cyprus on March 30, 1998. Formal accession negotiations were initiated with the Czech Republic, Estonia, Hungary, Poland, Slovakia, and Slovenia. Early official statements identified 2002 as a target date for the accession of these countries, but it seems unlikely that this will occur until the middle of the decade at the earliest.

It was recognized at an early stage that the expansion of the Union toward the east would create significant problems for the CAP. Collectively, the applicant countries are surplus producers of grain and livestock products, often at prices well below those in the Union. Agriculture is generally a larger part of the economy in these countries than in current EU members and accounts for a larger share of employment. An extension of the CAP to the new countries, coupled with the demands for assistance to address structural issues, such as small farm size and the need for rural economic diversification and development, could place large pressures on the Union's budget.

Attempts to address this issue began at the Berlin summit of the heads of state of the EU member countries in 1999. A package of measures, known as Agenda 2000, was agreed for the "reform" of the CAP. The changes agreed were weaker than those proposed by the EC Commission. The final agreement involved the following major provisions:

- A reduction in the support price for cereals by 15 percent phased in over two years (2000-2001), with an increase in area payments to compensate for half the price reduction
- A reduction in direct payments for oilseeds over a three year period
- A reduction in the support price for beef of 20 percent to be phased in over three years and to be partially offset through direct payments (of various types of premiums)
- The continuation of the dairy quota scheme with an increase in the quota of 1.2 percent in the first two years, allocated to specified deficit countries; an additional 1.2 percent over three years for the remaining countries from 2005; reduction in milk support prices of 15 percent over a period of three years to begin in 2005

The reduction in support prices, at least for cereals and beef, should make the Union more competitive internationally. However, the failure to address the problems in the dairy sector and the neglect of other key sectors such as sugar will limit the effectiveness of the policy changes. The increase in direct payments and in the milk quota mean that it is unlikely that the reforms will be effective in constraining total budgetary expenditures, although these are supposed to be capped in real terms. Most important, Agenda 2000 does not go far enough in dealing with the challenges to current agricultural policies created by enlargement. It is unclear whether the Union will be prepared to extend its generous system of direct payments to new members and will therefore be forced to provide even more funds for agriculture. Furthermore it is not clear whether the reductions in support prices will be sufficient to prevent the build-up of surpluses in the current or enlarged Union. If surpluses result, the Union will likely be forced to use supply controls (set-asides and quotas) to constrain output in order to meet its export subsidy commitments under the AoA.[6] The change in dairy policy will be unlikely to make the EU competitive internationally, and it is unclear how the new applicant countries would adapt to a quota system. Overall, the changes to the CAP in Agenda 2000 do not appear to prepare the Union for the expansion of its membership.

Agenda 2000 also does not appear to create the possibility for significant progress in the trade negotiations. The EU Commissioner for Agriculture has stated that the Agenda 2000 reforms will form the basis of the EU's negotiating position (Fischler, 1999). However, it seems clear that the changes agreed under Agenda 2000 do not add greatly to the Union's ability to negotiate significant increases in market access, through cuts in tariffs and increases in the quotas in the TRQs, or to negotiate a reduction in export subsidies. The Union may be able to make some concessions within current domestic policy parameters by continuing to eliminate some of the "water" in its tariffs (reductions in some of the excessively high bound tariffs agreed in the AoA) and through further concessions on levels of support—both of which could be achieved without affecting domestic markets. However, it seems likely that further concessions on export subsidies would be difficult to absorb; significant increases in the TRQ quantities could begin to create problems for internal market balance. In an analysis of the implications of Agenda 2000, Swinbank (1999) concludes that the changes made to the CAP are relatively minor. He argues that a further reform package will be necessary by mid-decade, if not before.

## EU PERSPECTIVES ON FUTURE AGRICULTURAL POLICY

In addition to the constant struggle with the costs of the Common Agricultural Policy and the problems that this creates for its trade relationships with other countries, the European Union is grappling with the future policy direction for its food and agricultural sector. The magnitude of the changes that have occurred in EU agriculture over recent decades is startling. Agriculture's contribution to national income has declined dramatically and is now similar to that in the United States at roughly 2 percent of gross domestic product (Table 4.1). In 1960, agriculture accounted for 21 percent of civilian employment in the countries that now make up the European Union,

TABLE 4.1. Agriculture's Economic Contribution

|                       | 1960 | 1970 | 1980 | 1990 | 1995 |
|-----------------------|------|------|------|------|------|
| *Percent of GDP*      |      |      |      |      |      |
| European Union (15)   | 9    | 6    | 4    | 3    | 2    |
| United States         | 4    | 3    | 3    | 2    | 2    |
| *Percent of employment* |    |      |      |      |      |
| European Union (15)   | 21   | 9    | 7    | 6    | 4    |
| United States         | 9    | 5    | 4    | 3    | 3    |

*Source:* Organization for Economic Cooperation and Development (1996). Economic Accounts for Agriculture Database, Data on CD. Paris.

compared to 9 percent in the United States. By the mid-1990s, the proportions were fairly similar at 3 to 4 percent. The out-migration of labor, fueled by rapid technical change and growth in the nonagricultural economy, has substantially decreased agriculture's economic importance. According to the OECD Secretariat, the number of full-time farmer equivalents in the EU-12 declined by roughly 25 percent between 1986 and 1996. The number of farms declined by 20 percent between 1985 and 1995. However, the number of farms greater than 100 hectares in size increased by 4 percent over the same period (OECD, 1998a).

Data for "main occupation farms" in the Union relating primarily to the late 1980s and early 1990s yield an average of 45 percent for the nonagricultural contribution to farm family income in nine current EU countries. The nonfarm share ranged from 25 percent in the Netherlands to roughly 70 percent in Italy (Blandford, 1996). Not only is agriculture a declining part of the overall economy, it is also a declining part of the rural economy. The contribution of agriculture to rural employment is largest in rural areas in Greece and Portugal, but even in those countries it averaged less than 40 percent in the early 1990s. In EU countries as diverse as Germany and Sweden, agriculture accounts for less than 10 percent of total employment in predominantly rural regions.

European policymakers have been attempting to promote the concept of a "European model of agriculture" as a means of establishing the future direction for agricultural policy. The EU's Agriculture Commissioner has spoken a great deal about the concept of the European model, recently defining the future objectives for EU agriculture as:

- A competitive agricultural sector, which can gradually face up to world markets without being oversubsidized
- Production methods that are sound and environmentally friendly, able to supply quality products that the public wants
- Diversity in the forms of agriculture, which maintain visual amenities and support for rural communities
- Simplicity and transparency in agricultural policy with shared responsibilities between the European Commission and EU member states
- Justification of farm support through the provision of services that the public expects farmers to provide (Fischler, 2000)

Several European countries both inside and outside the Union are promoting the concept of a "multifunctional" agriculture—one that provides a variety of desirable outputs, not simply food and fiber. References to the multiple roles or functions of agriculture have appeared in OECD ministerial communiqués for many years, although the term "multifunctional" did not surface publicly until the 1998 meeting of the Ministers of Agriculture (OECD, 1999). Commissioner Fischler observes, "Multifunctionality is the word we have found in Europe to describe the fundamental link . . . between sustainable agriculture, food safety, territorial balance, maintaining the landscape and the environment and what is particularly important for developing countries, food security" (Fischler, 2000).

The stress upon the multiple outputs of agriculture, and particularly its environmental implications, is an important theme in the debate on European agricultural policy. At the national level, considerable stress is laid upon environmental objectives for agriculture. Thus the OECD Secretariat in its 1998 evaluation of agricultural policies in the OECD member countries notes, "The EU Member countries have implemented more than a hundred agri-environmental programs at national and regional level, covering a substantial part of the farmland in some of the countries. The measures, which are on a voluntary and contractual basis, primarily relate to land management . . . " (OECD, 1998b, p. 37).

The concept of multifunctionality is and will continue to be controversial, both within the Union and outside it. Although space is not available for an exhaustive critique, a few key issues can be highlighted:

- Agricultural practices in Europe generate both positive and negative environmental effects. Most of the emphasis seems to be on the former as a rationale for providing government support to farmers. There is substantial evidence that many forms of support have negative environmental effects by stimulating intensive production.
- The prevailing view in Europe appears to be that current agricultural structure and production practices (with the correct incentives) will

provide an appropriate supply of environmental goods. Little consideration is given to allowing changes in the scale, intensity, or type of production that could also result in the desired supply of these goods and at the same time reduce trade distortions. The view appears to be that agricultural practices and the agricultural landscape should not be allowed to change significantly.

- Rural development is used as an argument for providing support to European agriculture. Given the limited and declining role that agriculture plays in the economies of many rural areas, and in rural employment, it is doubtful whether such support will have much of an impact on rural development.

Other issues, such as those relating to food safety, genetically modified organisms, and animal welfare are also important in the debate about the future of agricultural policy in the Union. There are substantial concerns about these issues in Europe. In survey results released by the European Commission, 34 percent of those questioned indicated that they did not consider fresh meat and frozen foods to be safe; 49 percent indicated that they did not feel that precooked meats were safe (European Commission, 1998a,b). Most important, the two leading priorities for the Common Agricultural Policy identified by those surveyed were guaranteeing food safety (89 percent) and guaranteeing that animals are well treated (84 percent).

## IMPLICATIONS FOR THE TRADE NEGOTIATIONS

The concept of multifunctionality will undoubtedly figure large in the Union's approach to future trade negotiations. Commissioner Fischler has been very clear on this point, arguing that the most fundamental nontrade concern that needs to be addressed in the WTO negotiations on agriculture is the multifunctional role of agriculture (Fischler, 2000).

Acceptance of the multifunctionality concept implies acceptance that agriculture is different from other industries in important ways and that it must be treated differently on a continuing basis, including the disciplines affecting international trade. If agriculture is a provider of multiple outputs, not simply food and fiber, then the value of these outputs needs to be recognized and a sufficient supply of them needs to be maintained. If the market will not reward farmers for supplying the outputs, then the public sector must do this. Thus multifunctionality is the mantra for the continuation of agricultural support. To the extent that the Union can use direct payments that fall into the green box category to achieve its aims, this should not create a barrier to progress in the trade negotiations. It would simply be necessary to permit the Union to use direct payments, coupled with practice requirements that would allow it to meet its environmental aims.

Unfortunately, there is a problem with this approach if one accepts the EU view that the supply of desired environmental and other goods depends on the continuation of existing systems of production and that a certain level of production is required to maintain the supply of these goods. Under this view, it is not sufficient that producers keep livestock, but that they keep certain types of livestock with a certain stocking density, and use certain animal husbandry practices. In order to achieve this, support must be linked to production, since the provision of the nonagricultural goods is directly linked to production. For this reason, the Union is likely to have difficulty accepting the decoupling argument that is embedded within the green box category of support for policies necessary for the supply of public goods.[7] The current blue box category, which links support to production limitations and compensation for lower support prices, may not be sufficiently flexible to meet EU needs in this area.

It is not clear what the Union's strategy will be on this issue in the negotiations, although Commissioner Fischler has stated recently, "a major review of the specific instruments provided in the Uruguay Round is neither necessary or desirable. This position does not rule out some updating of the blue and green boxes. But it stresses the continuation of the current distinction of policies according to their degree of trade-distortion as the essential element to move away from support linked to prices or products towards transparent and non-distorting support policies" (Fischler, 2000). It is difficult to see how the need for the Union to maintain flexibility in its policies to ensure that the level of joint production of goods, including agricultural output, can be reconciled with this view. Perhaps the Commission believes that the application of an environmental or other label to policies will be sufficient to exempt them from additional discipline regardless of the degree to which they affect production and trade. Alternatively it may believe that the negotiations will not force reductions in EU internal prices that are sufficiently large to make this an issue.

The Union is likely to place a lot of emphasis on other key domestic concerns, such as food safety and animal welfare in the negotiations. The ability of the Union to restrict imports of products that it believes to be a threat to human health is already an area of contention, as demonstrated by the dispute with the United States over imports of hormone-treated beef. It is likely to be an increasingly important issue in the treatment of imports of genetically modified organisms (GMOs) or food products made from these organisms. It is also important in regard to animal welfare standards. The Union has gradually increased the standards for the treatment of farm animals, particularly for poultry, pigs, and veal calves (Blandford et al., 2001). In the Union, there is pressure for even higher public standards to govern produc tion methods. In the past the Commission has been concerned about the impact of such standards on the competitive position of European agriculture. There may be pressure to legitimize the ability of countries to restrict imports of agricultural products that do not meet domestic standards (similar to pressures in other sectors relating to labor and environmental standards), or at the very minimum to allow countries to impose a mandatory labeling requirement on suppliers.

The areas of so-called non-trade concerns are likely to be the most difficult for the Union and its trading partners in making progress in the trade negotiations. However, they are not the only ones. As indicated earlier, increased market access and reductions in export subsidies would likely require further changes in the CAP, and in commodity sectors that are highly sensitive, such as dairy and sugar. It will not be easy for the member states to agree on major changes in these areas that would improve the competitive position of EU agriculture. The Commission has indicated that it will be prepared to negotiate further reductions in export subsidies, providing that all forms of export support (including export credits and food assistance) are considered (Fischler, 2000). Finally, the Union would like to see a continuation of the Special Safeguard Clause, which allows countries to take action in the event of sudden surges in imports or "abnormally low" import prices. The Union would argue that this provision enables it to ensure domestic market stability.

Regardless of the prospects in the negotiations, the Union may even have difficulty in meeting its current international obligations if European public concern over food safety and public health continues to grow. The Union apparently preferred to endure punitive tariffs on some of its exports to the United States rather than open its market to U.S. beef as a result of the WTO ruling in the beef case. Growing concerns in Europe about foodstuffs containing genetically modified organisms, well-publicized threats to public health posed by chemical contamination, and the continuing fallout from the Bovine Spongiform Encephalopathy (BSE) epidemic are likely to put increasing pressure on policymakers to control agricultural imports when there are questions, real or imagined, about food safety.

## CONCLUSION

The European Union's attitude toward international trade and trade negotiations has evolved substantially since the Uruguay Round negotiations were launched in 1986. While trade is still viewed as being secondary to the achievement of domestic objectives, there is greater willingness to take international implications into account in framing domestic agricultural policies. In some commodity sectors, most notably cereals, the Union has moved to a more market-oriented position through reductions in internal support prices. There is growing acceptance that such market orientation is necessary in order to avoid costly surpluses and the trade conflicts generated by subsidized exports.

Despite this, from a European perspective, there are two major obstacles to achieving progress in the agricultural trade negotiations. The first is the difficulty in reaching consensus among the member states on changes to the Common Agricultural Policy that would permit the Union to move to world price levels in all key commodity sectors and to absorb new members, without placing major additional demands on the budget. Although changes made to the CAP as part of its Agenda 2000 initiative were supposed to prepare the Union for a further expansion of its membership, it seems unlikely that they will achieve this. Furthermore, Agenda 2000 provides very limited flexibility for achieving further liberalization of EU agricultural trade in the WTO negotiations.

Second, there is the difficulty of reconciling the Union's view of agriculture as a multifunctional activity, yielding various public goods in addition to food and fiber, with further trade liberalization. A major issue is how to ensure the desired supply of public goods without affecting the volume of production and trade. If internal EU prices are reduced in order to liberalize trade, the desired supply of environmental services, for example, can only be guaranteed if EU direct payments are linked to particular production practices. It would be difficult to argue that such payments would be minimally trade distorting, as required by the green box criteria. It is not clear that production objectives can be met through current blue box policies nor whether the other parties to the negotiations will be willing to accept an indefinite continuation of the blue box exemption for the Union's compensatory payments.

Other domestic concerns associated with agriculture, in particular food safety and animal welfare, are likely to complicate the trade negotiations for the European Union, and make it increasingly difficult for the Union to meet its existing international trade obligations. Public opinion surveys suggest that food safety and the way that food is produced are at the forefront of public concern about the future of European agriculture and the CAP. The recent beef-hormone dispute with the United States and controversy over the introduction of genetically modified organisms into the food chain show that consumer concerns about the food system are highly emotional and politically charged issues in Europe. Reaching an accommodation with other trading partners on how these issues are treated in international law will be very difficult.

There is a long way to go before it will be possible to address the concerns that the European Union brings to the trade negotiations. This is partly due to the fact that the EU's thinking on its objectives and the options for achieving these does not appear to be sufficiently advanced. It will be necessary for EU policymakers to come to grips with these issues in order to conclude an agreement that will be acceptable to key domestic political constituencies. If the other participants in the negotiations insist on making significant progress in liberalizing agricultural trade, finding a balance between EU domestic objectives and international obligations is likely to take several years. The prospect of many all-night negotiating sessions in Brussels, Geneva, or elsewhere looms.

## NOTES

1. The terms European Economic Community, European Community (EC), and European Union (EU) relate to various stages in the evolution of the Union. In general, the current term, European Union, is employed.

2. The original members of the European Economic Community were Belgium, France, Germany (Federal Republic), Italy, Luxembourg, and the Netherlands.

3. As its name suggests, the variable levy was adjusted when world prices changed in order to maintain a roughly constant price for imported products in the Union.

4. The six new members were Denmark, Ireland, and the United Kingdom (1973); Greece (1981); and Portugal and Spain (1986). The Union expanded to fifteen members with the addition of Austria, Finland, and Sweden in 1995.

5. The fill rate for TRQs is often used as an indicator of the degree to which a TRQ is operated fairly. Many factors can affect the fill rate, some of which are totally unrelated to the administration of the system. The average fill rate for EU TRQs has tended to exceed that for US TRQs (IATRC, 2001).

6. These will become more severe since the new members do not have any significant export subsidy "credits" under the AoA.

7. From an economic perspective, "decoupled" policies are not appropriate in this case. Without corrective action to ensure the necessary supply of the unpriced public goods, free trade volumes would be distorted. Coupled support is necessary to correct the distortion.

## REFERENCES

Blandford, D. (1996). "Overview of Microeconomic Results in OECD Countries and Policy Interests: Characteristics of Incomes in Agriculture and the Identification of Households with Low Incomes." In Hill, B. (Ed.), *Income Statistics for the Agricultural Household Sector.* Luxembourg: Eurostat.

Blandford, D. (1999). "The Upcoming Round of Agricultural Trade Negotiations: Possible US Objectives and Constraints." Agricultural Policy Discussion Paper No. 15. Centre for Applied Economics and Policy Studies, Massey University, New Zealand.

Blandford, D., J.C. Bureau, L. Fulponi, and S. Henson. (2001). *Potential Implications of Animal Welfare Concerns and Public Policies in Industrialized Countries for International Trade.* New York: Kluwer Academic Press.

Blandford, D. and J. Dewbre. (1994). "Structural Adjustment and Learning to Live Without Subsidies in OECD Countries." *American Journal of Agricultural Economics* 76:1047-1052.

European Commission (1998a). *Eurobarometer,* No. 48, March, Brussels.

European Commission (1998b). *Eurobarometer,* No. 49, September, Brussels.

Fischler, F. (1999). "WTO Negotiations—Agricultural Aspects." Speech delivered to the Civil Society Consultation, Brussels, October 22.

Fischler, F. (2000). "Framework for World Agri-Food Trade." Speech delivered at the Dublin Castle Centenary Conference, Dublin, March 30.

Hasha, G. (1999). "Overview." In *The European Union's Common Agricultural Policy: Pressures for Change.* International Agriculture and Trade Report WRS-99-2. Washington, DC: Economic Research Service, USDA.

International Agricultural Trade Research Consortium (IATRC). (2001). *Issues in the Administration of Tariff-Rate Import Quotas in the Agreement on Agriculture in the WTO.* Commissioned Paper Number 13.

Josling, T.E. and S. Tangermann. (1999). "Implementation of the WTO Agreement on Agriculture and Developments for the Next Round of Negotiations." *European Review of Agricultural Economics* 26:371-388.

Josling, T.E., S. Tangermann, and T.K. Warley. (1996). *Agriculture in the GATT.* New York: St. Martin's Press.

Leetmaa, S.E. and K.Z. Ackerman. (1998). "Export Subsidy Commitment: Few Are Binding Yet, But Some Members Try to Evade Them." In *Agriculture in the WTO.* International Agriculture and Trade Report WRS-98-4. Washington, DC: Economic Research Service, USDA.

Organization for Economic Cooperation and Development (1998a). *Agricultural Policy Reform, Stocktaking of Achievements.* Paris: OECD.

Organization for Economic Cooperation and Development (1998b). *Agricultural Policies in OECD Countries: Monitoring and Evaluation.* Paris: OECD.

Organization for Economic Cooperation and Development (1999). *Agricultural Policies in OECD Countries: Monitoring and Evaluation.* Paris: OECD.

Ritson, C. and D.R. Harvey. (1997). *The Common Agricultural Policy* (Second edition). New York: CAB International.

Swinbank, A. (1999). "CAP Reform and the WTO: Compatibility and Developments." *European Review of Agricultural Economics* 26:389-407.

Tangermann, S. (1996). "Implementation of the Uruguay Round Agreement on Agriculture: Issues and Prospects." *Journal of Agricultural Economics* 47:315-337.

Young, E., P. Liapis, and R. Schnepf. (1998). "Domestic Support Commitments: A Preliminary Evaluation." In *Agriculture in the WTO*. International Agriculture and Trade Report WRS-98-4. Washington, DC: Economic Research Service, USDA.

Chapter 5

# One Perspective on U.S. Policy and Agricultural Progress in World Trade Organization Negotiations

Daniel A. Sumner

The first step in this chapter is to clarify my topic by defining terms and objectives. Progress is like beauty. Most economists behold open agricultural markets as beautiful as well as conducive of human well-being in the broadest sense, but such a view is not universal. To the eye of this beholder, "progress" in trade negotiations is defined as movement toward more open markets with fewer barriers and with trade conducted primarily on a commercial basis without government trade taxes or subsidies. This old-fashioned view (Johnson, 1950) is standard for economists and for much of the educated public (although we must be careful not to define the term "educated" so as to be circular in defining progress). This view of progress is also imbedded within the very definition or at least the mission statement of the World Trade Organization (WTO). The WTO Web site states,

> The WTO is the only international body dealing with the rules of trade between nations. At its heart are the WTO agreements, the legal ground-rules for international commerce and for trade policy. The agreements have three main objectives: to help trade flow as freely as possible, to achieve further liberalization gradually through negotiation, and to set up an impartial means of settling disputes. (WTO, <http://www.wto.org/english/thewto_e/whatis_e/whatise.htm>)

Not everyone would behold movement toward trade liberalization as progress. Especially after the WTO meeting held in Seattle in December 1999, it is not at all clear that the economic way of thinking on these issues has made much progress in capturing popular opinion in the last century or so. The reason for being explicit of my definition of progress is precisely because some may have a different perspective (Chapter 2, this volume).

Of course, even economists do not generally claim optimality for completely open borders with no trade restrictions. For example, even many otherwise liberal economists allow for a government role in regulating trade that may spread harmful plant, animal, or human pests and diseases or regulating trade when national defense risks are at stake. Interestingly, President Bush has regularly placed agricultural issues in the context of national security. This chapter does not discuss national defense, but I do discuss food security and sanitary and phytosanitary regulations.

This chapter was initially assigned the task of providing the U.S. perspective on obstacles to progress in the WTO negotiations. Even given my definition of progress, there are many U.S. perspectives on what constitutes "obstacles." There are probably even many U.S. government perspectives on this topic, depending on the agency or other situation of the viewer. Also, any official U.S. government perspective would probably indicate that obstacles to progress are mainly to be found in the negotiating stance of other parties.

This chapter does not represent a government perspective on obstacles to progress, and I do not examine obstacles created by other countries. Instead, the core of this chapter investigates the extent to which the WTO positions and farm policies of the United States may contribute to progress or barriers to progress in negotiations toward more liberal agricultural trade. But, before turning to U.S. trade and agricultural policies, I want to consider briefly some broader issues.

## GENERAL OBSTACLES TO PROGRESS IN CURRENT WTO NEGOTIATIONS IN AGRICULTURE

There are a host of potential barriers to progress in the WTO negotiations in agriculture. Some of these are traditional (Josling, Tangermann, and Warley, 1996). Some we might classify as "post-Seattle" in the sense that that event raised the general awareness of these issues (Chapter 3, this volume). A few comments on each of these broader potential barriers are appropriate before turning to a discussion of agricultural policy directly.

### Nature of the WTO and the Process of Negotiation and Agreement

The WTO is an organization of nations. WTO members negotiate, sign agreements, and resolve disputes using an agreed framework. The WTO does not help negotiate contracts between private firms or other nongovernment organizations and does not help settle disputes among such groups or between a nongovernmental organization and a WTO member. Concerns raised by firms or other organizations get into a WTO process only if they are raised by a member nation. Thus, many issues go through a multistep process of being raised within a nation and then being raised by that nation at the WTO. The purpose of the WTO is to aid members to move toward lower trade barriers and to establish, and work within, a rule of law with respect to commercial relationships across national boundaries. Some seem to want to the WTO to take on an expanded role, but it is not clear that the WTO is suited for such a role, and such pressure may slow progress on trade liberalization. Indeed, that may be exactly the objective of some nongovernmental organizations.

As noted, neither the Sierra Club nor the Farm Bureau has a place at the table at the WTO in Geneva. The two important ways these groups participate are indirect. First, they may provide direct information and advice to negotiators in formal and informal submissions and forums. Second, they may participate more generally in the political process, letting elected representatives know their views and working for (or funding) candidates that are more likely to have positions consistent with their own. This latter process then requires the elected representatives to communicate with the negotiators. This complicated process, like much of agricultural politics, encourages the effective efforts from those who have strong (special) interests in particular topics or narrow issues.

In the WTO, broad agreements require consensus. No agreement may be imposed on a sovereign nation. The WTO does not have tanks or planes or bombs. It does not even have "peacekeeping" forces like the United Nations. If a country does not comply with UN demands, it may be invaded or bombed into submission. A member that does not comply with its WTO obligations as agreed either pays compensation to those injured or, in extreme cases, may lose its privileges of membership. Assurance of national sovereignty probably allows the WTO to make more progress than otherwise. However, it seems from listening to public discussion during and after Seattle that this characteristic of the WTO is not widely appreciated. Some seem to oppose the WTO because they think it may threaten national environmental or labor policy. Others oppose the WTO because it does not impose their favored environmental or labor standards on other countries.

### *Global Antiglobalization Sentiment*

Post-Seattle protests make clear that an active collection of individuals and groups simply oppose international commercial relationships. Many seem to oppose foreigners in general, some oppose economic growth, some oppose market forces in general, and some seem to simply oppose economic reasoning. Many of these individuals and groups seem little interested in tinkering with reform of international economic organizations; they want to shut down such organizations. The WTO garners opposition both from those who think it represents oppressive world government dominated by foreigners and from others who think the WTO respects national sovereignty too much.

The power of these "anti" groups is hard to assess. They affect the WTO by affecting the positions taken by member governments. Protests since Seattle have been relatively ineffective, and even the press is losing interest. In the United States, the election of 2000 indicated relatively little influence of the antitrade lobby. Anti-WTO presidential candidates on the left and the right received less than 3 percent of the vote, and the U.S. Congress remained in the control of the Republicans who generally support trade liberalization compared to congressional Democrats who recently have opposed liberalization. Under this interpretation, the Seattle protest represents a high water mark for the antitrade tides and not the first wave of a tsunami large enough to flood Geneva.

### General Political Situations or Calendars in Certain Countries

U.S. presidential and congressional elections in 2000, and the slow resolution of the presidential election, slowed the ability of the United States to make strong commitments in the WTO negotiation, and it limited the willingness of other countries to table their own serious proposals. Clearly, there is no time window when no change in government is pending in any of the significant WTO members, and yet negotiations must go on.

The leaders of the new U.S. administration have very strong trade credentials and commitments to liberalization. Both Secretary Veneman and Ambassador Zoellick were leaders in the Uruguay Round negotiations and appreciate the benefits of open markets and the effort required to achieve further agricultural liberalization. They also have experience working with negotiating partners and the U.S. Congress on these issues. Veneman in particular has a long and distinguished record on agricultural trade, and her appointment signaled that the United States will pursue an aggressive liberalizing agenda in this round.

### Broader Issues in International Relations

In the past, trade negotiations have been a part of overall foreign policy and, in particular, the struggle of the cold war. This linkage is muted after the fall of communism, but it is still possible for trade concessions to be offered to gather support for nontrade initiatives. Certainly, agricultural interests in the United States often express concern over such concessions. For example, they point to agreements to maintain NATO access to military bases in Italy as a reason for little success in issues concerning subsidies for Mediterranean farm products. Agricultural negotiations are mainly concerned with details of tariff cuts and similar specifics. Nonetheless, at the highest level of government the tradeoffs inherently include nontrade relationships between nations.

### WTO Positions and Issues in Nonagricultural Areas

Just as nontrade concerns affect the tradeoffs in trade negotiations, generally within the negotiations there is natural spillover across otherwise separate areas. In the Uruguay Round, the consensus view is that progress in agricultural was aided by expected gains in nonagricultural negotiations to some countries which generally resist opening agricultural markets. The current round does not yet have such linkages, and agricultural progress may be limited as a consequence. It is interesting in this regard that governments in Japan and the EU currently support a broader round that will allow them to claim victory even if their initial positions are not adopted in the agricultural segment of the negotiations. However, this spillover across areas may be less helpful to progress toward agricultural liberalization this time because there is less left to gain from liberalization in the nonagricultural goods markets. Further, agricultural forces that support protection understand that they are also vulnerable to the general linkage across industries and issues.

### Agricultural Market Conditions

Extremely low global farm prices, which have continued for several years through early 2002, have at least four potential effects on trade negotiations. The first effect aids progress; the next three of these effects may act as obstacles to progress. First, low prices emphasize the urgency of removing barriers and allowing more trade to raise world prices. This idea is consistent with the hypothesis that low farm prices at the beginning of the Uruguay Round provided impetus for reform. Second, antitrade forces emphasize that the opening achieved in the Uruguay Round did not solve the problem of low world prices, and thus it is false to attempt to solve the problem with more market opening. Third, in the United States these low prices have spawned a massive infusion of popular farm subsidies that may seem vulnerable to international negotiations. This vulnerability is heightened by the recent arguments to subsume all farm support, including decoupled payments, in the category of policy to be controlled (see ABARE, 1998; Chapter 2, this volume; or for an alternative perspective, Sumner, 1995). Fourth, low market prices in global markets reinforce the concerns and resolve of those commodity interests in the United States and especially in other countries that do not want to be exposed to world market forces or want to be able to export their own commodity market problems.

We now turn to considering the effects of early negotiating positions and policies of the United States on progress toward agricultural trade liberalization.

## THE NEGOTIATING POSITION OF THE UNITED STATES AND PROGRESS IN THE CURRENT WTO ROUND

During the early stages of the current round of WTO negotiations, it is tempting to ignore the predictable early posturing of the major players in the negotiation. The current positions of major WTO participants seem to represent mainly the interests of leading producer groups in each country. This section does not catalog why the opening positions and policies of countries that *oppose* liberalization are inimical to progress in the WTO negotiations. Clearly, countries such as Korea, Japan, Norway, etc., have powerful agricultural interest groups that command sympathy from the rest of the population. These nations have generally defined progress in WTO negotiations as the exact opposite of the way I use the term in this chapter. The EU is at least mainly in this camp. Canada is more mixed. The grains industry wants open markets and the dairy industry wants to keep Canada's market closed. The government claims having both is feasible. Obviously, the biggest problem for a liberal agreement is that many governments of WTO member states do not want agricultural liberalization, and this is reflected in their negotiating positions. This section is devoted to considering the early position of the United States in the negotiations.

It was often argued that the high visibility of the strong opening position of the United States slowed progress in the early years of the Uruguay Round, and that position was restated in a recent history of that period (Orden, Paarlberg, and Roe, 1999), so it is worthwhile to discuss such a hypothesis applied to this round. So far at least, the current approach of the United States has been much less aggressive compared to the approach taken in 1986 to 1990 in the Uruguay Round, when the United States put forward the so-called zero option to eliminate all trade barriers and trade-distorting subsidies within ten years.

WTO Ambassador Hayes (2000) succinctly restated the broad U.S. agricultural position:

> Within the agricultural negotiations, we are focusing on substantive reform proposals, such as the elimination of export subsidies, reducing tariffs and reducing trade-distorting domestic support levels. We also want to ensure market access for biotechnology products, through the application of open, transparent, science-based regulatory regimes.

In testimony to the U.S. Congress, then Ambassador Barshefsky restated more fully the material that was submitted in Geneva in July 1999.

> While the work has just begun, our fundamental principles for agricultural trade reform in these talks are clear:
>
> - Eliminate Agricultural Export Subsidies—A principal goal of the U.S. will be to completely eliminate, and prohibit for the future, all agricultural export subsidies.
> - Lower tariff rates and bind them—this should include reduction and elimination of tariffs, elimination of tariff disparities, and simplification of tariff policies, for example in cases where WTO members use "compound" tariffs that include both ad valorem and cent-per-kilogram tariffs.
> - Substantially reduce trade-distorting domestic supports and strengthen rules that ensure all production-related support is subject to discipline, while preserving criteria-based "green box" policies. In addition, all trade-distorting supports should be more tightly disciplined.
> - Improve access for U.S. exports under tariff-rate-quotas (TRQ)—by increasing quantities eligible for low-duty treatment, reducing high out-of-quota duties, and improving disciplines on administration of TRQs to ensure that they offer real market access.
> - Strengthen disciplines on the operation of state trading enterprises—While state trading enterprises are subject to WTO limits on subsidized exports, there are a number of concerns about their operations, and in particular those of monopoly exporters. These include the possibility of disguised circumvention of export subsidy commitments, and anti-competitive practices such as predatory pricing.
> - Address disciplines to ensure trade in agricultural biotechnology products is based on transparent, predictable and timely processes. While WTO rules cover trade measures affecting biotechnology products, we are concerned about the utter collapse of the European Union's approval process for biotechnology. We continue to work with our industry, Congress and other interested groups in developing the best approach for dealing with this subject bilaterally and in WTO negotiations.

The United States elaborated on these positions in submissions to the WTO in June 2000 and November 2000 (WTO, 2000b,c). The June submission added the following points: more disciplines on import state trading enterprises, a proposed ban on export taxes, more specifics on internal support, and allowances for developing countries for additional internal support, especially related to food security. The November submission elaborated on ideas for reforming TRQ administration.

A few comments about this position are worth noting. First, the clear and direct positions are to eliminate export subsidies and taxes. As we will discuss, the United States has not been inclusive in the scope of export programs to be eliminated. We should also note that the export tax position implies no discipline on the United States itself, since export taxes are forbidden by the U.S. constitution. Second, the United States indicates a willingness that farm product tariffs remain into the foreseeable future. There is no "zero option" here. Third, the emphasis on trade-related domestic support and the proposal to clarify these definitions suggests that the U.S. government expects that its own domestic programs would pass muster. Finally, the biotech statement suggests, the United States still believes it has more to gain than lose by opening this topic in the WTO. With the year 2000 positions of the United States as background, let us compare these to the early days of the Uruguay Round and see if parallels exist.

First we need a bit of background. In the first days of the Uruguay Round, the United States presented an aggressive statement that its proposal was to gradually eliminate all trade-distorting agricultural policies, including both border measures and internal supports that more than minimally affect trade.

Orden, Paarlberg, and Roe (1999) state, "Little or no progress was made in the 1987 and 1988 phase of the Uruguay Round, mostly because of the extreme nature of the U.S. zero option proposal" (p. 97). On its face this statement is remarkable in that it attributes lack of evident progress not to those whose initial positions were to oppose any serious farm trade reform, but to the United States because it made the remarkable offer to discipline its own notorious farm programs and trade barriers. It seems just as likely to say that a lack of progress (if there was indeed a lack of progress) was due to the inflexible opposition to reform expressed by other major negotiating parties. Particularly in the face of a forty-year history of failure of farm trade negotiations, it seems odd to blame the strategy that clearly broke with the past rather than the strategy of status quo. Of course, such historical claims are impossible to verify one way or the other, but the question is directly relevant to the objective of the present paper, because the years in question correspond roughly to 2000-2001 in the current round.

Although it is impossible to refute the historical claim of Orden, Paarlberg, and Roe (1999), I think one could just as plausibly argue exactly the opposite. For example, one might make a parallel statement that little or no progress was made in the year 2000 phase of the current trade round, mostly because the United States failed to put forward any clear proposal with farm trade liberalization as a national objective. The list provided by Ambassador Barshefsky provides no straightforward declaration for what the United States stands for in this round. In that sense, it is much like the positions of Canada, though somewhat less bipolar or bilingual.

But before discussing the idea of interim progress any further, we should clarify briefly what evidence of "progress" may mean at the beginning of a negotiation. When parties start far apart, few want to give up what may become bargaining chips later. Early negotiations are therefore usually about positioning for the end game. In negotiations, one does not get to store up points as one goes along. Lack of evidence of progress to outsiders is natural at such early stages of any complex negotiation. Progress in the early stages of negotiation means building foundations for eventual success and is not the same as progress measured at the end. Thus, perhaps no early evidence of progress is in fact the optimal path to eventual success. The fact that Orden, Paarlberg, and Roe saw no progress in 1987 and 1988 may be due to their vantage point outside the actual negotiations.

In this vein, one might argue that whatever success the Uruguay Round achieved in the end was due mostly to the zero-option proposal, which signaled to the other WTO members that the United States was serious about more trade liberalization in agriculture. The United States opened the negotiations with a clear statement of its goals. That clarity created the opportunity for other nations to claim victory in the end when they had, in fact, set up a framework and moved a step or two along the path toward the very zero option that is now scorned. We cannot know if this scenario is actually true, but it seems as plausible, and just as irrefutable, as the one specified.

Orden, Paarlberg, and Roe (1999) also assert, even more confidently, that " . . . *most obviously* it [the zero option] went *far* beyond anything *remotely* acceptable to the Agricultural Committees of Congress" (p. 97, italics added). This is a very strong and confident assertion, especially when we remember that in the U.S. system, Congress is an integral part of negotiation. Then as now, the congressional committees were part of the process of setting the U.S. position. As Ambassador Barshevsky noted in her March 2000 testimony, "In preparation for more detailed proposals, we will extensively consult with stakeholders and Congress, and work in tandem with the drafting of a new Farm Bill. We are now continuing our work with American producer groups and other interested parties. . . . " Thus, whereas Congress and the administration have different roles and perspectives, one can be assured that the negotiators are not planning to bring back an agreement that is sure to fail in a final vote.

The same collaboration was as true in 1986 as in 2000. Congress was a part of the process, and the agricultural committees may have been more willing to support the zero option as a goal than some complex compilation of half measures. Frankly, it was silly to use such language as "obviously" and "far beyond" and "remotely" in discussing a negotiating position in which Congress was a participant! Under the unlikely hypothetical situation that the United States' initial position could have been achieved, it does not seem implausible that the Congress would have accepted the tradeoff of U.S. programs for enforceable elimination of all the other farm trade barriers and subsidies in the world. But, in fact, what is clear, and was clear at the time, is that no one expected the opening U.S. position to become the final outcome, so the whole question is moot and irrelevant. (Nor, obviously enough, was the opening position from the EU, the Cairns Group, or Japan or Korea, "not one grain of rice," likely to be the final outcome.) What is clear is that the U.S. Congress aggressively supported the idea that U.S. agriculture would be a winner if farm trade barriers and subsidies came down on a multilateral basis. Further, given their belief that U.S. barriers were already lower than those of other WTO members, the zero option for a final position removed the complaint that trading partners would be left with subsidies and barriers much higher than those in the United States.

So what does this mean for the current round of negotiations? Are we about to reap an early harvest of progress because there are no explicit zero options on the table? No, of course not. The real concern is that because the United States has been so timid and "realistic" in its early positions, negotiating partners suspect that the United States is no longer serious about progress toward reform. One suspects that this timidness is about to change with the change in administration. Indeed, for better or worse, the zero option continues to frame the debate and discussion. Perhaps if it had not been invented in 1986 it would have to be invented now.

## TRADE AND AGRICULTURAL POLICIES OF THE UNITED STATES AS OBSTACLES TO PROGRESS IN THE CURRENT WTO ROUND

One reason that GATT progress in agriculture was limited before the Uruguay Round was that opening borders was inconsistent with the policies that countries had established to protect and support agricultural interests (Josling, Tangermann, and Warley, 1996; Sumner and Tangermann, 2000). Open borders were clearly threatening to industries that relied on government protection and subsidy, and governments have been generally unwilling to shift to support policies that were more compatible with trade liberalization. Any country may choose unilateral free trade in agriculture. Most do not. Some policies are pursued as responses to policies of other countries that would be reversed in a multilateral reform, but many WTO member governments act as if opening agricultural markets is not in their interest. It is hard to see what is likely to change that policy behavior in the short run.

In the rest of this chapter I want to discuss how policies of the United States itself may be inconsistent with achieving the goal of liberalization.

The simplest case relates to policies by which the United States maintains high tariffs. The United States has made clear in the past that these policies are on the block and would have to change radically if the goal of liberal agricultural trade were to be achieved. As we have seen, the current negotiating position of the United States does not specify a zero option on tariffs. But it does suggest that the United States continues to urge more market access. Of course, there are many remaining tariffs for U.S. agricultural goods. While few industries volunteer for unilateral tariff reductions, many industries with tariffs do support *multilateral* tariff liberalization. Clearly, any reduction in border protection is a threat to some U.S. industries that produce products such as sugar, peanuts, or frozen concentrated orange juice. The political power of these commodities is the most obvious U.S. obstacle to broad-based liberalization, but it may not be the most important.

Two other policies in the area of market access get less attention than high tariffs. First is TRQ administration in the United States. Much of the benefits from U.S. TRQs go to historical exporters into the U.S. market. There is less rent seeking by the United States with respect to the distribution of those rents than there is by exporting nations. Some of these are odd cases. For example, New Zealand now enjoys rents from the U.S. TRQ for beef. If this tariff were eliminated the market would be more open to Argentina, which, because it entered the market late, has a small share of the current TRQ. New Zealand meat interests calculate that they would now lose from opening this market even though they pushed hard for further liberalization back in 1993, before Argentina had controlled foot and mouth disease.

Ironically, in 1999 the United States introduced a new temporary lamb TRQ, and this one directly harms New Zealand meat exporters. As part of the Uruguay Round agreement, countries may create new trade barriers if a "surge" of imports harms a domestic industry. The United States applied this option to limit imports of lamb from New Zealand and Australia, although no unfair trade practices were suggested. The U.S. government is in the awkward position of creating trade barriers to protect a failing U.S. industry simply because someone else has a comparative advantage. This application of U.S. trade law may have been within WTO rules (the United States lost the first round of the WTO case), but it surely sent an antitrade signal to the world.

The United States is also one of the most vigorous users of WTO provisions that allow countries to limit access when "unfair trade practices" are behind imports that harm a domestic industry (Chapter 17, this volume; Table 5.1). Antidumping and countervailing duty cases have high profile in the affected industry and a remedy for industries that lose standard protection. Both countervail and antidumping have economic underpinning in theory, but both have problematic aspects in practice, especially in agriculture. The

TABLE 5.1. Value of Export Programs

| Year | Export Enhancement Program | Credit Programs (Guarantee) | Concessional Programs |
|---|---|---|---|
| | $ Million | $ Million | $ Million |
| 1980 | 0.0 | 1,417.0 | 1,341.6 |
| 1981 | 0.0 | 1,873.7 | 1,333.0 |
| 1982 | 0.0 | 1,393.2 | 1,107.6 |
| 1983 | 0.0 | 4,069.1 | 1,194.7 |
| 1984 | 0.0 | 3,646.3 | 1,505.9 |
| 1985 | 86.5 | 2,761.1 | 1,905.9 |
| 1986 | 715.7 | 2,416.5 | 1,345.0 |
| 1987 | 1,684.4 | 2,784.4 | 1,077.2 |
| 1988 | 3,313.5 | 3,880.0 | 1,469.2 |
| 1989 | 2,826.7 | 5,057.0 | 1,311.4 |
| 1990 | 2,384.2 | 4,299.6 | 1,434.5 |
| 1991 | 2,009.3 | 4,111.3 | 1,323.9 |
| 1992 | 3,296.8 | 5,564.7 | 1,516.0 |
| 1993 | 3,733.5 | 3,831.4 | 2,363.7 |
| 1994 | 3,118.9 | 2,948.5 | 1,167.3 |
| 1995 | 2,408.0 | 2,547.1 | 892.4 |
| 1996 | 135.2 | 3,240.2 | 869.5 |
| 1997 | 254.9 | 2,894.5 | 673.0 |

*Source:* U.S. Department of Agriculture, Foreign Agricultural Service, "Commodity Exports by USDA Programs: Concessional, and Other Authorized Government Programs for Fiscal Years 1955-1997."

idea behind countervailing duties is the notion that if a government subsidizes the production of a product it is "unfair" to allow the product to hurt domestic producers of the same good. This idea is not necessarily based on domestic welfare if an exporting government wants to subsidize welfare increases for the importing country consumers. But it is accepted as a part of trade law and the GATT. The problem in agriculture is that many of the most traded commodities and products have some element of domestic subsidy. Thus, if all countries were to pursue countervail vigorously, much farm trade might be blocked. The peace clause in the Uruguay Round agreement was designed to avoid just this outcome (Chapter 3, this volume).

Dumping involves exporting below cost or below domestic market prices. The theory here is that of predatory pricing. A firm with market power may lower prices in the short run to drive out competition so as to raise the price later. As defined in the United States, however, costs are determined in such an arcane and biased way that dumping is found in just about every investigation. The two main problems with applying this concept in agriculture is that predatory pricing seldom makes much sense and farmers regularly sell below (ex post) cost in competitive and variable commodity markets (Chapter 17, this volume). The use of antidumping in agriculture looks to be protection to replace other lost trade barriers. And even when foreign industries expect to prevail, the cost of litigation may serve as a deterrent to serve the U.S. market.

In Seattle, the United States faced a majority of nations that wanted a review of these policies to be part of the current trade round. Unfortunately, one way the negotiators maintain support for liberalization is by assuring import competing industries that they will be protected in the case of import surges or unfair trade practices. This set of issues is one in which U.S. policy and its current negotiating position may be an obstacle to progress.

In the area of export competition, the United States has a long history of programs (Ackerman, Smith, and Suarez, 1995; Sumner, 1995). The U.S. government, through the Commodity Credit Corporation, a government owned and operated entity, sells commodities from its stocks, administers the international food aid program, finances export credit guarantees, finances export price subsidies (EEP), and funds export promotions. In its current position statement the United States ignores export programs other than explicit price subsidies, and this allows others, such as the EU, to downplay the U.S. seriousness to really discipline export programs.

Table 5.1 shows the extent of selected export programs. The value of commodities exported under the credit programs are for the GSM 5, GSM 101, GSM 102, GSM 103, GSM 201, GSM 301, and blended credit programs. The concessional programs include Public Law 480 (P.L. 480), Section 416 (b), and Food for Peace (FFP) programs. The export programs also built up during the 1980s, reaching a peak in the middle of the decade and showing a marked reduction in the middle 1990s. The years since 1995 indicate the effect of the Uruguay Round constraints as well as autonomous changes in U.S. domestic policy. The data for 1998 and 1999 show a jump in credit and concessional programs and a further drop in the EEP.

As we have seen, the United States has attempted to raise the profile of state trading exporters in the context of the current round of WTO. Many observers raise the issue of state importing in the context of TRQ administration, but the United States seems to be the major participant to express such concerns on the export side (Chapter 3, this volume; Ackerman and Dixit, 1999). This issue may be linked to antidumping and countervail, competition policy, or circumvention of export subsidy commitments. Here, however, I want to ask the United States whether the State Trading Enterprise (STE) is likely to be an obstacle of reform. The United States listed the Commodity Credit Corporation (CCC) as an STE along with several other organizations in its initial WTO notification. The question is, could the CCC withstand scrutiny as an STE in the WTO context? The existence of the CCC itself would not be a WTO issue. State trading as such is not contrary to the WTO articles. But there are issues. The definition of an STE as agreed in the Uruguay Round begins with the notion that some public or private enterprises have been given special rights and privileges in the market for agricultural products. As an explicit arm of the government, the CCC clearly has "special rights and privileges." The CCC is an agent of government policy; it cannot operate on a commercial basis because its role is to implement government policy. Thus the CCC violates GATT rules. In fact, my judgment is that there is a problem with the WTO definition or with the U.S. decision to list the CCC as an STE.

The CCC is a part of the USDA, which is a policy agency. The market activity of the CCC is explicitly an instrument of policy. The use of the CCC for these functions may be seen as an accident of history. Whether lending to farmers, guaranteeing credit to foreign buyers, shipping food aid, or financing direct payments, it simply implements policy. The policies that the CCC funds are already subject to disciplines under the Uruguay Round Agreement on Agriculture. There seems no more reason to apply additional disciplines to the CCC than to the customs authority or to government agencies that serve these same functions in the European Union. A WTO rule requiring more transparency for STEs would do little to affect the operation of U.S. farm programs, and in fact the CCC as a distinct institution does not seem particularly important to the operation of those programs.

The recent increase in outlays has caused the U.S. direct farm payment policy to attract more attention than other parts of farm policy. This was true also back at the beginning of the Uruguay Round (Miller, 1986). It is still true in trade policy circles, even though since 1996 the main program was supposed to be in the so-called green box of minimally distorting programs (Burfisher, Robinson, and Thiefelder, 2000; Young and Westcott, 1996, 2000; ABARE, 1998; Roberts, Podbury, and Hinchy, 1999). I have argued in several places that it is counterproductive to devote scarce negotiating resources attempting to discipline the great variety of domestic programs, all of which have some aspects of supply impact (Sumner, 1995). The Uruguay Round disciplines on domestic supports are widely accepted to have had little if any trade effects (Silvis and van der Hamsvoort, 1996; Konandreas and Greenfield, 1996).

Clearly there is no space in this chapter to describe U.S. programs in any detail (Young and Westcott, 1996). In recent years there has been a mix of policies, including about $6 billion in "contract" payments based on a history of production of certain crops; about $6 billion in ad hoc increases to these contract payments that are clearly linked to low market prices for some commodities; about $5 billion in marketing loan benefits tied directly to very low prices for wheat, soybeans, rice, and cotton; and $3 or 4 billion in crop insurance or ad hoc disaster benefits (Young and Westcott, 2000; Westcott, 1999). This is a lot of money even by U.S. farm program standards. And almost all are associated with the grains and cotton industries, which comprise less than one-third of all U.S. agriculture.

Our question here, however, is not whether this set of payments is good policy for the United States. Rather, the question is, What does this sort of policy mean for progress in the WTO negotiations? I argue not much unless the forces supporting liberal farm trade make a serious strategic error.

The idea underlying the colored boxes of the Uruguay Round was that various farm programs tended to have different effects on trading partners. Some farm support policies had minimal trade effects, and these would be accepted as inevitable policies to deal with externalities, public goods, and even political power by influential constituencies. The general idea that governments have a variety of reasons to transfer resources to agriculture underlies the notion of multifunctionality that has been getting attention in the context of the current WTO round (Bohman et al., 1999).

The current spate of payments to U.S. farmers resulted from the confluence of a budget surplus, relatively modest automatic new payments in the FAIR Act, and some unique political configurations. Do these payments affect trade? Of course they do. Do they block progress in the current WTO round? They do not if the round focuses on border measures. I see no chance that the United States would give up farm compensation programs outright in the short run and no chance that this is a feasible part of a WTO bargain. That is, I see no chance that the United States would eliminate farm payments in exchange for similar pledges elsewhere.

## CONCLUDING REMARKS

For this WTO round, deadlines are now quite loose. However, as Ambassador Barshefsky noted in March 2000, "While no deadline for the conclusion of negotiations has yet been established, the expiration of the peace clause at the end of 2003 should encourage countries to proceed expeditiously." Because it related to the relationship between border measures and internal supports, the peace clause focus may help negotiators along the lines that I have advocated.

I see a compromise of the following sort. The United States agrees with Korea, Quebec, and Norway that supporting farmers is the business of the sovereign governments. It agrees with the Cairns Group and western Canada that markets should be open. The United States then pushes to eliminate all border measures including credit and promotion subsidies, curb the use of new barriers in the case of import surges, eliminate the use of anti-dumping in agricultural cases, reform rules for countervailing duties to make them harder to apply in agriculture (say by showing direct and substantial effects on exports) and maintaining the line on the Sanitary and Phytosanitary Agreement.

Some may not call the result a pretty picture, but this plan could create beauty in the eyes of this beholder.

# REFERENCES

ABARE (Australian Bureau of Agricultural and Resource Economics). (1998) "Farm Income Support." *ABARE Current Issues.* 98:4(August) <http://www.abare.gov.au/pdf/CI98_4.pdf>.

Ackerman Karen Z. and Praveen M. Dixit. (1999) *An Introduction to State Trading in Agriculture.* Market and Trade Economics Division, Economic Research Service, U.S. Department of Agriculture. Agricultural Economic Report No. 783 (AER-783).

Ackerman, Karen Z., Mark E. Smith, and Nydia R. Suarez. (1995) *Agricultural Export Programs: Background to the 1995 Farm Legislation.* Agricultural Economic Report No. 716, U.S. Department of Agriculture, June.

Barshefsky, Charlene. (2000) "U.S. agricultural agenda at the WTO." Testimony of Ambassador U.S. Trade Representative, Trade Subcommittee of the Senate Committee on Finance, Washington, DC, March 7.

Bohman, Mary, Josephy Cooper, Daniel Mullarkey, Mary Anne Normile, David Skully, Stephen Vogel, and Edwin Young. (1999) "The Use and Abuse of Multifunctionality." United States Department of Agriculture, Economic Research Service, Online Publication <http://www.ers.usda.gov/briefing/WTO/PDF/multifunc1119.pdf>.

Burfisher, Mary E., Sherman Robinson, and Karen Thiefelder. (2000) "North America Farm Programs and the WTO." *American Journal of Agricultural Economics,* 82(3):768-774.

Hayes, Rita Derrick. (2000) "The World Trade Organization: Next Steps." Remarks of the U.S. Ambassador to the WTO, April 14, Malta.

Johnson, D. Gale. (1950) *Trade and Agriculture: A Study of Inconsistent Policies.* New York: John Wiley and Sons, Inc.

Josling, Timothy E., Stefan Tangermann, and Thorald K. Warley. (1996) *Agriculture in the GATT.* Houndmills and London: Macmillan Press.

Konandreas, P. and J. Greenfield. (1996) "Uruguay Round Commitments on Domestic Support: Their Implications for Developing Countries." *Food Policy.* 21(4,5):433-445.

Miller, Geoff. (1986) *The Political Economy of International Agricultural Policy Reform.* Department of Primary Industry Australia. Canberra.

Orden, David, Robert Paarlberg, and Terry Roe. (1999) *Policy Reform in American Agriculture: Analysis and Prognosis.* Chicago, IL: University of Chicago Press.

Roberts, Ivan, Troy Podbury, and Mike Hinchy. (1999) *Reforming World Agricultural Trade Policies.* Australian Bureau of Agricultural Economics, Research Report 99.12, September.

Silvis, H.J. and C.P.C.M. van der Hamsvoort. (1996) "The AMS in Agricultural Trade Negotiations: A Review." *Food Policy.* 21(4,5):527-539.

Sumner, Daniel A. (1995) *Agricultural Trade Policy: Letting Markets Work.* American Enterprise Institute Press, Washington, DC.

Sumner, Daniel A. and Stefan Tangermann. (2000) "International Trade Policy and Negotiations." In Bruce Gardner and Gordon Rausser (Eds.), *Handbook of Agricultural Economics,* North Holland Press, Amsterdam (forthcoming).

USDA, Foreign Agricultural Service. <http://www.fas.usda.gov/itp/wto/>.

Westcott, Paul. (1999) "Ag Policy: Marketing Loan Benefits Supplement Market Revenues for Farmers." United States Department of Agriculture, Economic Research Service, *Agricultural Outlook,* p. 4.

World Trade Organization. (1995) "United States' Notification Concerning State Trading Enterprises." World Trade G/STR/N/1/USA, p. 140.

World Trade Organization. (2000a) <http://www.wto.org/wto/about/>.

World Trade Organization. (2000b) *Proposal for Tariff Rate Quota Reform, Submission from the United States.* Committee on Agriculture, Geneva, Switzerland. G/AG/NG/W/58. November 14, 2000.

World Trade Organization. (2000c) *Second Special Session of the Committee on Agriculture, Statement by the United States.* Committee on Agriculture, Geneva, Switzerland. G/AG/NG/W/32. July 12, 2000.

Young, C. Edwin and Paul Westcott. (1996) *The 1996 Farm Act Increases Market Orientation.* Agricultural Information Bulletin No. 726, Economic Research Service, USDA, Washington, DC.

Young, C. Edwin and Paul Westcott. (2000) "How Decoupled Is U.S. Agricultural Support for Major Crops?" *American Journal of Agricultural Economics* 82 (August): 762-767.

Chapter 6

# Agricultural Negotiations in the Context of a Broader Round: A Developing Country Perspective

Thomas W. Hertel
Bernard M. Hoekman
Will Martin

The success of the WTO negotiations on agriculture depends heavily on whether a broader agenda can be assembled. Only with such an agenda can the tradeoffs be made that will allow for the hard decisions to reduce agricultural protection.

The Uruguay Round (UR) of multilateral trade talks, concluded in 1994 after eight years of often confrontational negotiations, was a landmark in the history of the trading system. Agriculture, and textiles and clothing, two sectors that for all intents and purposes had been removed from the ambit of the General Agreement on Tariffs and Trade (GATT), were brought back into the fold. The system of multilateral rules was extended to include intellectual property rights and services.

Reflecting the very limited liberalization that had occurred in agriculture and services, the two agreements on these subjects included provisions calling for new negotiations within five years of the entry into force of the WTO. Other WTO agreements contained review provisions. In order to increase the scope for beneficial trade-offs across issues, the 1998 WTO Ministerial meeting called for the development of an agenda for "further liberalization sufficiently broad-based to respond to the range of interests and concerns of all members" (WTO, 1998). In the lead-up to the subsequent Ministerial meeting that was expected to launch a new round, numerous proposals were submitted by WTO members regarding the issues that should be included on a negotiating agenda. In the event, the November 1999 Ministerial meeting in Seattle turned out to be a fiasco, failing to launch a round.

Domestic U.S. politics played a key role in the failure to attain consensus on a broad negotiating agenda, greatly reducing the willingness of the U.S.

administration to agree to put items on the table that were opposed by domestic lobbies. Strong differences on the scope of agricultural liberalization between the EU on the one hand and the United States and other agricultural exporters on the other were also important.[1] Another major factor was the active and full-fledged participation by developing countries, many of which refused to accept the agenda being pushed by a number of high-income countries in some areas—most notably the United States on labor standards. Many also expressed general dissatisfaction concerning the process through which a negotiating agenda was being set. Small countries in particular perceived themselves to be left completely in the cold, not having access to the forums where potential agenda-setting compromises were being crafted.

This chapter summarizes some of the results emerging from a collaborative research and capacity-building project involving scholars in developing countries,[2] international experts, and World Bank staff. The aim of the project is to generate analysis, both cross-country and country-specific, on the costs and benefits of further multilateral rule-making and liberalization. The chapter focuses both on market access issues (the potential gains from further liberalization of trade in agriculture, manufactures, and services) and topics that are of particular concern to developing countries—implementation of Uruguay Round agreements and the operation of the WTO, and attempts to expand WTO disciplines on national regulatory policies.

## AGRICULTURAL LIBERALIZATION

Barriers to trade tend to be highest in agriculture and services. Average tariffs on agricultural imports are in the 15 to 20 percent range, with peaks for some commodities exceeding 100 percent in many countries, both developing and developed. In contrast, average manufacturing barriers are quite low in OECD (Organization for Economic Cooperation and Development) countries, but significantly higher in developing nations. However, certain manufactures (such as clothing) continue to confront high tariffs in many high-income countries. Tariff barriers faced by developing countries on their exports of agricultural products are estimated to average 15.6 percent in high income countries, and 20.1 percent in developing countries (Table 6.1). The rates for the industrial countries are much lower. Estimates of the implied tariffs paid (constructed by multiplying the marginal tariffs levied on the relevant trade flows by the value of the corresponding trade flow) suggest that more than half of the levies charged on developing country exports are associated with their exports to industrial countries.

These prevailing patterns of protection imply that many developing countries have a large stake in achieving significant agricultural liberaliza-

TABLE 6.1. Patterns of Protection in Agriculture, 1995

| | Importing Region | |
|---|---|---|
| Exporting Region | High Income | Developing |
| Implied tariffs paid (US $ bn) | | |
| High Income | 37 | 20 |
| Developing | 16 | 14 |
| World | 53 | 34 |
| Import-weighted average tariffs (%) | | |
| High Income | 15.9 | 21.5 |
| Developing | 15.1 | 18.3 |
| World | 15.6 | 20.1 |

*Source:* Hertel and Martin (2000)

tion. Hertel et al. (1999) build a model of the world economy in 2005—at which time UR commitments will have been fully phased in. They estimate that a 40 percent reduction in post-UR agricultural tariffs and export subsidies will cause an increase in global real income of about $60 billion per year. This figure increases by $10 billion if domestic support is also reduced by 40 percent, although the uncertainty in the degree to which such producer payments are linked to production decisions makes such analysis difficult (ABARE, 1999).

Measured in dollar amounts, developed countries capture the largest gains from liberalization, reflecting the reduction in the cost of agricultural support policies for OECD consumers. However, the percentage real income gains—reported in the first set of bars in Figure 6.1—are largest in developing regions such as South Asia (other than India) and Southeast Asia (other than Indonesia). Virtually all developing regions except the net food importing Other Middle East region experience overall gains from these multilateral reductions in agricultural protection. The bulk of these gains derive from efficiency improvements generated in the developing countries themselves (second set of bars in Figure 6.1).

### Modalities for Agricultural Negotiations

The precise outcome of the agricultural negotiations will depend heavily on the specific modalities used in the negotiations. Anderson, Hoekman, and Strutt (1999) identify the priorities in further progress on agriculture as: reducing import barriers, disciplining domestic support, and elimination of

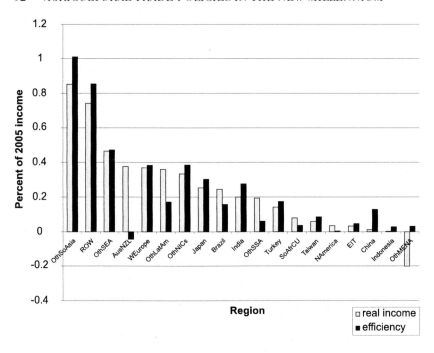

FIGURE 6.1. Implications of a 40 percent reduction in agricultural trade barriers. (*Source:* Hertel et al., 1999.)

export subsidies. As they make clear, substantial reductions in import barriers would be required to even begin to approach parity with the treatment of manufactures trade. Nothing short of elimination of export subsidies would be sufficient to do so. They also note that there is a large overlap between the agricultural reform agenda and the "second generation" regulatory issues that have been proposed for negotiation by a number of countries—competition policy, procurement, product standards, environmental regulation, and investment regimes.

The UR led to virtually complete tariffication of agricultural border protection. Unfortunately, the process that achieved this allowed substantial "dirty tariffication" in developed countries—setting tariff bindings far above the tariff implied by prevailing nontariff barriers—and very high ceiling bindings in developing countries (Hathaway and Ingco, 1995). The gap between applied tariff rates and tariff bindings in agriculture is particularly large in many developing countries (Abbott and Morse, 1999), implying that substantial reductions in tariff bindings are required to achieve any liberalization in applied rates (Francois, 1999). One approach to dealing with the gap between bound and applied rates is to make applied rates the basis for future negotiations, in effect requiring all countries to bind at applied rates. This is unlikely to be feasible. It would also create perverse incentives for countries to keep applied tariff rates high in order to conserve bargaining chips for future negotiations. A better approach is probably to devise a formula that imposes the largest reductions in the highest tariff bindings. Josling and Rae (1999) suggest the use of a "cocktail" approach in which the very highest tariffs are reduced using a formula approach, moderate tariffs are subject to a uniform percentage cut, and nuisance tariffs are abolished.

Understanding the impact of tariff-rate quotas (TRQs) is critical to predicting the outcome of attempts to liberalize trade. For example, reducing out-of-quota tariffs will increase imports only if the current demand for imports exceeds the quota amount such that the out-of-quota tariff is operational. If imports are less than the quota level, reductions in out-of-quota tariffs will be ineffective. On the other hand, marginal expansion of the TRQs will be ineffective if imports are greater than the TRQ—the only effect will be to increase the volume of imports on which scarcity rents are earned. If imports are less than the TRQ, expanding the quota will be also be ineffective. Only reductions in in-quota tariffs will stimulate greater imports in this case. The conclusion drawn is that reductions in out-of-quota tariffs would be the most effective instrument for achieving market liberalization in the majority of cases. However, it could be desirable to accompany such cuts with expansion of the quotas.[3]

Agricultural liberalization, especially moves toward elimination of export subsidies, may increase world prices of food products, and thus have a negative effect on net food-importing developing countries. However, any such impact will be offset to some degree by the increase in domestic supply that will be stimulated by higher prices. Current policies result in large global price swings that are highly detrimental to developing countries, and farmers in many developing nations suffer from a significant antiagriculture policy bias. Even if the prices of imports rise, complementary reforms at home can make net food importers better off: they are initially losing welfare by unnecessarily stimulating food imports, and the price rise curtails that stimulus (Wang and Winters, 2000). However, mechanisms are needed to ensure that any price-increasing effects of reforms do not reduce the real income and consumption of the poorest in society.

## INDUSTRIAL TARIFFS

There has been a sweeping change in the structure of international trade in the past two decades. In the mid-1960s, manufactures exports accounted for only around a quarter of developing country exports. By the early 1980s, it had only risen to around a third. Since then, growth has accelerated, and as of the mid-1990s the share had risen to around three-quarters, and is projected to go on rising (Figure 6.2). Much of the increase in the exports of developing countries during the past three decades has not followed a north-south pattern. The share of exports of developing countries going to other developing countries has risen sharply as the importance of developing countries in the world economy has risen, and barriers to trade in these countries have declined. Developing countries therefore have a strong interest in including industrial products in WTO negotiations. Although industrial countries impose low average tariffs on their imports of manufactures, the average tariff on imports from developing countries is four times higher than those originating in the OECD (Table 6.2). This is primarily because of the relatively high tariffs on products such as textiles and clothing. Estimates of the implied tariffs paid suggest that barriers that developing countries face in other developing countries account for over 70 percent of the total tariffs levied on their industrial exports. This contrasts sharply with the situation in agriculture (Table 6.1).

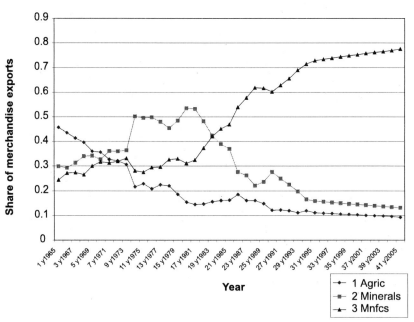

**Share of merchandise exports from developing countries**

FIGURE 6.2. The increasing share of merchandise in developing country exports. (*Source:* Hertel and Martin, 2000.)

TABLE 6.2. Patterns of Protection in Manufacturing, 1995

| | Importing Region | |
|---|---|---|
| Exporting Region | High Income | Developing |
| Import-weighted average tariffs (%) | | |
|    High Income | 0.8 | 10.9 |
|    Developing | 3.4 | 12.8 |
|    World | 1.5 | 11.5 |
| Implied tariff paid (US $ bn) | | |
|    High Income | 16 | 93 |
|    Developing | 23 | 57 |
|    World | 40 | 150 |

*Source:* Hertel and Martin (2000)

A Computable General Equilibrium (CGE) analysis of the impact of a 40 percent cut in applied tariffs on manufactures by all countries suggests that global trade volume would expand by some $380 billion in 2005—or about 4.7 percent of projected merchandise and nonfactor service trade. This increase is reflected in almost all products, including nonmanufactures. The largest increase is for wearing apparel. Even after the phase-out of multi-fiber arrangement (MFA) quotas, trade volume in this sector rises by more than 20 percent, reflecting the heavy tariff protection in high-income countries. Textiles and autos follow in importance. Real income and efficiency gains, by region, as a share of 2005 income are reported in Figure 6.3. The difference between these two variables reflects terms of trade effects. (If the real income gain exceeds the efficiency gain, then the terms of trade effect are positive, and vice versa.) Efficiency gains depend on the degree to which a country liberalizes its markets. Sharp tariff cuts give rise to increased access to cheaper imported goods, and generate gains in consumption as well as improvements in the efficiency with which domestic resources are used.

The largest efficiency gains (as a share of income) occur in developing economies, with countries or regions where tariffs are highest in the 2005 base gaining the most (China, Other South Asia, and India). China's greater gains, relative to India (which is projected to have higher protection levels in 2005), are due to the fact that the manufacturing sector in China is larger and more trade oriented. Tariff cuts in the industrialized economies of Japan, Western Europe, Australia/New Zealand, and North America generate almost no efficiency gains as tariffs are already extremely low. While their real income gains are somewhat larger due to positive terms-of-trade effects

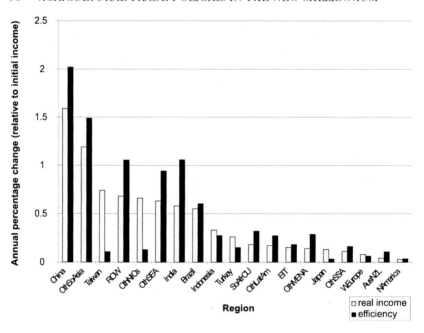

FIGURE 6.3. Welfare impact of a 40 percent cut in manufactures tariffs. (*Source:* Hertel and Martin, 2000.)

following increased exports from developing countries, the bulk of the gains go to the developing countries. The latter are estimated to receive three-quarters of the total gains from liberalizing manufactures trade.

These results suggest that there are strong economic and political-economy reasons for developing countries to support the inclusion of industrial products in any multilateral round of negotiations. From a political perspective, industrial products make up a very large share of exports, and frequently these products are produced by a relatively small number of producers who can provide active support for the politically difficult reforms required by a trade negotiation. From an economic perspective, the substantial static welfare gains outlined above are a good reason to support their inclusion, as are the potential dynamic gains associated with moving to a more outward-looking manufacturing sector.

The quantitative analysis of liberalization of trade in manufactures and agriculture that has been surveyed here is highly stylized and simplified. The use of a uniform percentage cut in applied rates of protection provides only a rough guide to the potential benefits from a broad-based liberalization. In practice, the policy instruments on which negotiations focus are tariff bindings—which may be higher than applied rates. The actual outcome will depend heavily on the precise approach to liberalization chosen. Theory predicts that the gains are likely to be larger than those indicated if negotiators choose a top-down approach that reduces the variance of protection more than a uniform cut. The gains will be less if less reduction in the variance of protection is achieved—i.e., if politically sensitive tariff peaks are preserved. Further work that takes these potential differences into account is necessary.

The research summarized uses static models and does not consider the dynamic effects of liberalization or the fact that many industries are imperfectly competitive. Current CGE techniques allow such factors to be incorporated into analyses. However, while this will affect the magnitude of the predicted net gains, it will not affect the basic message that emerges: developing countries have a major stake in the attainment of further reductions in barriers to trade in both agriculture and manufactures.

## SERVICES

In contrast to agriculture and industrial tariffs, it is much more difficult to employ numerical general equilibrium techniques to assess the potential gains from alternative liberalization options. The required information on prevailing barriers to trade and investment simply does not exist. In the case of merchandise trade the main barrier is the tariff. Although differences between tariff bindings and applied rates and accounting for preferential trade agreements and subsidies certainly complicates analysis, the prevailing policies are relatively straightforward to characterize. This is not the case with services. Frequently market access barriers are enforced "behind the border" and are embodied in regulations that control entry and/or operations, impose limitations on foreign equity holdings or nationality constraints, or require professionals to recertify as a condition for operating on a market.

Work is ongoing to improve tariff equivalent estimates of the effect of services policies  so as to be able to use this information in CGE modeling (e.g., Brown and Stern, 1999). This also involves efforts to construct openness indicators for modes of supply, especially foreign direct investment (FDI), and for specific sectors, using qualitative assessments of the extent to which actual policies raise the costs of entry and/or operation postentry.[4] One noteworthy attempt has been made by the staff of the Australian Productivity Commission, who identify existing policies affecting FDI, assign each a weight, and sum across weights to obtain an overall restrictiveness index. Their results suggest that across Asian-Pacific Economic Cooperation (APEC) countries, communications, financial services, and transport are subject to the greatest barriers to FDI, reflecting the existence of ownership limits or outright bans on foreign ownership. The most restrictive countries include Korea, Indonesia, Thailand, and China—all countries that appear to have restrictive service sectors using a variety of other measures (Francois and Hoekman, 2000).

Developing countries have a large stake in enhancing the efficiency of domestic service providers and improving their ability to contest foreign service markets. Although the large OECD countries dominate global trade in services, developing countries dominate the list of countries that are most specialized in (dependent on) services exports as a source of foreign exchange. Often this reflects the importance of tourism and/or transportation services. But developing countries have also become large exporters of transactions-processing, back-office services (Jamaica), and information and software development services (India). The potential to exploit recent and emerging technological developments—such as E-commerce—that facilitate cross-border trade in services and provide firms with incentives to slice up the value chain geographically is enormous. Recent research suggests the emphasis in the next set of General Agreement on Trade in Services (GATS) negotiations should be on three issues: (1) expanding the coverage of specific commitments; (2) increasing transparency of prevailing policies; and (3) improving multilateral disciplines.[5]

### Expanding the Coverage of the GATS

The sectoral coverage of specific commitments on national treatment and market access is limited for many countries. By one measure, high-income countries only made commitments on about half of all services, of which only half involved commitments of "free access." That is, governments committed to imposing no restrictions on market access or national treatment for only 25 percent of all service activities. Developing countries made even fewer commitments. In the case of major developing countries, on average "free access" commitments were made for only 15 percent of the service sector (Hoekman, 1996). Subsequently, successful negotiations expanded the coverage of specific commitments for basic telecoms and financial services. These negotiations were important both for keeping momentum going, and because the services involved are vital intermediate inputs. Despite the success in concluding these agreements, they have not led to a significant increase in the coverage of the GATS, as the Hoekman (1996) compilation included commitments made as of 1994 in both sectors.[6]

A strong case can be made that the GATS should cover *all* services. There is no rationale for excluding certain sectors or modes of supply from the national-treatment and market-access disciplines, given that the GATS allows for derogations of both principles. One way of moving toward this is to apply a formula approach to expanded coverage in the next round of negotiations, setting minimum coverage targets for GATS members, to be attained by a specified date (which may vary depending on per capita income level to allow for a transition period). This could include agreement that a specified share of all commitments involve full binding of status quo policies. A more ambitious approach would be to seek agreement on a deadline for full coverage to be reached. This should include politically sensitive but economically important sectors such as air and maritime transport. Weak (managed) competition on many international transport routes—both sea and air—imposes large costs on developing countries (Francois and Wooton, 2000).

From a market access perspective, developing countries have a great interest in ensuring that substantially more commitments are made with respect to supply of services through temporary entry by services providers and cross-border trade, which is of great importance for e-commerce. Although the tradability of services has been increasing rapidly due to technological developments, in many cases it remains imperative that service providers be able to work on the premises of their clients. Currently, virtually all GATS members maintain restrictions on such trade, usually through the application of economic needs tests and other requirements imposed on requests for entry visas. Achieving concrete agreements to liberalize access to services markets through mode 4 would go far toward making the GATS a more balanced market access liberalization instrument.

### Toward Greater Transparency and Better Rules

A major weakness of the GATS is that it does not force members to come clean regarding the measures that are used to restrict the ability of foreigners to contest domestic service markets. It is very unlikely that negotiators will be willing to reopen the issue of scheduling commitments, and efforts to adopt a "negative list" approach are likely to be counterproductive. But a negative list reporting exercise for transparency purposes deserves serious consideration. This would involve all members reporting information on all measures that affect market access and national treatment in all services sectors. It would result in a comprehensive database of status quo policies and provide a focal point for reform efforts.

Ideally, scheduling of liberalization commitments should shift from the sectoral (specific) to the horizontal (general). This would allow negotiating efforts to center more on developing disciplines that make sense from a long-term growth and economic development perspective. In general, these are likely to focus on safeguarding the contestability of markets while maintaining national sovereignty to regulate activities to attain health, safety, prudential and related objectives. In this perspective, it may be useful to consider "generalizing" the appropriate parts of the so-called Reference Paper for telecoms to other infrastructure network services in order to establish a "horizontal" set of pro-competitive disciplines. Work could also usefully be done to strengthen the reach of the most favored nation (MFN) principle and extend it to the area of standards and certification to ensure that (mutual) recognition agreements minimize discrimination.

A number of issues were left open after the Uruguay Round, including whether to adopt rules on procurement, subsidies, and safeguards. Evenett and Hoekman (2000) argue GATS-specific disciplines on procurement should not be sought, as any disciplines should cover both goods and services. Moreover, what really matters for foreign firms is to have access to services procurement markets, and frequently this can only be achieved if they have a commercial presence in a country. In such cases the binding constraint is not a policy of discrimination, but the ability of foreign firms to establish (enter). This suggests the focus of attention should be on expanding market access commitments under the GATS.

Multilateral disciplines on subsidies might help avoid mutually destructive policies from the viewpoint of developing countries—e.g., seeking to attract FDI via the use of incentives. Subsidies are an important source of distortions in OECD markets for some services (e.g., transport). However, to be effective in disciplining the use of firm-specific fiscal incentives, subsidy rules will have to be quite comprehensive to ensure that countries cannot sidestep them through the use of alternative policies. Here again the same conclusion arises as for procurement—any disciplines should be general, not sector-specific.

One area where a more compelling case can be made for service-specific rules concerns safeguards. The limited nature of liberalization commitments on mode 4—temporary movement of service providers—may in part be due to the nonexistence of safeguards instruments. Given that this mode of supply is of major interest to developing countries, one could envisage a safeguard instrument that is tied to mode 4 liberalization commitments. These safeguards could be explicitly aimed at providing country governments with an insurance mechanism that can be invoked if liberalization should have unexpected detrimental impacts on their societies (Hoekman, 2000).

## *INDUSTRIAL, INVESTMENT, AND EXPORT DEVELOPMENT POLICIES*

Many developing countries pursue a variety of industrial development, agricultural extension, and export promotion programs. These may involve assistance with adopting new technologies, penetrating new markets, and general advertising campaigns that aim at "selling" the country and enhancing the visibility of export products. During the 1990s an increasing number of countries also implemented so-called matching grant schemes that subsidize a proportion of the cost of improving production facilities, obtaining ISO 9000 certification of management systems, and exploring new export markets.

In almost all these cases, the rationale for activist policies is the existence of distortions created by market failures or other government policies. It is well known that if the source of the problem is policy-induced, the case for a subsidy is very much a second-best one (Bora, Lloyd, and Pangestu, 2000). A good case can be made that the stricter disciplines on export subsidies and Trade Related Investment Measures (TRIMs) are generally likely to be beneficial. Export subsidies are distortionary for the world as a whole and can easily be captured by private interests seeking rents. In practice they are very difficult to justify on the basis of distortions or market failure. In contrast, production subsidies (and taxes) can be an efficient way to offset externalities, and are allowed under WTO rules (although the effect of direct subsidies may be countervailed by importing countries if they can be shown to materially injure domestic competitors). The adoption of a green box approach toward subsidies in the Uruguay Round allows substantial freedom for governments to use subsidy instruments in cases where this is called for on economic grounds, and reduces the scope for other countries to second guess the motivation underlying the use of such instruments.

Neither the economics nor the political economy of seeking WTO disciplines on FDI are straightforward (Markusen, 1999). Hoekman and Saggi (2000) argue that there are potential payoffs, but that these may be difficult to realize. They also argue that in the area where FDI matters most as a mechanism to contest markets—services—a WTO instrument already exists. The GATS extends to FDI policies as countries can make specific market access and national treatment commitments for this mode of supply for any or all services. Thus, a lot can already be achieved using existing structures.

As far as more traditional TRIMs are concerned, the available empirical evidence suggests that local content and related policies are generally ineffective or costly to the economy—they often do not achieve the desired backward and forward linkages, encourage inefficient foreign entry, and create potential problems for future liberalization as those who enter lobby against a change in regime. The major policy question is implementing the agreement—that is, phasing out illegal TRIMs. Governments may be constrained in eliminating costly status quo TRIMs because protected industries are politically powerful. The UR agreement incorporated transition periods, but these were not based on economic criteria. Some countries may need extensions of transition periods as well as assistance in designing effective and credible transition paths.

## ACHIEVING BALANCE: RULE MAKING
## AND IMPLEMENTATION

Resource constraints impede the ability of many developing countries to identify and defend their interests in multilateral negotiations and WTO activities (Blackhurst, Lyakurwa, and Oyejide, 2000). Even if countries are able to influence the set of subjects to be negotiated so that notional symmetry prevails in terms of defining the agenda, outcomes can easily be asymmetric, reflecting differences in negotiating power. Under GATT this asymmetry was exemplified by the exclusion of agriculture and textiles and clothing from many multilateral disciplines, and the use of various instruments of contingent protection, including some that were illegal under GATT—e.g., voluntary export restraint agreements (VERs). In the Uruguay Round, negotiating power asymmetries were illustrated by the MFA/VER abolition for Trade Related Intellectual Property Rights (TRIPS) deal—the extraction of "payment" for elimination of practices that violated the spirit, if not the letter, of the GATT—and by the fact that very little, if anything, was done to significantly enhance developing country opportunities to export services (e.g., through the movement of natural persons).

Asymmetry under the Uruguay Round also was reflected in the fact that developing countries became subject to a large number of disciplines in areas that under GATT were voluntary—including customs valuation, antidumping, subsidies, technical product standards, and sanitary and phytosanitary measures—as well as new rules in areas such as intellectual property. In these areas it is difficult, if not impossible, to trade "concessions." Negotiators focused instead on the identification of specific rules that should be adopted by all. In practice the norms chosen were those that were (are) applied in industrialized countries. In contrast to traditional trade liberalization, a "one size fits all" approach may not be optimal (Finger and Hoekman, 1999). Nonetheless, "one size fits all" was a central pillar of the Uruguay Round—developing countries were only granted additional time in which to implement obligations. In many areas these periods were five years and expired at the end of 1999.

Implementation became an issue in part because the costs associated with complying with some WTO agreements can be significant. As noted by Finger and Schuler (2000), such costs can easily exceed the entire development budget of a least developed country.[7] It is not at all clear from a development perspective that the resources required for implementation of WTO agreements, whatever the amount might be, would not be better used to build schools or improve infrastructure. Ensuring that WTO agreements are conducive to (consistent with) attainment of development objectives should be a major objective of the next round. This requires development of mechanisms and methods to determine whether and when disciplines in specific areas should be implemented, and flexibility to allow countries to experiment with domestic regulatory regimes and mechanisms that maximize national welfare.

At the time of the Uruguay Round, there was only limited developing country experience in the "new areas" on which the negotiators could draw. Poor countries have yet to attempt to create intellectual property regimes that make traditional knowledge or cultural products into a negotiable and defensible asset. Nor have they identified the alternative options that can be used to upgrade and enforce national product, health, and safety standards, or to regulate service sectors that are subject to market failures. In many of these areas, the trial and error experience—the assessments of the real-world impacts of alternative policy options—that can inform the effective incorporation of the development dimension into multilateral rules does not exist. The implication is that WTO rules should allow for experimentation and learning, and that implementation of UR commitments should be made conditional on this being done in a way that makes economic (development) sense (Finger and Hoekman, 1999).

### *Intellectual Property Rights and Competition Policy*

The Uruguay Round TRIPS agreement obliged all WTO members to enforce intellectual property rights (IPRs), though with transition periods for developing countries. Whether developing countries will gain from stronger protection of IPRs is a matter of vigorous debate. Those in favor argue that dynamic benefits—operating through FDI, technology transfers and licensing, and innovation within and for the domestic market—will more than offset any static losses. Those against note that dynamic benefits are uncertain, while on balance the short-run impact of the TRIPS regime—which will operate through higher prices, lower domestic output, and more imports—is likely to cause a transfer of income from poor to rich countries, with at best marginal impacts on economic efficiency, resulting in net transfers to firms in high-income countries.[8]

The scale of the transfer very much depends on the market structures that prevail and the closeness of available substitutes. Estimates for Lebanon by Maskus (2000b) suggest that because Lebanon is a net importer of pharmaceutical products and technologies and currently has relatively little inventive capability in the sector, the static impacts of stronger patents are likely to be negative, increasing average prices by some 10 percent and resulting in lower output and fewer firms. But in other sectors where Lebanon produces IPR-sensitive products (such as printing and publishing, music, and film/video), stronger protection would be beneficial. The static net impact is therefore unclear. Dynamic effects are even more uncertain.

An important factor determining the impact of the TRIPS agreement is the ability of governments to intervene to offset socially detrimental outcomes. The agreement has a number of provisions that authorize the use of policy measures against abuses of IPRs. Competition law has an important role to play in this connection. For example, as right holders will frequently use their IPRs to segment markets, developing countries may have a strong interest in applying an international exhaustion rule (which implies allowing parallel imports). This would imply that domestic buyers could purchase patented and branded products wherever they find the most favorable prices. This is fully compatible with the TRIPS agreement, even though both the EU and the United States are active proponents of a national/regional approach to exhaustion. Whether to adopt an international exhaustion rule should be a matter for national authorities to decide independently. Of course, this is just one aspect of the inter-relationship between IPR and competition regimes, but it serves to illustrate that here again a "one size fits all" rule should be avoided (Maskus, 2000a).

Developing countries have an interest in adopting strong competition policies, the main pillar of which should be a liberal trade and FDI policy stance. Competition law is required to ensure markets are contestable, especially in nontradable sectors. Antitrust legislation may also be required to maximize the benefits (or minimize the costs) of certain WTO agreements, the TRIPS agreement being one example. Hoekman and Holmes (1999) argue that developing countries should use the occasion of a trade round to put their interests on the table, recognizing that the quid pro quo they can expect will depend importantly on what they are willing to offer. Seeking modifications in antidumping law and commitments by OECD competition authorities to provide assistance to developing country competition authorities are examples of the type of quid pro quo that could be sought. Realism suggests, however, that the primary focus should be on the design of appropriate national policies. Once more experience has been obtained with the design and implementation of national regulations, countries will be better able to judge what type of multilateral agreement in this area is appropriate.

## *LABOR, ENVIRONMENTAL, AND RELATED STANDARDS ISSUES*

The collapse of the Seattle Ministerial meeting was in part due to differences in views between developed and developing countries on whether the WTO should regulate and enforce labor and environmental standards. Determining the extent to which multilateral disciplines should extend "behind the border" and cover more areas of domestic regulation is one of the key challenges facing WTO members. While the case for domestic regulation to address market failures or pursue noneconomic objectives is indisputable, the case for international harmonization is not. Efforts to impose standards on all WTO members and enforce them via the threat of trade sanctions threaten to embroil the WTO in issues in which it has neither the technical ability nor the political legitimacy to act effectively (Rollo and Winters, 2000).

The TRIPS agreement reflects the first clear attempt at harmonizing domestic policy and setting certain minimum absolute standards of IP protection to which all members must adhere. The GATT was an instrument of negative integration—it revolved around agreements *not* to do certain things (discriminate in trade policy, raise tariffs above bound levels, etc.). TRIPS is the first major example of what has been called "positive" integration (Tinbergen, 1954). One result is that for the first time failure to implement a certain type of regulatory regime (in this case prevent the use of production processes by domestic firms that violate IPRs) can give rise to dispute settlement and possibly trade sanctions.

A question that has arisen concerns the implications of TRIPS on the scope of the WTO. If IPRs can be brought into the WTO, why not other areas of domestic regulations as well? A useful approach to answering this question is offered by Maskus (2000a), who identifies a series of screens or criteria that should be used to determine if there is a good economic case for bringing a regulatory area under the WTO. These are that the issue: (1) be strongly trade-related; (2) gives rise to international externalities; (3) is associated with policy coordination failures that can be addressed effectively through WTO dispute settlement; and (4) has the potential to strengthen the trading system. In the case of labor standards, Maskus surveys the literature and concludes that, first, there is very little credible evidence that deficient enforcement of core labor standards has an impact on trade. Second, he finds that both the theoretical and empirical bases for arguing that lax labor standards in developing countries suppress wages of low-skilled workers in OECD nations are very weak. Third, he argues that WTO-type enforcement (relying on trade sanctions) will worsen labor outcomes, and finally, he points out that the linkages between core labor standards and existing WTO disciplines are nonexistent (Maskus, 1997, 2000a).

There is considerable international agreement that certain core labor rights should be globally recognized and protected. Development and encouragement of implementation of such rights is the task of the ILO. One of the principal arguments for inclusion of labor standards in the WTO is to provide an enforcement mechanism for ILO conventions. However, trade remedies to enforce labor standards, as proposed by Rodrik (1997) and de Wet (1995) should be resisted: they would worsen the problems at which they are aimed (by forcing workers in targeted countries into informal or illegal activities) and burden the trading system (by increasing the likelihood of controversial disputes). Account should also be taken of the nonnegligible danger that such instruments will be captured by protectionist interests, seeking to limit imports from labor-abundant developing countries. Attainment of core labor standards can be pursued more effectively through instruments that are targeted directly at improving outcomes.

In the case of the environment, the argument for adopting substantive rules in the WTO is also weak. However, one important difference as compared to labor standards is that trade policies may have adverse environmental consequences. An example would be subsidies and trade protection for the coal industry—this may discourage a shift to cleaner-burning fuels. There may also be potentially important cross-border environmental spillovers. Some multilateral environmental agreements (MEAs) include trade sanctions as enforcement instruments. As this can give rise to WTO dispute settlement if a nonsignatory is targeted by signatories, the WTO membership may need to devise procedures that set out the conditions under which they would permit such sanctions. Rollo and Winters (2000) suggest the following as necessary conditions to ensure legitimacy and reduce the chances of protectionist capture:

- Sanctions must be genuinely a last resort.
- Decisions to sanction a transgressor should be collective, with a very substantial majority of WTO membership.
- Provision must be made for frequent review to determine if sanctions can be removed.
- The sanctions should be applied by all signatories of the MEA.
- WTO members should have no discretion about what products to restrict.

## DEVELOPING COUNTRY PARTICIPATION

Many of the contributions that have emerged from the ongoing research on which we have drawn emphasize that participation constraints are a general problem, in particular for least developed countries; that lack of information and limited cross-country experience greatly constrains the ability of countries to exploit the "wiggle room" that is embodied in many WTO agreements; that there is a need to ensure that governments have the scope to pursue policies in a manner that makes sense from a development perspective; that fulfillment of offers of financial and technical assistance by high-income countries have proven to be disappointing; and that provisions requiring such countries to take into account the interests of developing countries have proven to be meaningless (Blackhurst, Lyakurwa, and Oyejide, 2000;

Finger and Schuler, 2000; Finger and Hoekman, 1999; Michalopoulos, 1999; Hoekman and Mavroidis, 2000).

The end result has been an absence of developing country "ownership" of many agreements, and a general suspicion of the WTO in large segments of civil society. This can only be remedied if the next round results in a more balanced outcome (Stiglitz, 2000), one that addresses the selective liberalization and rule making that has been a characteristic of the system to date, and allows greater flexibility regarding the specific rules that are imposed on WTO members. The preconditions for achieving greater balance appear to be there—developing countries have demonstrated a willingness to participate actively and constructively in the WTO. This was reflected in the run-up to Seattle and the role played in the process of defining a negotiating agenda. The inability (unwillingness) of the industrial countries to accept the necessary compromises helped scuttle the talks, but arguably this has helped set the stage for a more balanced agenda to be crafted in the future. That said, to paraphrase Wang and Winters (2000), much will have to be done to "put Humpty Dumpty back together again" and repair the damage manifested in Seattle.

## *CONCLUDING REMARKS*

This chapter surveys some recent research regarding the interests of developing countries going into a new round of WTO negotiations. It is impossible to do full justice to the complexity of the issues that arise in many of the areas that could figure on the WTO negotiating agenda. Accordingly, readers are referred to the papers provided in the references for a more thorough treatment of these issues.[9] It is also impossible to generalize regarding the interests of developing countries. Nations are very diverse, reflecting differences in per capita incomes, initial conditions, and endowments.

Starting in the late 1970s, assessments of trade rounds have relied heavily on computable general equilibrium (CGE) modeling techniques. These allow the economy-wide effects of policy changes to be simulated. Such models are particularly well suited to assessing the impact across industries and countries of reductions in tariffs, quotas, and subsidies. We have summarized some key findings from one such model with respect to agricultural and manufacturing liberalization in the wake of the Uruguay Round. We conclude that the biggest winners from further liberalization are likely to be developing countries. Indeed, given their heavy reliance on manufacturing exports and the relatively high tariffs projected to remain after the Uruguay Round, developing countries are expected to reap the majority of the benefits from further tariff cuts in this sector.

It is important to separate initiatives that aim at directly reducing barriers to trade from initiatives to further expand and deepen WTO provisions pertaining to regulatory regimes. The available research strongly suggests that the potential benefits from vigorous pursuit of a market access agenda are significant. This conclusion spans agriculture, trade in manufactures, and services. Future efforts in the WTO should center on further reductions in traditional barriers to trade, which in the case of services include policies that restrict the ability of foreign firms to contest markets through a variety of entry modes, including FDI. Much of what an investment agreement might do can be achieved via the GATS. Significant scope exists for mutually beneficial quid pro quos within the services area. Although industrial products are notably absent from the "built-in" negotiating agenda, developing countries have a strong interest in taking up the banner of manufacturing tariff cuts given that manufactures account for some three-quarters of their merchandise exports. Efforts to further discipline the ability of governments to abuse instruments of contingent protection, minimize the trade-restricting impact of product standards, especially sanitary and phyto-sanitary measures, and facilitate trade more generally—while not discussed in this chapter due to space constraints—are also of great importance (see UNCTAD, 1999; Messerlin and Zarrouk, 2000).

A good case can be made for shifting away from attempts to introduce substantive rules on domestic regulatory and legal regimes that imply harmonization to high-income country standards. Instead, the focus could more fruitfully be on procedural disciplines that aim at ensuring transparency of policies, and facilitating the adoption of national policies that assist governments to implement development programs and attain sustainable economic growth. Concretely, this implies a need to revisit the issue of transition periods for certain UR agreements, rejection of attempts to negotiate substantive multilateral disciplines in areas such as investment and competition law until more experience has been obtained at the national level, and refusal to consider the inclusion of rules on labor standards and the environment in the WTO. Doubts can be expressed regarding the payoffs associated with the introduction of substantive disciplines in the WTO on domestic regulatory regimes if this entails harmonization to OECD norms that have not been determined to be in the interests of least-developed countries. However, such doubts certainly do not extend to the bread and butter of the multilateral trading system—the progressive liberalization of barriers to trade in goods and services on a nondiscriminatory basis. This is an area where there still remains much to be done, and where traditional GATT negotiating modalities can be an effective mechanism to overcome resistance to reform, in both developing and developed countries.

## NOTES

1. For a concise report see the December 11, 1999, issue of *The Economist*. Wang and Winters (2000) discuss the implications of the Seattle failure for developing countries and for policies toward such countries.

2. In addition to researchers based in national think tanks, the project draws on the work of a number of research networks, including the Latin American Trade Network (LATN); the Economic Research Forum for the Arab Countries, Iran, and Turkey (ERF); the African Economic Research Consortium (AERC); the Coordinated African Program of Assistance on Services (CAPAS); and the Trade Policy Forum of the Pacific Economic Cooperation Council.

3. Elbheri et al. (1999) find that reducing over-quota tariffs on sugar imported into the United States and the EU by one-third results in losses for almost half of the exporting countries, a consequence of the reductions in quota rents. When the out-of-quota tariff reduction is paired with a 50 percent increase in the TRQ, most countries experience gains, and the remaining losses fall to negligible levels. Thus there may be some grounds for liberalizing on both price and quantity margins simultaneously in order to secure acceptance of the overall liberalization program by importing and exporting nations alike.

4. See Warren and Findlay (2000) for an excellent survey of recent work.

5. For a comprehensive discussion, see the contributions in Sauvé and Stern (2000).

6. However, the quality of the commitments, especially in basic telecom, improved substantially. Of particular importance was the adoption of the so-called Reference paper on regulatory principles.

7. While not costing high-income countries anything, given that the rules basically codified existing practices in these countries.

8. See Primo Braga (1996) for a survey of the literature.

9. Many of these papers have been posted on the Web site: <www.worldbank.org/trade>.

## REFERENCES

ABARE. 1999. *Reforming Agricultural Trade Policies,* Research Report No. 99-12, Canberra, Australia.

Abbott, P. and Morse, A. 1999. "TRQ Implementation in Developing Countries." Paper presented to the Conference on Agriculture in the WTO 2000 Negotiations, October 1-2, Geneva.

Anderson, K., Hoekman, B., and Strutt, A. 1999. "Agriculture and the WTO: Next steps." <www.worldbank.org/trade>.

Blackhurst, R., W. Lyakurwa, and A. Oyejide. 2000. "Options for Improving Africa's Participation in the WTO." *The World Economy,* 23:491-510.

Bora, B., P.J. Lloyd, and M. Pangestu. 2000. "Industrial Policy and the WTO." *The World Economy,* 23:543-560.

Brown, D. and R.M. Stern. 1999. "Measurement and Modeling of the Economic Effects of Trade and Investment Barriers in Services." World Bank Mimeo. Washington, DC.

de Wet, E. 1995. "Labor Standards in the Globalized Economy: The Inclusion of a Social Clause in the GATT/WTO." *Human Rights Quarterly* 17:443-462.

Elbheri, A., Ingco, M., Hertel, T., and Pearson, K. 1999. "Agriculture and WTO 2000: Quantitative Assessment of Multilateral Liberalization of Agricultural Policies." Paper presented at the Conference on Agriculture and the New Trade Agenda in the WTO 2000 Negotiations, WTO, Geneva, October 1-2.

Evenett, S. and B. Hoekman. 2000. "Government Procurement of Services and Multilateral Disciplines." In P. Sauve and R. M. Stern (Eds.), *GATS 2000—New Directions in Services Trade Liberalization* (pp. 151-175). Washington, DC: Brookings Institution.

Finger, J.M. and B. Hoekman. 1999. "Developing Countries and a New Trade Round: Lessons from Recent Research." World Bank Mimeo. Washington, DC.

Finger, J.M. and P. Schuler. 2000. "Implementation of Uruguay Round Commitments: The Development Challenge." *The World Economy,* 23:511-526.

Francois, J. 1999. "The Ghost of Rounds Past: The Uruguay Round and the Shape of the Next Multilateral Trade Round." Paper presented at the Conference on Agriculture and the New Trade Agenda in the WTO 2000 Negotiations, WTO, Geneva, October 1-2.

Francois, J. and B. Hoekman. 2000. "Estimates of Barriers to Trade in Services." World Bank Mimeo. Washington, DC.

Francois, J. and I. Wooton. 2000. "Trade in International Transport Services: The Case of Competition." Mimeo <www.intereconomics.com/francois>.

Hathaway, D. and M. Ingco. 1995. "Agricultural Liberalization and the Uruguay Round." In Martin, W. and Winters, L.A. (Eds.), *The Uruguay Round and the Developing Countries.* Cambridge: Cambridge University Press.

Hertel, T., K. Anderson, J. Francois, and W. Martin. 1999. "Agriculture and Non-Agricultural Liberalization in the Millennium Round." Paper presented at the Conference on Agriculture and the New Trade Agenda in the WTO 2000 Negotiations, WTO, Geneva, October 1-2.

Hertel, T. and W. Martin. 2000. "Liberalizing Agriculture and Manufactures in a Millennium Round: Implications for Developing Countries." *The World Economy,* 23:455-470.

Hoekman, B. 1996. "Assessing the General Agreement on Trade in Services." In W. Martin and L.A. Winters (Eds.), *The Uruguay Round and the Developing Countries* (pp. 117-124). Cambridge: Cambridge University Press.

Hoekman, B. 2000. "Towards a More Balanced and Comprehensive Services Agreement." In J. Schott (Ed.), *Preparing for the Seattle WTO Ministerial* (pp. 119-136). Washington, DC: Institute for International Economics.

Hoekman, B. and P. Holmes. 1999. "Competition Policy, Developing Countries and the WTO." *The World Economy,* 22:875-893.

Hoekman, B. and P. Mavroidis. 2000. "WTO Dispute Settlement, Transparency and Surveillance." *The World Economy,* 23:527-542.

Hoekman, B. and K. Saggi. 2000. "Assessing the Case for Multilateral Disciplines on Investment-related Policies." Policy Research Working Paper 2138 <www.worldbank.org/trade>.

Josling, T. and Rae, A. 1999. "Multilateral Approaches to Market Access Negotiations in Agriculture." Paper presented at the Conference on Agriculture and the New Trade Agenda in the WTO 2000 Negotiations, WTO, Geneva, October 1-2.

Markusen, J. 1999. "Commitment to Multilateral Rules on Investment: The Developing Countries' Stake." <www.worldbank.org/trade>.

Maskus, K. 1997. "Should Core Labor Standards Be Imposed Through International Trade Policy?" Policy Research Working Paper 1817, World Bank. Washington, DC.

Maskus, K. 2000a. "Regulatory Standards in the WTO: Comparing Intellectual Property Rights with Competition Policy, Environmental Protection and Core Labor Standards" World Bank Mimeo. Washington, DC.

Maskus, K. 2000b. "Strengthening Intellectual Property Rights in Lebanon." In B. Hoekman and J. Zarrouk (Eds.), *Catching Up with the Competition: Trade Opportunities and Challenges for Arab Countries* (pp. 251-284). Ann Arbor: University of Michigan Press.

Messerlin, P. and J. Zarrouk. 2000. "Trade Facilitation: Technical Regulations and Customs Procedures." *The World Economy* 23:577-594.

Michalopoulos, C. 1999. "Trade Policies and Market Access Issues for Developing Countries." Policy Research Working Paper 2214, World Bank. Washington, DC.

Primo Braga, C. 1996. "Trade-Related Intellectual Property Issues: The Uruguay Round Agreement and Its Economic Implications." In W. Martin and L.A. Winters (Eds.), *The Uruguay Round and the Developing Countries* (pp. 381-412). Cambridge: Cambridge University Press.

Rodrik, D. 1997. *Has Globalization Gone Too Far?* Washington, DC: Institute for International Economics.

Rollo, J. and L.A. Winters. 2000. "Subsidiarity and Governance Challenges for the WTO: The Examples of Environmental and Labour Standards." *The World Economy,* 23(4)(April):455-470.

Sauvé, P. and R. Stern (Eds.), 2000. *Services 2000: New Directions in Services Trade Liberalization.* Washington, DC: Brookings Institution.

Stiglitz, J. 2000. "Two Principles for the Next Round, or, How to Bring Developing Countries in from the Cold." *The World Economy,* 23:437-454.

Tinbergen, J. 1954. *International Economic Integration.* Amsterdam: Elsevier.

UNCTAD. 1999. *A Positive Agenda for Developing Countries: Issues for Future Trade Negotiations.* <www.unctad.org/en/posagen>.

Wang, Z.K. and L.A. Winters. 2000. "Putting 'Humpty' Together Again: Including Developing Countries in a Pro-WTO Consensus." World Bank Mimeo. Washington, DC.

Warren, T. and C. Findlay. 2000. "How Significant are the Barriers? Measuring Impediments to Trade in Services." In P. Sauvé and R. Stern (Eds.), *Services 2000: New Directions in Services Trade Liberalization* (pp. 58-83). Washington, DC: Brookings Institution.

Chapter 7

# Agricultural Trade Liberalization and the Environment: Issues and Policies

Wesley Nimon
Utpal Vasavada

## *INTRODUCTION*

Although economists have long argued that a country serves its own interest by adopting free trade, recent attempts at further trade liberalization by the World Trade Organization (WTO) have come under heavy criticism. Some environmental, consumer, labor, and religious groups formed a broad "civil society" coalition, which blamed trade liberalization for numerous maladies including domestic job losses, environmental damage, low wages, and child labor in less developed countries (LDCs). The main objective of this chapter is to summarize the key issues raised by fourteen major environmental nongovernmental organizations (NGOs) at the Seattle Ministerial meeting and to review concrete policy proposals these specific NGOs made for reform of the WTO. Although the Ministerial meeting failed to agree on objectives for a new round of negotiations, the built-in agenda of the Uruguay Round Agreement on Agriculture has allowed WTO member countries to launch new agricultural negotiations. Environmental NGOs will likely continue to attempt to affect the outcome of these negotiations.

The plan of this chapter is as follows. To frame the recent debate between environmentalists, as characterized by fourteen major environmental groups, and free traders, as characterized by economists Paul Krugman and Jagdish Bhagwati, the next section briefly surveys the theoretical arguments and the empirical evidence regarding the environmental impact of trade liberalization. This cursory survey provides a context to assess the economic merit of recent demands made by environmental NGOs for WTO reform. The following section discusses the demands that were made by fourteen major environmental NGOs at the recent Seattle WTO Ministerial meeting. These demands were predicated on a belief that further trade liberalization is inimical to the environment and can contribute to an unsustainable pattern of natural resource use. The next section, Countries' Agricultural Negotiating Positions, examines whether attempts by some environmental NGOs to influence the negotiating positions adopted by WTO member countries were successful. Negotiating positions of principal trading nations on agri-environmental issues are reviewed in this section. While there were common elements in the negotiating positions adopted by some WTO member countries, numerous unresolved differences remain to be addressed. The final section explores some unresolved agri-environmental issues that continue to divide WTO member nations and will likely slow the pace of further agricultural trade liberalization.

## THE IMPACT OF TRADE LIBERALIZATION ON THE ENVIRONMENT

Proponents of free trade argue that, with a few notable exceptions, a nation serves its own economic interest by unilaterally adopting free trade, regardless of whether other countries erect trade barriers. A frustrating reality, however, is that trade negotiations are usually predicated on a mercantilist philosophy in which trade is assumed to be a zero-sum game. No matter how great the opportunity cost of domestic production, all imports are viewed as a necessary evil to be tolerated only in exchange for exports. Free traders are left to reiterate their arguments for free trade to each other or join the larger debate, indulge in a bit of mercantilist language, and attempt to salvage a second-best outcome from negotiations based on a fallacious premise.

Environmental groups, such as the Sierra Club and Greenpeace USA, often urge negotiators to call for upward harmonization of environmental and labor standards to the more stringent Organization for Economic Cooperation and Development (OECD) criteria (Downs, 1999). While this may appeal to some OECD nations, many LDCs, who in part derive their comparative advantage from less stringent regulations, may object to this proposal. Free traders point out that as long as domestic and world prices differ (assuming domestic prices accurately reflect the opportunity cost), there are efficiency gains from trade. Furthermore, the gains from trade are not contingent on the source of these price differences. Even if the source of a nation's comparative advantage is exploitation of child labor or operating factories that pollute the environment, the economic gains from trade remain undiminished and harmonization should be rejected (Krugman, 1997).

While the economic benefits of trade may create welfare gains, adoption of lower environmental and labor standards may violate first-world sensibilities, resulting in calls for "upward harmonization" of these standards. A diverse set of standards across countries, however, should be expected to emerge due to different incomes and/or initial conditions (Bhagwati, 1996). For example, suppose the United States has clean water and dirty air and Canada has clean air and dirty water. The United States is likely to prefer devoting resources to air pollution abatement and Canada to water pollution abatement. If the United States forces Canada to alter its air quality standards upward to reflect its preferences, then the United States gains but Canadians lose because upward harmonization forces them to move away from an optimal resource allocation consistent with Canadian public preferences. Similarly, because of its higher income, the U.S.'s optimal regulatory framework may involve more stringent environmental and labor standards than Mexico's.

Trade liberalization, however, offers the prospect for improved environmental outcomes. Through exploitation of comparative advantages, international trade improves the allocative efficiency of resources worldwide. This creates greater incomes in LDCs and increases their demand not just for material goods but for environmental amenities such as cleaner air and water as well. To the extent that government policies reasonably reflect the underlying preferences of the population they represent, stricter environmental regulations will be voluntarily enacted and enforced. As long as per capita incomes vary across countries, however, different environmental preferences are likely to persist. Nonetheless, increasing worldwide incomes create conditions under which it becomes incentive compatible for all countries, including LDCs, to raise their environmental standards.

Many free trade economists argue that trade is welfare enhancing even if it tends to temporarily increase pollution. Chichilnisky (1994), however, shows that LDCs' lack of institutional capacity to define property rights may create the illusion of comparative advantage when one does not actually exist. The result is overexploitation of the environment as a sink for waste. Resources are then underpriced and overproduction ensues. Trade liberalization then transmits this distortionary effect to the rest of the world. Bhagwati (1996) points out that there are many imperfections in the world economy, and market failures abound. That being the case, he argues that regardless of what others do, free trade is still in a nation's best interest.

The effects of trade liberalization on the environment can be decomposed into several effects (Cole, Rayner, and Bates, 1998). These effects are the technique effect, the scale effect, the composition effect, and the transportation effect.

### Technique Effect

Increasing per capita income, other things being equal, tends to result in calls for increased regulation mandating cleaner technologies. Trade liberalization may have a technique effect as producers alter production methods to adopt either cleaner or dirtier production technologies. This change can occur for several reasons.

1. Increases in consumer income may precipitate stricter environmental restrictions (Grossman and Krueger, 1995).
2. Changes in relative prices may create incentives to alter production technologies (Anderson, 1992).
3. International diffusion of clean technologies, perhaps through foreign direct investment, could reduce polluting emissions (Leonard, 1998; Wheeler and Martin, 1992).
4. Lowered environmental standards to attract industry could tend to increase polluting emissions (Barrett, 1994).

### Scale Effect

Empirical evidence has long linked open economies to economic growth (Edwards, 1992; Harrison, 1996). Increased output and scale of production, however, may generate additional pollution emissions and accelerate natural resource depletion. This is known as the scale effect. The opposite is possible as well if an economy contracts. Vukina, Beghin, and Solakoglu (1999) find that, during the transition of twelve formerly centrally planned economies to more open, market-oriented economies, large reductions in pollution emissions followed when the manufacturing sector of these economies collapsed.

### Composition Effect

Trade liberalization may also impact the composition of output produced in an economy. Resources devoted to protected inefficient industries will be utilized elsewhere. For instance, Cole, Rayner, and Bates (1998) note that the phasing out of the Multi-Fibre Agreement under the Uruguay Round will cause textile production in LDCs to increase and their heavy manufacturing sectors to contract. Since textile production is less pollution intensive than heavy industry there may be a positive composition effect, but it may be swamped by a negative scale effect.

### Transportation Effect

Increased trade tends to increase the movements of goods across countries and change the way in which they are transported. Since different methods of transportation entail different levels of pollution intensities, the net effect depends on which types of transportation dominate the posttrade liberalization world economy. The OECD estimates that the net effect will be positive but small as a switch from trucks to ships and trains mitigates the environmental impact of increased volumes of trade (OECD, 2000).

The technique, scale, composition, and transportation effects may interact to create an inverted-U relationship between income and pollution. The argument is that when a nation develops from an initial low level of income, the scale effect dominates as there is an increase in the demand for all inputs, including the environment as a sink for waste. If trade liberalization is the engine of growth, then there may be a relatively small adverse transportation effect as well (OECD, 2000). Rising incomes, however, increase the willingness to pay for environmental amenities. Regulations are enacted forcing a shift to cleaner production processes, and the "technique effect" tends to reduce harmful emissions and environmental damage. Furthermore, trade liberalization may facilitate the international diffusion of clean technologies as Wheeler and Martin (1992) found in the case of the wood pulp production industry. As resources are shifted out of protected polluting industries and/or rising incomes shift preferences to cleaner goods, then the composition and technique effects eventually dominate the scale and transportation effects. Grossman and Krueger (1995) give empirical support to this hypothesis and find that for most pollutants emissions decline before per capita income reaches $8000, or about that of the Republic of Korea.

### Pollution Haven and Pollution Migration

Copeland and Taylor (1994, 1995) make the case that lower incomes give LDCs a comparative advantage in pollution-intensive goods because a higher marginal utility of income translates into lower environmental valuations and less stringent optimal regulations. This implies that the composition effect will tend to create "pollution havens" as trade liberalization induces developed countries to shift production to less pollution-intensive goods and LDCs to shift production to more pollution-intensive goods. Despite pollution migration, trade nonetheless increases welfare because environmental regulations are assumed to be at their optimal levels. After initial trade liberalization induces pollution migration, the favorable environmental outcome becomes more likely as incomes and political pressure for environmental regulation increase. In other words, the faster economic growth caused by trade liberalization moves LDCs over the inverted U sooner. Rock (1996), however, challenges this conventional wisdom that trade liberalization creates a "win-win" scenario because outward-oriented countries tend to have greater pollution intensities, i.e., toxic chemical intensity, of GDP.

### Trade and Environmental Policy Coordination

A number of papers (Copeland, 1994; Beghin, Roland-Holst, and van der Mensbruggle, 1995; and Beghin and Potier, 1997) make the case for environmental and trade policy coordination. The incentive for firms to develop and adopt "green" goods and production methods tends to be regulation driven, and improvements would likely be much slower without regulatory inducements. Coordination of trade and environmental policies improves the probability that the adverse environmental scale effects will be soon dominated by the positive environmental effects of trade liberalization. Applying a computable general equilibrium model to Mexico, Beghin, Roland-Holst, and van der Mensbruggle (1995) find that the composition effect under joint trade liberalization and increased pollution taxes is significantly larger than either policy individually. This occurs in part as the reforms induce greater substitution of inputs for dirty domestic output. Thus, trade liberalization may reduce the cost of environmental reforms.

### Trade Liberalization and Agriculture: Empirical Evidence

There are only a limited number of empirical studies specifically focused on the environmental effects of agricultural trade liberalization, and they are often specific to a particular country. While drawing general conclusions is difficult and dangerous, the existing research provides some insights.

One study finds that multilateral trade liberalization will shift food production away from the developed countries toward LDCs that use more labor and less potentially polluting chemicals. This shift will not, however, accelerate tropical deforestation because land use is not very responsive to price changes. Because of increased worldwide allocative efficiency, trade liberalization will increase income, which will further lessen environmental degradation from farming (Anderson, 1992).

A more recent OECD study indicates that trade liberalization would cause agricultural prices and production intensity to decrease in those countries that have historically had chemical-intensive production practices. In those countries where pesticide and fertilizer usage has been historically low, and hence better able to accommodate increased agricultural intensity, there would be increased application rates. Corroborating Anderson's earlier work, this study concludes that trade liberalization will have only modest impacts on agricultural land use. Although the effect is projected to be small, increases in the total ruminant livestock herd might lead to some increases in greenhouse gas emissions (OECD, 2000).

A detailed general equilibrium study of twenty-two agricultural subsectors in Mexico indicates that unilateral trade liberalization would decrease both agricultural output and pollution, as measured by thirteen indicators of water, air, and soil effluents. Overall Mexican real gross domestic product, however, increases significantly (Beghin et al., 1997).

More recently, the impact of NAFTA, economic growth, research investment, and farm policy was examined (Williams and Shumway, 2000). Real farm income is projected to increase in both the United States and Mexico, dramatically so in Mexico. Unlike previous studies, it predicts that both fertilizer and pesticide usage in the United States will increase substantially and, although pesticide usage will decrease in Mexico, there will be substantial increases in fertilizer usage.

## THE POSITIONS OF FOURTEEN MAJOR ENVIRONMENTAL NGOs

Based on anecdotal and other evidence that a more open economy can accelerate environmental degradation, fourteen environmental NGOs made several specific proposals to alter the institutional framework governing trade. Along with other members of civil society, these fourteen environmental NGOs argue that the environmental and labor costs of trade liberalization have been ignored or underestimated. They fear a "race to the bottom" in which nations compete for industries and jobs by lowering environmental and labor standards. These concerns have mobilized environmental and labor groups against further trade liberalization without fundamental revisions of the terms under which trade is conducted.

Among environmental groups there is a fair amount of diversity of interests, but fourteen major environmental NGOs communicated specific common demands to the Office of the U.S. Trade Representative (USTR) and the U.S. Environmental Protection Agency. These fourteen environmental NGOs were the National Wildlife Federation, the Sierra Club, the World Wildlife Fund, Friends of the Earth, the Natural Resources Defense Council, Greenpeace USA, Defenders of Wildlife, and Community Nutrition Institute, American Lands Alliance, Consumer's Choice Council, Earthjustice Legal Defense Fund, Institute for Agricultural and Trade Policy, the Center for International Environmental Law, and the Pacific Environment and Resource Center (Downs, 1999).

### Suspend Trade Negotiations in Environmentally Sensitive Sectors and Investment

At the root of these fourteen environmental NGOs' call for reform is the underlying belief that trade, at least on the terms it is currently conducted, degrades the environment. In particular, they have expressed concerns over trade liberalization in areas such as forest products which they deem particularly environmentally sensitive.

These environmental NGOs petitioning the USTR advocate suspending WTO negotiations on investment in order to preempt regulatory takings rules similar to those under NAFTA. They refer to the methyl tertiary-butyl ether (MTBE) gas additive case, in which under NAFTA investment rules a Canadian producer of MTBE, Methanex, is suing the U.S. government for $970 million dollars for lost future profits due to California's proposed phase out of MTBE. MTBE is commonly believed to cause groundwater contamination. The aforementioned NGOs are concerned that giving firms the right to sue for lost profits as "regulatory takings" may create a chilling effect in which countries do not raise environmental standards for fear of violating their WTO commitments. They are also concerned that with investment liberalization may come "industrial flight" as firms are more easily able to move across borders to avoid compliance with labor and environmental regulations (Downs, 1999).

### Deference to National Regulatory Bodies and MEAs

The fourteen environmental NGOs want the WTO to defer judgment on environmental regulations to national regulatory bodies and multilateral environmental agreements (MEA) if those regulations conflict with WTO rules. MEAs include the recent Cartagena Protocol on Biosafety, the Basel Convention on transboundary movements of hazardous wastes and their disposal, and the Convention on International Trade in Endangered Species. They argue that the WTO lacks the environmental expertise to overrule national regulatory bodies and want to avoid a situation in which national governments are inhibited from setting strict environmental, health, and labor standards. They often cast this issue in terms of a loss of domestic sovereignty that undermines democracy by granting veto power to a secret international institution (Downs, 1999).

### Deference to National Regulatory Bodies— PPM-Based Regulations and Mandatory Labels

Granting deference to national regulatory bodies entails allowing member nations to impose import restrictions based on production processes and methods (PPM). The environmental NGOs argue that allowing distinctions based on the environment, human rights, and labor standards is a needed reform to promote sustainable commerce. As a case in point they refer to the shrimp/turtle ruling in which the WTO Dispute Settlement Body found that the U.S. import ban on shrimp caught without turtle exclusion devices violated its WTO commitments. The law was subsequently altered to comply with the ruling (Downs, 1999).

Similarly, these environmental NGOs want to give nations the indisputable right to impose mandatory labels on goods based on PPMs, i.e., non–performance-based criteria. The EU's mandatory labeling of products of biotechnology is largely driving this concern. Although there have been no WTO rulings against labels for biotechnology, there has long been concern that they could be viewed as "creating unnecessary obstacles to international trade" that are "more trade-restrictive than necessary to fulfill a legitimate objective" (Article 2, Section 2 of the TBT Agreement). The petitioning environmental NGOs want to ensure that the TBT Agreement is never found to be inconsistent with mandatory labeling programs such as the EU mandatory labeling of products with GMOs (Downs, 1999).

### WTO Deference to MEAs

Similarly, these fourteen environmental groups call for the WTO to defer to MEAs when environmental concerns are raised. This requires that the MEAs' role be explicitly recognized in the WTO such that any trade-related environmental measures allowable under them be exempt from WTO challenges. Further, on environmental matters they would like to see cooperative agreements requiring the WTO to defer to the secretariats of MEAs and other organizations such as the United Nations Environment Programme (UNEP) (Downs, 1999).

### Enshrine the Precautionary Principle in the WTO

In addition to this revised hierarchy allowing governments to impose strict environmental standards based on PPMs, the fourteen environmental NGOs petitioning the USTR want to allow these restrictions to be based on the precautionary principle (Downs, 1999). The idea is that member nations could without challenge impose restrictions or even bans on products that are feared, although not proven, to have adverse environmental or health effects. The argument for the precautionary principle enters most prominently into discussions about biotechnology. Indeed, the Cartagena Protocol on Biosafety adopted on January 29, 2000, incorporates this concept. At the same time, it also contains a "savings clause" stating that the protocol is not to be interpreted as altering the rights or obligations of any of the parties under existing international agreements. As such the United States maintains that the language of the existing WTO Sanitary and Phytosanitary (SPS) Agreement and its somewhat more limited expression of the precautionary principle takes precedence (U.S. Department of State, 2000).

### *Amend the Agreement on Government Procurement (AGP)*

The listed fourteen major environmental NGOs advocate an amendment to the AGP in order to allow governments to discriminate on the basis of social and environmental criteria in their procurement decisions (Downs, 1999). For example, they cite the Maryland/Nigeria case in which the Maryland Senate Economic and Environmental Affairs Committee in 1998 voted down a bill that would have limited state business with Shell Oil Corporation, whose activities in Nigeria are alleged to have caused environmental degradation. One reason given for the vote was that it would violate the U.S.'s WTO commitments, and some environmental NGOs view the AGP as inhibiting government attempts to improve environmental outcomes via its purchasing decisions (Seligman, 1999).

### *Conduct Environmental Impact Assessments*

Among environmental NGOs weighing in on trade matters there appears to be a sense that environmental concerns are not well incorporated within WTO decision making, either in the dispute settlement process or the negotiating positions of the United States and other member countries. To more intimately incorporate environmental concerns, the fourteen environmental NGOs petitioning the USTR advocate reforming the Committee on Trade and the Environment, requiring all WTO councils, committees, and working bodies to periodically consult environmental NGOs. These fourteen NGOs also want the United States to conduct environmental impact assessments of all current and proposed trade policies. The environmental impact assessments should use the principles of the National Environmental Policy Act as a basis for establishing the procedures and methodologies, as elaborated through the Council on Environmental Quality (Downs, 1999).

### *Improve Transparency*

One of the most common criticisms of the WTO is that it is an unaccountable organization shrouded in secrecy. To increase its transparency, these fourteen major environmental NGOs propose to open the dispute settlement and appellate body proceedings to the public; increase the participation of NGOs in discussions of environment-related issues; include NGOs in the U.S. delegation to high-level WTO meetings; and provide faster access to working documents.

### Protect Voluntary Ecolabeling from TBT Objections

Protection of ecolabeling from potential objections realted to a Technical Barriers to Trade Agreement (TBT) ranks high on the fourteen environmental NGOs' list of demands. Voluntary third-party ecolabels are viewed as indispensable tools to bring markets closer to full information and improve their ability to allocate resources in a socially optimal manner (Downs, 1999). Ecolabels are labels indicating that a product is more environmentally friendly in terms of its production, use, and/or disposal than similar products. Agricultural examples include organic labels and Food Alliance Labels indicating that fruits and vegetables were produced in a way they deem sustainable. Although there have been no WTO rulings against voluntary ecolabels, there has long been concern that they, like mandatory labels, might be viewed as "creating unnecessary obstacles to international trade" that are "more trade-restrictive than necessary to fulfill a legitimate objective" (Article 2, Section 2 of the Technical Barriers to Trade Agreement). It is not clear, however, whether the TBT would even apply to voluntary ecolabels, but many environmental NGOs want assurances that ecolabels are exempt from WTO rules.

### Eliminate Environmentally Damaging Subsidies

These fourteen environmental NGOs are pressing for the elimination of environmentally damaging subsidies. They view fisheries subsidies as a large contributor to overfishing. Other harmful subsidies include those for forestry, fossil fuels, nuclear energy, and below-market pricing for water on U.S. government lands (Downs, 1999).

### Permit Environmentally Beneficial Subsidies

The petitioning fourteen environmental NGOs view environmentally beneficial subsidies as positive outcomes for trade and the environment because they decrease trade distortions by internalizing externalities (i.e., they compensate farmers who minimize negative impacts on others) and they improve the environment. Although these fourteen NGOs are concerned about the expansion and intensification of agriculture, they sympathize with agricultural production subsidies because of agriculture's multiple social functions—i.e., the multifunctionality of agriculture. These external benefits include cultural heritage, rural beautification, and food security as well as environmental benefits. The environmental benefits are rarely specified, but agriculture may combat desertification and promote biodiversity (at least to the extent that "sprawl" would replace unsubsidized farmland). Although non–trade-distorting agricultural subsidies are green box compliant under the URAA, the listed environmental NGOs do not want agri-environmental policies constrained by the non–trade-distorting criterion (Downs, 1999).

These fourteen major environmental NGOs undoubtedly have to some degree affected positions that individual nations assume. Although no country recites their demands verbatim, many countries go to great lengths to portray their proposals as environmentally friendly. Although far from realizing all their goals, these environmental NGOs have certainly raised public awareness of the trade and environment debate and the potential conflict between trade rules and environmental policies. To assess the influence of these environmental NGOs, WTO member nations' positions are discussed in the next section.

## COUNTRIES' AGRICULTURAL NEGOTIATING POSITIONS WITH ENVIRONMENTAL IMPLICATIONS

These fourteen environmental NGOs fundamentally believe WTO-driven trade liberalization poses substantial environmental risks and advocate extensive WTO reforms to avoid any damage. The bargaining positions of most countries, however, seem to reflect a core belief that further trade liberalization is beneficial but that WTO reforms could protect the environment and domestic industries of particular national or political interest (Barshefsky, 2000).

### The United States

Since the collapse of the Seattle Ministerial meeting the United States has not fundamentally altered its trade liberalization objectives. In short, the goals are to increase market access, increase export competition, and reduce trade-distorting domestic agricultural supports. In particular, U.S. Trade Representative Charlene Barshefsky (2000) has directed attention toward identifying areas where progress can be made in the built-in agenda on agriculture and services.

#### Eliminate Agricultural Export Subsidies

One major objective the United States advocates is the elimination of export subsidies, including those of the EU's Common Agricultural Policy. In addition to hurting U.S. agricultural exports, the United States argues that this policy only serves to increase environmental degradation through the intensification of agriculture and encouragement of monoculture (WTO, 1999a).

## Eliminate Trade-Distorting Production-Linked Agricultural Subsidies

While open to allowing minimally trade-distorting agricultural support policies to be exempt from further reductions by placing them in the green box, the United States favors elimination of production-linked domestic agricultural support policies. In general, however, the United States remains skeptical of arguments for production-linked supports made on the conjecture that agriculture is multifunctional and provides external benefits ranging from cultural heritage and rural beautification to biodiversity and other environmental goods (Esserman, 1999).

Similarly, the United States advocates eliminating subsidies to create "win-win" scenarios in which both trade distortions and environmental degradation are reduced. Elimination of fisheries subsidies that lead to overfishing is cited as a promising example.

## Require Biotechnology Restrictions to Be Based on Sound Science

Another major U.S. objective is to obtain a guarantee that decisions on new technologies, especially biotechnology, will be made on scientific grounds through transparent regulatory processes. The United States is particularly concerned that the EU may restrict market access to genetically modified crops based on fears but not scientific evidence of adverse environmental or health effects. Although the Catagena Protocol on Biosafety adopted on January 29, 2000, addressed these concerns, it is unlikely to resolve the debate between the fourteen environmental NGOs, the EU, and others who favor restrictions on biotechnology and the United States, which emphasizes that any discipline be transparent, timely, and predictable to ensure market access. Furthermore, the United States believes that the Sanitary and Phytosanitary Agreement provides a sufficient framework in which to address environmental concerns about biotechnology and argues that reopening the agreement risks destroying a successful system that is fundamentally functioning as intended.

## The United States to Conduct Environmental Reviews of Trade Agreements

On November 16, 1999, President Clinton committed the United States to conduct environmental reviews of comprehensive multilateral trade rounds, bilateral or plurilateral free trade agreements, and major new trade liberalization agreements in natural resource sectors. Furthermore, the United States will factor environmental considerations into the development of its trade negotiating objectives (Executive Order 13141, 1999).

*Improve Transparency of the WTO and State Trading Enterprises*

Although the United States rejects many of the criticisms leveled at the WTO by the environmental NGOs, it advocates a number of reform measures to restore public confidence in the organization, such as increasing transparency. The United States proposes opening the dispute settlement processes to the public, a position supported by the fourteen environmental NGOs but not present in the EU's proposal. The United States advocates expediting the release of panel reports and looking for ways for civil society to participate more fully in WTO procedures (USTR, 1999).

In addition to enhancing WTO transparency, the United States advocates increased transparency of state trading enterprises as well. In less specific objectives, the United States advocates improving the administration of the expanding tariff rate quotas, improving market access for least developed countries, and pursuing trade liberalization in a way that supports high environmental standards (Scher, 1999).

### The European Union

Although the EU has expressed disappointment at the outcome of the Seattle Ministerial meeting, they continue to make the case for the launch of a new round of trade negotiations with a broad agenda. In general, the EU supports major reductions in tariff and nontariff trade barriers on a large range of products from agriculture to services as well as negotiations on trade facilitation and investment. The argument for a broad round of negotiations is that it increases the flexibility of nations to make multilaterally beneficial tradeoffs. As the EU advocates the launch of a new round, there is no indication that its core Seattle objectives have changed (Lamy, 2000).

*Recognize the Multifunctionality of Agricultural Production*

One of the EU's top priorities remains to achieve a recognition of the multifunctional role of agriculture as fundamentally linked to food safety, territorial balance, maintenance of the landscape, food security, cultural heritage, and environmental amenities. The EU appears to assume that these public goods are largely inseparable from food production and hence justify production-linked supports. While the EU favors substantial liberalization in other sectors, their agricultural proposals are more modest as they believe positive agricultural externalities must be considered in order to take a more "future oriented perspective than mechanical calls for total liberalisation of farm trade" (Fischler, 2000; WTO, 1999b).

### Ensure that the SPS Agreement Is Compatible with the Precautionary Principle

Although provisional measures may be taken if pertinent scientific certainty is unavailable, as written the SPS Agreement states that a risk assessment based on "techniques developed by the relevant international organizations" must be used to impose permanent SPS-based restrictions. The EU emphasizes "strengthening" that criterion to ensure that regulatory actions may be taken in the absence of scientific certainty. Though they maintain that the precautionary principle is already contained in the SPS Agreement and need only be "clarified," this likely means interpreting the SPS Agreement as requiring a less stringent science-based risk assessment. For example, in their proposal they argue that "when the available data are inadequate or non-conclusive, a prudent and cautious approach to environmental protection, health or safety could be to opt for the worst-case hypothesis. When such hypotheses are accumulated, this will lead to an exaggeration of the real risk but gives a certain assurance that it will not be underestimated" (WTO, 2000a). This proposal in large part reflects the EU's concern about the environmental and health impacts of biotechnology (WTO, 1999c).

### Reform the TBT Agreement to Permit Ecolabels Involving PPM-Based Criteria

The EU also proposes to clarify the relationship between WTO rules and restrictions based on production processes and methods. In particular, they are concerned about ensuring the WTO compatibility of ecolabeling schemes that indicate a particular product is more environmentally friendly than similar products. The EU argues that ecolabels can be designed in a nondiscriminatory, nonprotectionist manner, and those that are should be considered legitimate market-based means of achieving environmental objectives under WTO rules. To that end, the EU advocates clarifying the role of the TBT Agreement to ensure its compatibility with ecolabeling initiatives. More generally the EU wants to safeguard compulsory labeling schemes based on "non-product related process and production methods" (WTO, 2000a). This concern largely emanates from the EU's skepticism over the health and environmental effects of biotechnology products.

### Enhance MEAs' Status in WTO Rules

The EU advocates clarifying the relationship between MEAs and WTO rules to ensure that environmental measures with trade impacts permissible under MEA rules may be implemented without WTO challenges. The EU feels that MEAs represent the best way to tackle global environmental problems and want to avoid situations in which MEA measures conflict with

WTO rules (WTO, 1999b). For example, the Cartegena Protocol on Bio-safety contains the clearest expression of the EU-advocated precautionary principle.

### Japan

With an export-driven economy, Japan approached the 1999 Seattle Ministerial meeting proposing rapid trade liberalization in most goods. On the issue of agriculture, however, its strong free trade stance is qualified somewhat by its heavy emphasis on the multifunctional role of agriculture. In general, Japan expresses concerns that agriculture provides multiple social functions by contributing "to the preservation of land and environment, to the creation of a good landscape and to the maintenance of the local community" (WTO, 1999e). As a net food-importing country, its greatest concern is food security, and domestic production is viewed as a way to reduce the risk of unexpected events causing future food shortages. Environmental benefits such as increased biodiversity are often cited as positive externalities of agriculture as well. Japan views domestic agricultural supports as an important policy designed to secure a stable food supply. Food security concerns largely explain their resistance to increasing market access for agricultural imports.

Japan believes that it is appropriate to maintain the green, blue, and amber box classification system for domestic policies. At the same time their position is that a certain level of intervention is required for the fulfillment of the multifunctionality of agriculture. In accordance with their situations and needs, Japan supports giving special consideration to developing countries (WTO, 1999e).

Japan advocates establishing new rules concerning the production, labeling, and export or import of products of biotechnology. Like the EU, Japan's position reflects concern about the health and environmental impacts of biotechnology. Although Japan offers no specific proposals, they call for discussions on issues that are of increasing importance to consumers such as food safety, recycling, and organic agricultural practices (WTO, 1999e).

Japan wants to strengthen existing rules on export taxes, export subsidies, and export state trading. They view these practices as trade distorting and harmful. Since more lenient rules apply to state trading enterprises, Japan views them as a means through which a nation's export subsidy reduction commitment can be circumvented (WTO, 1999e).

With regard to fishery subsidies, Japan argues that they should be addressed in a larger context that considers all factors hindering the sustainable utilization of fisheries. They argue that not only do other factors such as ineffective fisheries management contribute to the overexploitation of fisheries, but there are some "positive aspects" of fisheries subsidies (WTO, 1999f).

## Norway

Like Japan, Norway advocates substantial trade liberalization, but takes a far more limited approach with respect to agriculture. Norway places much emphasis on the multifunctional role of agriculture to provide not only food but "food security, . . . the viability of rural areas, the maintenance of agricultural landscapes and the cultural heritage, the preservation of agri-biological diversity and the maintenance of a good plant, animal, and public health" (WTO, 2001). In their view, these special environmental and social externalities of agriculture justify intervention, and Norway anticipates that agricultural issues "will have to be treated separately within the multilateral trading system" (WTO, 1999f).

## Cairns Group

The Cairns Group represents the most influential group of nations coordinating their negotiations. The Cairns Group consists of eighteen nations with a common interest in freeing agricultural markets. Members include nations as diverse as Brazil, Australia, South Africa, the Philippines, and Fiji, which are largely net agricultural exporters. They call for the complete elimination of all trade-distorting subsidies and substantial improvement in market access. This includes an end to all export subsidies, trade-distorting domestic subsidies, and deep cuts in tariffs. They argue that liberalization will increase the stability of worldwide food supplies and hence food security as well as generate environmental improvements. They argue that agricultural subsidies and access restrictions have stimulated farm practices that are harmful to the environment. At the same time, they propose differential treatment and technical assistance to the least developed countries (WTO, 1999g).

## Developing Countries

While there is a diversity of objectives among developing countries, including some Cairns members, there are a few concerns common to many. Although they are less concerned about the multifunctional role of agriculture to provide rural amenities such as environmental benefits, many are concerned about how trade liberalization might affect food security. They advocate removal of any WTO rules that might hinder food production for domestic consumption, favor support and import restraints designed to protect poor farmers, and oppose export subsidies (Varma, 1999; McKinnell, 2000). In general, they are adamantly opposed to imposing labor standards and environmental restrictions because they view them as thinly veiled protectionist nontariff trade barriers. For example, they fear that restrictions on environmentally sensitive goods, e.g., forest products, might further limit their market access in developed countries.

## RELATIONSHIP BETWEEN FOURTEEN ENVIRONMENTAL NGOs AND COUNTRIES' WTO PROPOSALS

Numerous country positions are in part justified on the basis of their positive environmental impacts, and a number of positions reflect specific proposals advocated by the fourteen major environmental NGOs. For instance, the U.S. position on enhanced WTO transparency, the elimination of fisheries subsidies, and environmental reviews closely resemble the fourteen environmental NGOs' demands. The EU call to enshrine the precautionary principle and enhance the status of MEAs in WTO rules, as well as its advocacy of agriculture's multifunctional role in providing environmental amenities, skepticism of biotechnology, and support for ecolabels all closely parallel their positions. At least in democratic states government positions reflect the preferences of the entire population, whereas environmental NGOs, even the fourteen major ones discussed, represent only a small subset of that population. Nonetheless, that subset may offer insights into the burgeoning agri-environmental trade issues that free traders and policymakers must increasingly consider. Indeed, environmental NGOs may have already influenced the terms of the trade liberalization debate. In his remarks to the WTO ministers in Seattle on December 1, 1999, President Clinton repeatedly refered to the protestors, the millions more they represent, and the U.S. environmental objectives (White House, 1999).

### Unresolved Agri-Environmental Trade Issues

Many of the most contentious debates involve agriculture and the environment, and often the policies proposed to address these concerns involve trade effects. Individual countries approach negotiations with both commercial and environmental objectives but attach different weights to each. The challenge will be to minimize the conflicts between these sometimes competing objectives. In the agricultural negotiations, resolving issues related to biotechnology and multifunctionality arguments poses the greatest future challenges.

## Biotechnology: PPM, Labeling, SPS, TBT, and the Precautionary Principle

Much of the debate over agriculture and the environment centers on biotechnology and multifunctionality arguments. The advent of the prominence of biotechnology has raised some of the most contentious disagreements. Those skeptical of the environmental and health impacts of products of biotechnology, especially the EU and many, though not all, environmental NGOs, have called for WTO compatibility of compulsory labeling based on PPMs (e.g., genetically modified foods). To the extent that such biotechnology labeling schemes might raise challenges under the TBT Agreement, the agreement would have to be modified. Although the United States accepts labeling, and indeed is considering a GMO-free label itself, market access for products of biotechnology remains a concern (Weiss, 2000).

On March 15 and 16 of 2000 the EU presented a paper to the SPS Committee clarifying what it believes is the proper role for the precautionary principle. The EU advocates ensuring that nations have the right to adopt regulations to achieve the level of protection, notably as regards environmental protection and human, animal, and plant health, it regards as appropriate. It sees the determination of the appropriate response as inherently political. The precautionary principle, or "safety first," is the central plank of the EU's policy, and it advocates "clarifying" the SPS Agreement to ensure that the precautionary principle may be invoked when it deems appropriate (WTO, 2000a).

The United States and like-minded governments expressed two primary concerns about the proposal. First, the precautionary principle already exists in Article 5, paragraph 7 of the SPS Agreement. Reexamination of the agreement would risk destroying an agreement the United States views as functioning successfully and as intended. Second, it fears that the proposed revision "might weaken WTO rules by reducing the certainty and predictability of the rules, upset the balance of 'rights and obligations' struck in Uruguay, and could allow every country to use precaution as an excuse for protectionism" (WTO, 2000b). Although the concerns are not likely to be resolved any time soon, the potential implications for the U.S. agricultural exports are substantial.

## THE MULTIFUNCTIONALITY OF AGRICULTURE AND PRODUCTION-LINKED SUPPORTS

The debate over the role governments should take in encouraging what has come to be known as the multifunctional role of agriculture has generated friction between those favoring extensive agricultural trade liberalization and others advocating domestic support to agriculture as a way to jointly provide environmental goods, cultural amenities, and food. The EU, Japan, and Norway are vocal proponents of the argument that decoupling agriculture from its positive externalities is often impossible. In that case, production-linked agricultural supports are the only way to achieve legitimate policy objectives. Since the goal is to achieve justifiable environmental and social policy objectives, they argue that WTO rules should accommodate such supports as green box compatible and hence not subject to reductions. Currently, however, green box policies must have "no, or at least minimal, trade-distorting effects or effects on production," which greatly limits what can qualify for placement in the green box (GATT, 1994).

The United States, Brazil, and others argue that countries can and should address their multifunctionality concerns in ways that meet the green box criteria as stated in Annex 2 of the Agreement on Agriculture. This would ensure that their objectives are met in ways that did not distort production or trade (CTE, 1999). Ways in which this could be accomplished include public food stocks to improve food security, the purchase or transfer of land development rights to secure the environmental benefits of farming, and improved rural infrastructure to promote viable rural communities that preserve cultural heritage. Perhaps not surprisingly, Bohman et al. (1999) find that the nations supportive of the multifunctional agenda are those that are mostly tightly bound by its amber box commitments, i.e., those policies that are subject to careful review and reduction over time under the terms of the Uruguay Round Agreement on Agriculture (URAA). The divisions over the environmental benefits of agriculture and to what extent they are jointly produced and inseparable from food production are likely to continue to future agricultural negotiations.

Although little progress on agri-environmental issues was made in Seattle, the built-in agenda on agriculture offers an opportunity for future progress. In their first meeting on March 23 and 24, 2000, countries made general statements that largely echoed their Seattle positions. Although failing to agree on a chairman, delegates did agree to conduct technical work on subsidies and protection and set a timetable for future negotiations. At the second meeting in June, the United States offered a new agricultural proposal, and negotiating sessions were to be held in September and November of 2000 and in March 2001. There was, however, disagreement as to whether these would be stand-alone negotiations or only considered in the context of broader negotiations. The EU favored more comprehensive negotiations, but the Cairns Group felt that they had obtained these negotiations in exchange for moderate concessions in the Uruguay Round and wanted them to be considered free standing (WTO, 2000c).

## CONCLUSION

Although Williams and Shumway (2000) provide a notable exception, the totality of the limited number of studies of the environmental impact of agricultural trade liberalization suggest only modest adverse environmental impacts. At least the fourteen major environmental NGOs, however, remain convinced of trade's insidious environmental consequences, and some of their ideas are reflected in the negotiating positions of individual nations. For example, the U.S. position on WTO transparency and the EU position on the precautionary principle closely resemble that of the aforementioned environmental NGOs. The United States and other nations emphasize not only the commercial benefits of trade but couch many of their WTO proposals in terms of the environmental and health benefits they would generate.

Many of the most contentious debates, such as those related to the precautionary principle and the extent to which the multifunctional nature of agriculture legitimizes placement of production-linked supports in the green box, involve agriculture and the environment. Although unlikely to be completely resolved quickly, there may yet be room for compromise. Recently the United States announced plans to increase regulatory oversight of genetically modified foods. The Food and Drug Administration intends to create a "free of gene-altered ingredients" label that would be validated by tests administered by the Department of Agriculture (Weiss, 2000). This may indicate a small convergence of philosophies toward biotechnology that could translate into changed trade policy objectives, but significant differences remain.

## REFERENCES

Anderson, K. (1992). "Effects on the Environment and Welfare of Liberalizing World Trade: The Cases of Coal and Food." In Anderson, K. and R. Blackhurst, *The Greening of World Trade Issues* (pp. 145-172). Ann Arbor: The University of Michigan Press.

Barrett, S. (1994). "Strategic Environmental Policy and International Trade." *Journal of Public Economics*. 54: 325-338.

Barshefsky, C. (2000). "Testimony of Ambassador Charlene Barshefsky Before the House Ways and Means Subcommittee on Trade, February 8, 2000." Reprinted by the U.S. Department of State. <http://www.ustr.gov/speech-test/barshefsky/barshefsky_t33.pdf>.

Beghin, J., S. Dessus, D. Roland-Holst, and D. van der Mensbrugghe. (1997). "The Trade and Environment Nexus in Mexican Agriculture: A General Equilibrium Analysis." *Agricultural Economics*. 17: 115-131.

Beghin, J. and M. Potier. (1997). "Effects of Trade Liberalization on the Environment in the Manufacturing Sector." *The World Economy*, 20(4): 435-456.

Beghin, J., D. Roland-Holst, and D. van der Mensbruggle. (1995). "Trade Liberalization and the Environment in the Pacific Basin: Coordinated Approaches to Mexican Trade and the Environment Policy." *American Journal of Agricultural Economics*. 77: 778-785.

Bhagwati, J. (1996). "Trade and the Environment: Exploring the Critical Linkages." in M. Bredahl, N. Ballenger, J. Dunmore, and T. Roe. (Eds.), *Agriculture, Trade, and the Environment: Discovering and Measuring the Critical Linkages.* Boulder, CO: Westview Press Inc., pp. 13-22.

Bohman, M., J. Cooper, D. Mullarkey, M. A. Normile, D. Skully, S. Vogel, and E. Young. (1999). "The Use and Abuse of Multifunctionality." Economic Research Service/USDA. November. WTO Briefing Room Web site. <http://www.ers.usda.gov/whatsnew/issues/multifunction/index.htm>.

Chichilnisky, G. (1994). "North-South Trade and the Environment." *Quarterly Journal of Economics*. 109: 755-787.

Cole, M., A. Rayner, and J. Bates. (1998). "Trade Liberalisation and the Environment: The Case of the Uruguay Round." *The World Economy*, 21(4): 337-347.

Committee on Trade and the Environment (CTE). (1999). "WTO Committee on Trade and Environment Discusses Services; Relations Between IGOs and NGOs; The Two Thematic Clusters; and Adopts 1999 Report and Schedule Meetings for 2000." Press Release November 8. <http://www.wto.org/english/tratop_e/envir_e/ te031_e.htm>.

Copeland, B. (1994). "International Trade and the Environment: Policy Reform in a Polluted Small Open Economy." *Journal of Environmental Economics and Management*. 26: 44-65.

Copeland, B. R. and M. S. Taylor. (1994). "North-South Trade and the Environment." *The Quarterly Journal of Economics*. August: 755-787.

Copeland, B. and S. Taylor. (1995). "Trade and the Environment: A Partial Synthesis." *American Journal of Agricultural Economics*. 77(August): 765-771.

Downs, D. (1999). "Technical Statement by the United States Environmental Organizations." June 16. Letter and report sent to Deputy U.S. Trade Representative, Ambassador Susan G. Esserman, and Acting Deputy Administrator, U.S. Environmental Protection Agency, Peter D. Robertson by David Downs on behalf of 14 Environmental NGOs. Reprinted by the Sierra Club. <http://www.sierraclub.org/trade/summit/letter.asp> and <http://www.sierraclub.org/trade/summit/report. asp>.

Edwards, S. (1992). "Trade Orientation, Distortions, and Growth in Developing Countries." *Journal of Development Economics*. 39: 31-57.

Esserman, S. Ambassador and Deputy Trade Representative. (1999). A speech given before the WTO General Council Session on July 29 in Geneva, Switzerland. Reprinted by the Center for International Development at Harvard University. <http://www.cid.harvard.edu/cidtrade/gov/usgov.html>.

Executive Order 13141. (1999). Issued by President Clinton on November 16. White House, Office of the Press Secretary. Accessed at (http://frwebgate.access. gpo.gov/cgi-bin/getdoc.cgi?dbname=2000_register&docid=00-4151-filed>.

Fischler, F. Agricultural Commissioner. (2000). "A New Common Agricultural Policy for a New Century—New Challenges Arising from Globalisation of Agricultural Markets, the New Round of World Trade Negotiations and EU Enlargement." March 9-10. London. Reprinted by the European Commission. <http:// europa. eu.int/comm/trade/whats_new/fisch001.htm>.

GATT. (1994). *The Results of the Uruguay Round of Multilateral Trade Negotiations. The Legal Text.* Geneva: GATT Secretariate.

Grossman, G. and A. Krueger. (1995). "Economic Growth and the Environment." *Quarterly Journal of Economics.* May: 353-377.

Harrison, A. (1996). "Openness and Growth: A Time-Series, Cross-Country Analysis for Developing Countries." *Journal of Development Economics.* 48: 419-447.

Krugman, P. (1997). "What Should Trade Negotiators Negotiate About?" *Journal of Economic Literature.* 35(March): 113-120.

Lamy, P. (2000). Member of the European Commission and Director General for Trade. "What Are the Options After Seattle?" Speech before the European Parliament, Brussels, January 25. Reprinted by the European Commission. <http:// europa.eu.int/comm/trade/2000_round/index_en.htm>.

Leonard, J. (1998). *Pollution and the Struggle for the World Product.* Cambridge: Cambridge University Press.

McKinnell, C. (2000). "Agricultural Negotiations—Setting the Table for Further Reform in Agriculture." ERS Commodity Roundtable presentation. Washington, DC, February 28.

Organization for Economic Cooperation and Development (OECD). (2000). "Domestic and International Environmental Impacts of Agricultural Trade Liberalization." Document # COM/AGR/CA/ENV/EPOC(99)72/REV2.

Rock, M. (1996). "Pollution Intensity of GDP and Trade Policy: Can the World Bank Be Wrong?" *World Development.* 24(3): 471-479.

Scher, P. L., Ambassador and Special Trade Negotiator. (1999). Testimony before the U.S. Senate Finance Committee Subcommittee on International Trade, Washington, DC, March 15, 1999 <http://www.ustr.gov/speech-test/scher/ scher4.pdf>.

Seligman, D. (1999). "Comments to the Trade Policy Staff Committee, United States Trade Representative." Washington, DC, May 20. <http://www.sierraclub.org/ trade/summit/testimony2.asp>.

U.S. Department of State. (2000). "The Cartagena Protocol on Biosafety." Fact sheet released by the Office of the Spokesman. February 16. <http://www.state. gov/www/global/oes/fs-cart_prot_biosaf_000216.html>.

United States Trade Representative (USTR). (1999). "Declaration of Principles on Trade and Environment." November. <www.ustr.gov/environment/finpol.pdf>.

Varma, S. (1999). "Workshop on Agriculture, Trade, and the WTO." Geneva, June 21-23. <www.cid.harvard.edu/cidtrade>.

Vukina, T., J. Beghin, and E. Solakoglu. (1999). "Transition to Markets and the Environment: Effects of the Change in the Composition of Manufacturing Output." *Environment and Development Economics.* 4(4): 582-598.

Weiss, R. (2000). "U.S. to Add Oversight on Biotech Food." *The Washington Post.* May 3, p. A01.

Wheeler, W. and P. Martin. (1992). "Prices, Policies, and the International Diffusion of Clean Technology: The Case of Wood Pulp Production." *International Trade and the Environment.* P. Low, (Ed.) World Bank Discussion Paper. World Bank, Washington, DC.

White House. (1999). "Remarks by the President to the Luncheon in Honor of the Ministers Attending the Meetings of the World Trade Organization." Office of the Press Secretary. December 1. <http://frwebgate.access.gpo.gov/cgi-bin/multidb.cgi>.

Williams, S. and R. Shumway. (2000). "Trade Liberalization and Agricultural Chemical Use: United States and Mexico." *American Journal of Agricultural Economics.* 82(February): 183-199.

WTO. (1999a). "Agriculture and the Environment—The Case of Export Subsidies." Submission to the Committee on Trade and Environment by Argentina, Australia, Brazil, Canada, Chile, Columbia, Indonesia, Malaysia, New Zealand, Paraguay, the Philippines, Thailand, the United States, and Uruguay. February 11. WT/CTE/W/106.

WTO. (1999b). "Preparations for the 1999 Ministerial Conference—EC Approach to Agriculture —Communication from the European Communities." Communication, dated July 23, received from the Permanent Delegation of the European Commission. WT/GC/W/273.

WTO. (1999c). "Preparations for the 1999 Ministerial Conference—EC Approach to Trade and the Environment in the New WTO Round—Communication from the European Communities." Communication, dated May 28, received from the Permanent Delegation of the European Commission. WT/GC/W/194.

WTO. (1999d). "Preparations for the 1999 Ministerial Conference—Proposal for a Joint ILO/WTO Standing Working Forum on Trade, Globalization and Labour Issues—Communication from the European Communities." Communication, dated October 30, received by the Director General from the EC Commissioner for Trade, Pascal Lamy. WT/GC/W/383.

WTO. (1999e). "Preparations for the 1999 Ministerial Conference—Negotiations on Agriculture—Communication from Japan." Communication, dated June 25, received by the World Trade Organization from the Permanent Mission of Japan. WT/GC/W/220.

WTO. (1999f). "Preparations for the 1999 Ministerial Conference—Agriculture—Communication from Norway." Communication, dated July 6, received by the WTO from the Permanent Mission of Norway. WT/GC/W/238.

WTO. (1999g). "Preparations for the 1999 Ministerial Conference—Communication from Australia." Communication, dated March 23, received by the WTO from the Permanent Mission of Australia. WT/GC/W/156.

WTO. (2000a). "Communication from the Commission on the Precautionary Principle." Communication received by the WTO from the European Communities on March 8. G/SPS/GEN/168.

WTO. (2000b). "SPS Completes Draft on Risk Consistency." SPS Committee March 15-16. G/SPS/R/18.

WTO. (2000c). "Talks Reach Swift Agreement on Phase 1." Committee on Agriculture. Press Release 172. March 27. Geneva, Switzerland. <http://www.wto.org/english/news_e/pres00_e/ pr172_e.htm>.

WTO (2001). *WTO Agriculture negotiations, Proposal by Norway,* World Trade Organization, Committee on Agriculture, Special Session, January 16. G/AG/NG/W/101.

# PART II:
# COMMODITY TRADE ISSUES

# Chapter 8

# Trade Liberalization in Rice

## Eric J. Wailes

Rice is one of the most important food grains in the world, accounting for more than 20 percent of global calories consumed. Rice trade accounts for only 5 percent of world consumption compared to wheat trade at 20 percent, coarse grains at 12 percent, and soybeans at 25 percent. The thinness of trade for rice is primarily a result of a variety of protectionist mechanisms based on national policy objectives in major rice-producing countries of domestic food security and producer support (Table 8.1).

In addition to the thinness of rice trade, another structural characteristic important for understanding the impacts of trade liberalization for the global rice market is the geographic concentration of production and consumption in Asia. More than 90 percent of rice production and consumption occur in Asia, with nearly two-thirds in just three countries—China, India, and Indonesia. Given that as much as 40 percent of Asian rice is cultivated under rain-fed systems, the monsoon weather effects on rice trade are magnified.

Finally, there is substantial market segmentation by rice type and quality (Wailes, 1996). One of the key structural dimensions is the high degree of end-use differentiation in rice. Substitution among rice types and qualities is limited by differences in cooking and taste characteristics.

An important end-use characteristic is stickiness. This is a particularly important characteristic of the medium- and short-grain rices. Long-grain rice is typically nonsticky, and cooks to a fluffier consistency compared to medium/short grain. Low substitutability for rice, however, exists both on the demand (mill and end-use) and supply sides. Different rice varieties require different climatic conditions, production, and milling technologies. This limits the ability of producers to use price incentives as a guideline in selecting which type of rice to produce.

The combination of high levels of domestic protection, geographically concentrated and erratic weather effects, inelastic price responses in production and end-use markets, and relatively thinly traded volume results in relatively volatile rice prices and trade. It is in this framework that liberalization of rice trade has been pursued (Wailes, 1999; Siamwalla and Haykin, 1983; Cramer, Wailes, and Shui, 1993).

TABLE 8.1. WTO Tariff Schedules of Rice Import by Country and Rice Types: Base Period 1986-88

| Country | Rough 1986-88 Base Period | Rough WTO Tariff Agreement | Brown 1986-88 Base Period | Brown WTO Tariff Agreement | Milled 1986-88 Base Period | Milled WTO Tariff Agreement | Broken 1986-88 Base Period | Broken WTO Tariff Agreement |
|---|---|---|---|---|---|---|---|---|
| Bangladesh | 50% | 50% | | | | | | |
| Brazil | 45% | 55% | 55% | 55% | 55% | 55% | 45% | 55% |
| Canada | 0% | 0% | 0% | 0% | $C5.51/MT | $C3.53/MT | $C5.51/MT | $C3.53/MT |
| China | 150% | 114% | 150% | 114% | 150% | 114% | 150% | 40% |
| Colombia | 210% | 189% | 210% | 189% | 210% | 189% | 210% | 189% |
| Costa Rica | 56% | 46% | 56% | 36% | 56% | 36% | 56% | 36% |
| Cuba | 40% | 40% | 40% | 40% | 40% | 40% | 40% | 40% |
| European Union | 330 ecu/mt | 211 ecu/mt | 413 ecu/mt | 264 ecu/t | 650 ecu/mt | 416 ecu/mt | 200 ecu/mt | 128 ecu/mt |
| Ghana | 125% | 99% | 125% | 99% | 125% | 99% | 125% | 99% |
| Guinea | 40% | 40% | 40% | 40% | 40% | 40% | 40% | 40% |
| Haiti | 66% | 66% | 66% | 66% | 66% | 66% | 66% | 66% |
| Hong Kong | 0% | 0% | 0% | 0% | 0% | 0% | 0% | 0% |
| Indonesia | 180% | 160% | 180% | 160% | 180% | 160% | 180% | 160% |
| Ivory Coast | 215% | 215% | 215% | 215% | 215% | 215% | 215% | 215% |
| Jamaica | 115% | 115% | 115% | 115% | 115% | 115% | 115% | 115% |
| Kenya | 100% | 100% | 100% | 100% | 100% | 100% | 100% | 100% |
| Malaysia | 45% | 40% | 45% | 40% | 45% | 40% | 45% | 40% |

| Country | Rough | | Brown | | Milled | | Broken | |
|---|---|---|---|---|---|---|---|---|
| | 1986-88 Base Period | WTO Tariff Agreement | 1986-88 Base Period | WTO Tariff Agreement | 1986-88 Base Period | WTO Tariff Agreement | 1986-88 Base Period | WTO Tariff Agreement |
| Mexico | 10% | 9% | 50% | 45% | 50% | 45% | 50% | 45% |
| Nigeria | 230% | 230% | 230% | 230% | 230% | 230% | 230% | 230% |
| Peru | 185% | 68% | 185% | 68% | 185% | 68% | 185% | 68% |
| South Africa | 5% | 0% | 5% | 0% | 5% | 0% | 5% | 0% |
| Senegal | 180% | 180% | 180% | 180% | 180% | 180% | 180% | 180% |
| Sierra Leone | 60% | | 60% | | 60% | | 60% | |
| Singapore | 27% | 10% | 27% | 10% | 27% | 10% | 27% | 10% |
| Sri Lanka | 66% | 50% | 66% | 50% | 66% | 50% | 66% | 50% |
| Turkey | 50% | 45% | 50% | 45% | 50% | 45% | 50% | 45% |
| Australia | 2% | 1% | 2% | 1% | 2% | 1% | 2% | 1% |
| United States | $28/MT | $18/MT | $33/MT | $21/MT | $22/MT | $14/MT | $6.90/MT | $4.40/MT |

Source: WTO, 2001.

143

## *PROGRESS IN RICE TRADE LIBERALIZATION*

Trade liberalization is having a profound impact on the international rice market because of the fact that rice trade has been highly restricted in both industrialized and developing nations. The relatively modest terms of agreement in the Uruguay Round Agreement on Agriculture (URAA) have contributed to global rice trade growth experienced in the latter half of the 1990s (Figure 8.1).

The URAA for the international rice markets has eliminated trade barriers by reducing import tariffs, increasing market access, reducing export subsidies, and lowering domestic support. The following discussion focuses specifically on the World Trade Organization (WTO) multilateral trade liberalization. However, the URAA has not been the only trade-liberalizing mechanism for rice. It can be reasonably argued that regional trade agreements, specifically the North American Free Trade Agreement (NAFTA) and the Southern Cone Common Market (MERCOSUR), also contributed significantly to the expansion in total world rice trade in the 1990s (Bierlen, Wailes, and Cramer, 1997; Hoffman et al., 2000). Growth of rice imports in Mexico and Brazil has been directly related to their respective regional free-trade agreements rather than the URAA.

The most significant impact of the URAA on rice has been the implementation of minimum access (MA) commitments for Japan and South Korea (Lee, Wailes, and Hansen, 1998). Prior to the URAA, Japan and South Korea maintained virtual bans on rice imports. To initiate the opening of

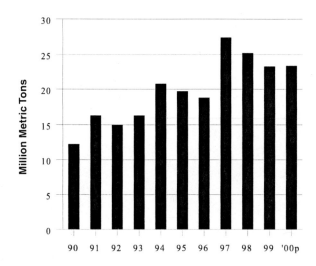

FIGURE 8.1. World rice trade in the 1990s

these markets, Japan and South Korea agreed, in lieu of tariffication, to a so-called rice clause requiring gradual increases in MA quotas. A grace period for tariffication was set at five years for Japan with "developed country" status and ten years for South Korea with "developing country" status. The MA quota for Japan in 1995 was set at 4 percent of base year consumption, increasing annually by 0.8 percent to reach 8 percent by the year 2000. The MA quota for South Korea required imports of only 1 percent of base period use in 1995 (51,000 metric tons [mt]), increasing to 4 percent by 2004 (205,000 mt). Japan also negotiated an option for early tariffication. This option was exercised without serious objections from other WTO members and was initiated on April, 1, 1999.

The minimum access quotas in Japan have been implemented using two mechanisms: (1) state trade purchases under the auspices of the Food Agency of the Ministry of Agriculture, Forestry, and Fisheries (MAFF), and (2) the simultaneous buy and sell (SBS) auctions for private trade. Food Agency purchases of rice imports have been largely isolated from direct food markets, with purchases going for food aid, industrial use, animal feed, and stocks. Food Agency data for the 1995-1996 fiscal year showed the following allocation of rice imports: 38 percent industrial use, 31 percent feed, 13 percent food aid, 11 percent stocks, and 7 percent table use. The share of the quota allocated to the two mechanisms has largely been an internal policy matter. Recently, however, burdensome stocks and external pressures have motivated MAFF to increase the share of the MA quota through the SBS from an initial 3 percent to 19 percent.

The MA commitment by Japan translated into rice imports of 379 million metric tons (mmt) in 1995. According to the scheduled increases, it would have reached 758 mmt by 2000 had the tariffication option not been implemented in 1999. Japan's tariffication agreement has three components. (1) A secondary tariff applied to rice imports above the MA import levels (where the markup associated with the quota rent (tariff equivalent) of the MA is considered the primary tariff). The initial secondary tariff rate is set at 351.17 yen per kilogram (kg) for 1999 (approximately $3,000/mt) and a 2.5 percent reduction to 341 yen in 2000. (2) In agreeing to the possibility for over-quota imports, Japan was allowed to reduce the annual increase in MA imports from 0.8 percent to only 0.4 percent. (3) The new policy adds a safeguard tariff of an additional 33 percent, triggered when imports are greater than 125 percent of the previous three-year moving rice import volume. This volume is adjusted by the addition of the volume change in domestic rice consumption in the most recent year from the previous year for which data are available. The safeguard tariff is equivalent to an additional 117.6 yen/kg (approximately $1,000/mt). The impact of these tariffication rates on Japan's import demand curve (ED) is shown in Figure 8.2.

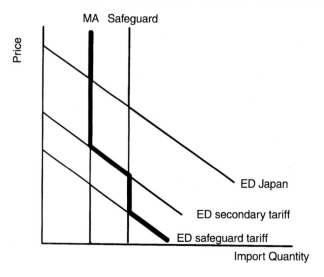

FIGURE 8.2. Japanese import demand under tariffication with safeguard

The secondary tariff is imposed when imports are greater than MA quantities. The safeguard tariff is imposed when imports are greater than the safeguard quantity. The combination of MA and the two tariffs result in a kinked import demand curve (Cramer, Hansen, and Wailes, 1999). The result of tariffication is a lower import volume than what would have occurred with the original MA quota, 38,000 mt less in 1999 and 76,000 mt less in 2000. The prohibitive tariff essentially results in a capping of Japanese rice imports until reductions in the tariff rate are negotiated in the next round. Japan's MA quota will remain at the 7.2 percent of base period use until another agreement is reached. Despite the negative effect on trade volume of Japan's recent tariffication, the MA quotas have resulted in an expansion of total world rice trade of approximately 5 percent and more than 30 percent more trade of high-quality medium-grain rice. Opening of the import market, along with domestic market reforms in Japan, has resulted in lower domestic market prices by 10 to 15 percent (MAFF, 2000). Japan's imports have stimulated resource adjustments in the rest of the world, such as Australia, China, and the United States, countries that have been the primary beneficiaries of the Japanese and South Korean market access. Relative price effects of Japanese rice imports on the medium-grain rice to long-grain price have been estimated by Sumner and Lee (2000). They suggest that a significant relative price effect has occurred as a result of Japan's market access imports.

South Korea's rice imports have also met their MA agreement requirements. Their program is administered through two government agencies, the Supply Administration of the Republic of Korea (SAROK) and the Agricultural and Fishery Marketing Cooperation (AFMC). Import purchases are made through a bidding process. In the most recent years, approximately 20 percent has been tenders for long grain and 80 percent for medium/short grain. Exporting companies in Thailand and Vietnam have won the bids for long grain while COFCO, China's state grain-trading company, has dominated the supply of medium/short grain. No MA imports by South Korea are allowed in the direct food market channels. The rice is resold for industrial food-processing purposes. As a consequence, retail rice prices in South Korea have actually risen.

South Korea's domestic support for rice production through the official procurement price is bound by WTO commitments. In 1995 South Korea had a rice AMS (aggregate measure of support) base of 2,034 billion won. The bound AMS declines to 1,360 billion won by 2004. In 2000, an election year, the South Korean National Assembly increased the official procurement price by 5.5 percent to 2,016,000 won (approximately $1,694) per mt. Such an increase in the procurement price, in the face of a declining AMS, has reduced the total quantity that the government can support.

The United States agreed in the URAA to reductions in rice tariffs, export subsidies, and domestic support programs that were trade distorting. Base period rice tariffs in the United States were already low, e.g., $22/mt for milled rice. The United States committed to reduce milled rice tariffs to $14/mt by 2000. Rice imports in the United States have been increasing steadily for the past twenty years and currently account for approximately 8 percent of domestic use. Nearly all U.S. rice imports are the higher priced, milled fragrant rices (jasmine and basmati) imported from Thailand and India. Prices for these rice types typically average $500-600/mt. The tariff reductions lowered import prices by slightly more than 1 percent by 2000.

No export subsidies for rice have been used by the United States since 1995. The export enhancement program (EEP) for rice was used aggressively during the base period to compete with EU exports, primarily into Mediterranean and Middle Eastern markets. Export credit guarantee programs have continued to be used to assist U.S. rice exports to developing countries.

The 1996 FAIR Act decoupled government income support payments from individual crop production decisions, replacing expenditures on the deficiency payment program with the flexibility contract payment program. This effectively moved most of the domestic support programs out from under "amber box" discipline. Only the marketing loan and marketing loan gain payments remain as trade-distorting program mechanisms for rice. Although marketing loan payments for rice as well as all program crops were sizable in the 1999 and 2000 fiscal years as a result of depressed farm prices, the AMS outlays were well below the bound level. Estimates using the Arkansas Global Rice Model (AGRM) suggest that marketing loan payments and marketing loan gains would account for approximately 27 percent of producer income in the 1999-2000 marketing year (Wailes et al., 2000).

On balance, the impact of the URAA on the U.S. rice industry has been favorable. The major benefit has been the access to the Japanese market. The U.S. share of this market has been about 50 percent. WTO restrictions on U.S. import tariffs, export subsidies, and farm program support have had little or no impact on world rice trade.

The European Union agreed to convert variable levies to fixed tariffs and reduce tariffs 36 percent by 2000. Different tariff levels apply to rough, brown, milled, and broken rice (Table 8.1). Although there have been several subtle changes to the rice import regime in the EU (Childs and Hoffman, 1999), imports have not been greatly impacted by the WTO commitments. One reason for this is that special preferences to Egypt, ACP (African, Caribbean, and Pacific) and Overseas Territories and Countries apply to nearly 30 percent of the EU's rice imports. The most recent EU accessions of Austria, Finland, and Sweden did result in special concessions to exporters including duty-free and reduced tariff rate quotas (TRQ). Duty-free allocations were given a quota of 63,000 mt; an allocation of 20,000 mt of brown rice was set at a tariff of 88 Euro/mt and 80,000 mt of broken rice at a tariff of 28 Euro/mt, below the bound tariff rate for brokens.

The EU also made export subsidy and AMS reduction commitments. Due to higher internal prices, EU exports are shipped with export subsidies or as food aid. The recent decline in world prices has resulted in a large rice stock accumulation through intervention buying as the commitment on export subsidies became binding. As for producer support programs, the rice regime has become harmonized with other cereals in terms of the compensatory area payment program. The compensatory payments were accompanied with an agreement to reduce intervention prices 15 percent, by 5 percent annually for the past three years. Since the compensatory payments are based on fixed yields and area (average of 1993 through 1995 marketing years, except Spain and Portugal 1992 through 1994), they are temporarily exempt (blue box) from WTO discipline.

URAA commitments by other countries such as the Philippines and Indonesia on tariff rate quotas have not been binding. In the case of these two countries, chronic production shortfalls have resulted in imports substantially higher than their TRQs. Developing countries agreed to not use export subsidies. None of the major rice exporters have used export subsidies. Both the "special and differential" and "de minimus" exemptions apply to most developing countries who support rice production.

Central American countries and Mexico use sanitary and phytosanitary measures to restrict rice imports. This has been used most aggressively against Asian rice exports, giving U.S. exports a marked advantage. Nevertheless, arbitrary application of the SPS measures is a concern for exporters into this region.

The Uruguay Round has expanded rice trade. The volume of global rice trade has averaged 22.8 mmt after implementation of the Agreement on Agriculture (AoA). That compares with an average of only 16.1 mmt in the five years preceding implementation of the AoA. The difference is 6.7 mmt; of that amount, it is reasonable to attribute at least 1 to 2 mmt to the AoA. The remaining growth can be reasonably attributed to regional trade agreements, unilateral policy reforms, and production shortfalls in key importing countries due to El Niño (Hansen, Wailes, and Cramer, 1998). Price effects are more difficult to identify, but have clearly been more important for the medium grain markets that have supplied most of Japan's and South Korea's imports.

## ISSUES FOR RICE IN THE CURRENT WTO ROUND

In 2000, WTO member countries submitted proposals on how they intend to further liberalize agricultural trade (WTO, 2001). Issues raised by various proposals that will affect rice trade liberalization in the current negotiation round include: market access, domestic support, export competition, state trading, special and differential treatment for developing countries, and nontrade objectives (including food security, rural conditions, and environmental preservation—issues grouped by some in the concept of multifunctionality). In addition to this list, the accession of China and Taiwan to the WTO are and will be important for agricultural trade liberalization in general and for rice in particular.

Expanded market access remains one of the most important issues for rice trade. The URAA can claim credit for part of the expansion in rice trade in the 1990s, but many rice markets remain highly protected. As indicated in Table 8.1, bound tariff levels in a number of countries are higher than 150 percent. By 2000, average applied tariffs on milled rice were 68 percent and were as high as 756 percent in Japan (Gibson et al., 2001). The TRQ for Japan will remain at 682,000 mt until a new agreement is negotiated. Tariffication for Japan established a prohibitive tariff and reduced the quantity of minimum access quota that would have resulted without tariffication. Both the tariff level and quota will receive considerable attention in the next round. Similar pressure will be on expanding the minimum access quota for South Korea along with a push for tariffication. Proposals by Japan and

South Korea have explicitly linked market access to nontrade or multi-functionality concerns.

The accession negotiations with China and Taiwan give considerable attention to market access. China has agreed in the U.S. accession agreement to reduce all agricultural tariffs to an average of 17 percent. China agreed to a rice TRQ of 2.6 mmt in 2000, increasing to 5.32 mmt by 2004. The tariff level within the TRQ will be 1 percent for milled rice. Current domestic use in China is 137 mmt, and current import levels are 0.2 mmt. China rice trade has been controlled by COFCO, the state grain trading enterprise. Imports of 2.6 mmt would increase world trade by 10 percent. The TRQ is only a market access opportunity and low, competitive costs of rice production in China suggest that the TRQ would not be fully traded. In fact, China has emerged over the past several years as a major net rice exporter. The China market access would most likely expand high-quality long-grain and fragrant rice imports.

WTO accession for Taiwan has been made conditional upon China's WTO accession. Taiwan's accession agreement with the United States was completed in 1998. The terms of that agreement are favorable for rice. Taiwan has agreed to remove its import ban and provide a minimum access for rice imports of 8 percent of domestic consumption. The initial quota will be approximately 72,000 mt ($25 million) but that will double in two years to 145,000 mt ($50 million). Part of the quota will be allocated for the private grain trade sector, increasing from an initial share of 21 percent to 35 percent in two years. The remaining share will be imported by a state trading company with the condition that it be made available for table use, unlike the situation in Japan and South Korea where most of the minimum access imports are sold for industrial or feed purposes. Taiwan's imports would be expected to favor the same medium-grain markets that supply to Japan and South Korea.

Another market access issue important to rice is the tariff distortion that discriminates among rice types, particularly by the degree of processing, i.e., rough, brown, and milled rice tariffs (Table 8.1). The expansion in rough rice trade into Mexico and Central America is based on tariff differentials that discriminate against milled rice. U.S. rice farmers complain that the opposite is true for rough rice in the EU rice import regime, which favors brown rice.

State trading enterprises (STEs) are important participants in major rice exporting and importing nations. Australia, China, Myanmar, and Vietnam rely on state trading or single-desk sellers for rice exports. On the import side, Japan, South Korea, Malaysia, China, the Philippines, and Indonesia have state trading organizations for rice. An effort will be applied to bring discipline for greater transparency to pricing of rice to determine if implicit export subsidies or tariffs are within the country's WTO commitment.

Domestic support concerns that will affect rice trade include the capping, binding, and reduction in trade-distorting domestic assistance programs for agriculture. A major controversy has emerged with the China accession agreement on domestic support. China has insisted on developing country status, which allows production subsidies up to 10 percent of producer returns, compared to 5 percent for developed country status. While China's current domestic support levels are low at approximately 2 percent, Chinese leaders worry about not having the ability to provide production support to their rural population, which accounts for 67 percent of their 1.3 billion total population.

An additional domestic support issue is whether to maintain the blue box exemptions for direct payments not tied directly to price supports or output. The EU is a major proponent of keeping the blue box as their internal reforms have shifted away from price supports to area payments for rice and other cereals. They are supported by Japan and South Korea but opposed by the United States and the Cairns Group in seeking to maintain this exemption.

Export competition issues relate to export subsidies and taxes, state trading, and export credit programs. Substantial progress in export subsidy reduction has been achieved. Only the EU continues to rely heavily upon export subsidies in rice trade. Their negotiating proposal therefore deflects attention to state trading, food aid, and export credit programs. State trading reform will focus on the need for greater transparency, cost and price accounting, and governmental financial assistance to single-desk sellers. With respect to food aid, Japan has used food aid aggressively to help balance the minimum access import requirements. The EU proposal on food aid would require assistance in the form of grants allowing the recipient to purchase food from the lowest-cost provider. Reform of export credit programs has been pursued in negotiations hosted by the Organization for Economic Cooperation and Development (OECD) as agreed upon in the WTO AoA. The primary issue on export credits relates to whether they should be treated as export subsidies. The U.S. uses credit programs for rice and other agricultural exports and has led the effort to keep export credit programs from being considered export subsidies. Negotiations within the OECD have focused on discipline on maximum days of tenor of credit guarantees, reporting, and limits on the use of direct government credits.

Rice is on the verge of commercialization of transgenic rice varieties. Biotechnology developments have focused on agronomic (primarily herbicide resistance), abiotic stresses (salinity, drought, etc.), and nutritional improvements (vitamin and nutrient fortification). Trade disputes are already important for other genetically modified (GMO) grains and oilseeds. Some of the major rice importers have developed or are in the process of developing regulations on GMO product trade. The primary challenge will be to negotiate science-based regulatory and labeling rules. A number of country proposals to the current WTO agricultural negotiations recommend the development of appropriate rules to address the GMO food safety concerns.

Consideration of the multifunctionality of agriculture has become an integral part of the negotiating proposals by the more protectionist group of countries, including Japan, South Korea, and the EU. Multifunctionality of agriculture refers to the broad array of services that the agricultural sector provides in addition to food production. The various proposals identify food security, environmental protection, rural economic vitality, and cultural heritage as multifunctional issues. The concern for rice exporters will be the attempt by importing nations to use these multifunctional outputs to pursue trade-distorting rice production support policies. Rice is a major crop in most Asian countries, and the food security and rural economic viability concerns are legitimate. Special and differential treatment on market access and domestic support for developing countries was built into the Uruguay Round agreement and is expected to continue. The use of government assistance to achieve environmental protection by protecting rice production is more questionable. Rice production is resource intensive, especially with respect to water use and emission of methane gases from the flooded rice paddy fields. Serious costs are imposed on the environment by rice production, including water depletion and contribution to global warming gases. It is unclear how multifunctional criteria will be quantified and built into a system of exemptions for domestic support, but it is important to note that the leading proponents, Japan, South Korea, and the EU, are all major rice importers.

The broad sets of issues under consideration in the current agricultural negotiations are important for rice trade liberalization. Rice remains as one of the most protected agricultural commodities. Reforms of the Uruguay Round, limited as they were, resulted in a significant increase in global rice trade. Completion of the next round holds promise that additional reforms will expand rice trade, improving the lives of producers who are cost competitive and the lives of consumers who will benefit from lower costs of rice imports. A recent USDA study reports that full trade-distorting policy elimination would result in a 10.1 percent increase in world rice prices and that tariff reforms alone would lift prices by 5.9 percent (Burfisher, 2001). These estimates reflect the degree to which rice production and consumption decisions are distorted by a broad array of protectionist policies in rice. Moving toward greater market orientation and an orderly international agricultural trading system is important for rice producers in the United States, who stand to benefit from greater market access.

# REFERENCES

Bierlen, Ralph, Eric J. Wailes, and Gail L. Cramer. (1997). *The Mercosur Rice Economy.* Bulletin 954, Arkansas Agricultural Experiment Station, University of Agriculture, Fayetteville, AR.

Burfisher, Mary E., ed. (2001). *The Road Ahead: Agricultural Policy Reform in the WTO–Summary Report.* USDA, ERS. Agricultural Economic Report No. 797, Washington, DC.

Childs, Nathan and Linwood Hoffman. (1999). "Upcoming World Trade Organization Negotiations: Issues for the U.S. Rice Sector." In *Rice Situation and Outlook Yearbook,* USDA, ERS, RCS-1999.

Cramer, Gail L., James M. Hansen, and Eric J. Wailes. (1999). "Impact of Rice Tariffication on Japan and the World Rice Market." *American Journal of Agricultural Economics* 81:1149-1156.

Cramer, Gail L., Eric J. Wailes, and Shangnan Shui. (1993). "Impacts of Liberalizing Trade in the World Rice Market." *American Journal of Agricultural Economics* 75:219-226.

Gibson, Paul, John Wainio, Daniel Whitley, and Mary Bohman. (2001). *Profiles of Tariffs in Global Agricultural Markets.* USDA, ERS. Agricultural Economic Report No. 796, Washington, DC.

Hansen, James M. , Eric J. Wailes, and Gail L. Cramer. (1998). "The Impact of El Niño on World Rice Production, Consumption, and Trade." Paper presented at the Southern Agricultural Economics Association Annual Meeting, Little Rock, Arkansas.

Hoffman, Linwood, Harjanto Djunaidi, Nathan Childs, Eric J. Wailes and Gail L. Cramer. (2000). "Assessing Impacts of the North American Free Trade Agreement on Market Integration: Evidence from Rice Markets in the United States and Mexico." Paper presented at the Southern Agricultural Economics Association Annual Meeting, Lexington, Kentucky.

Lee, Dae-Seob, Eric J. Wailes, and James M. Hansen. (1998). "From Minimum Access to Tariffication of Rice Imports in Japan and South Korea." Paper presented at the American Association of Agricultural Economics Annual Meeting. Salt Lake City, Utah.

MAFF (Ministry of Agriculture, Forestry and Fisheries, Government of Japan). (2000). *Statistics on Agriculture, Forestry and Fisheries.* Preliminary Statistical Report No. 12-75, Tokyo, Japan.

Siamwalla, Ammar and Stephen Haykin. (1983). "The World Rice Market: Structure, Conduct and Performance." Research Report No. 39. International Food Policy Research Institute, Washington, DC.

Sumner, Daniel A. and Hyunok Lee. (2000). "Assessing the Effects of the WTO Agreement on Rice Markets: What Can We Learn from the First Five Years?" *American Journal of Agricultural Economics* 80:709-17.

Wailes, Eric J. (1996). "Rice." In *Quality of U.S. Agricultural Products,* L. Hill (Ed.), Council for Agricultural Science and Technology (CAST) Task Force Report No. 126.

Wailes, Eric J. (1999). "Trade Liberalization in Rice." Paper presented at the International Agricultural Trade Research Consortium Annual Meeting. New Orleans, Louisiana.

Wailes, Eric J., Gail L. Cramer, Eddie Chavez, and James M. Hansen. (2000). *Arkansas Global Rice Model: International Baseline Projections for 2000-2010.* Arkansas Agricultural Experiment Station, University of Arkansas, Special Report 200, Fayetteville, AR.

WTO, World Trade Organization. (2001). Agriculture: Work in the WTO, Current Negotiations. At <http://www.wto.org/english/tratop_e/agric_e/negoti_e.htm>.

Chapter 9

# Major Issues in the U.S. Sugar Industry Under 2000 WTO Negotiations and NAFTA

Won W. Koo
P. Lynn Kennedy

## *INTRODUCTION*

In most years, less than 30 percent of the world sugar production is traded internationally. A significant share of this trade takes place under bilateral long-term agreements or preferential terms such as the U.S. sugar quota or the European Union's Lome Convention (Borremans, 1999), indicating that only a small proportion of world sugar is traded freely. In addition, most sugar-producing countries use various trade barriers to protect their own sugar industries or use export subsidy programs to influence their world market shares.

Under the Uruguay Round Agreement (URA) for agricultural goods, most countries made commitments to reduce their subsidies for sugar (WTO, 1998). However, the basic structure of protection for sugar in most countries remains unchanged. A new round of the WTO negotiations started in the spring of 2000 for further liberalization of agricultural trade. In addition, the North American Free Trade Agreement (NAFTA) may substantially impact the U.S. sugar industry. Mexico's sugar exports may increase up to 250,000 mt by 2008, with the potential for unlimited market access thereafter. Liberalization of the world sugar industry through the successful conclusion of the WTO negotiation and NAFTA would certainly alter the U.S. sugar industry.

The objective of this chapter is to analyze the major issues facing the U.S. sugar industry and ascertain their impacts on the U.S. sugar industry. Special attention is given to the impacts of bilateral sugar trade between the United States and Mexico under NAFTA.

*155*

## OVERVIEW OF THE WORLD SUGAR INDUSTRY

Sugar is produced in over 100 countries worldwide. During the 1994-1998 period, annual global sugar production was approximately 119 million tons (USDA-ERS, 1996). Only 30 percent of global production is exported from its country of origin. The largest sugar-producing region is the EU, followed by India and Brazil (Table 9.1). The EU primarily produces beet sugar, while Brazil and India produce cane sugar.

Per capita sugar consumption is highest in Cuba (58.89 kg), followed by Brazil and Australia. Per capita sugar consumption in the United States is 32.46 kg, which is above the world average (20.03 kg). Per capita sugar consumption is lowest in China (6.65 kg). China's per capita sugar consumption may increase substantially as its per capita income increases. Global sugar consumption for the 1994-1998 period was 117 million mt. Figure 9.1 shows world sugar production and consumption for the 1970-1998 period. In most years, total sugar production has been larger than sugar consumption. This has led to downward pressure on the world price of sugar. The major sugar-exporting countries are the European Union, Brazil, Australia, Thailand, Cuba, and Ukraine, accounting for 73 percent of global exports from 1990 to 1995 (Table 9.1). While relatively few countries dominate world sugar exports, imports are less concentrated. Major importing countries are the European Union, Russia, China, the United States, Japan, Korea, and Canada. Their imports accounted for about 46 percent from 1994 to 1998. Under the Lome Convention, the EU is required to import sugar under preferential terms from certain African, Caribbean, and Pacific countries. Figure 9.2 shows export to production ratios. The ratios fluctuate widely with a gradual downward trend, indicating that a smaller portion of production was traded in the global market.

The Caribbean raw sugar price is usually considered to be the world market price for sugar. The U.S. import price is the price paid by U.S. refineries for imported raw sugar, which includes import duties. Except for years with high world market prices, there is a significant wedge between the U.S. import price of raw sugar and the world market price (USDA-ERS, various issues). Over the last decade, U.S. import prices fluctuated between $0.20 per pound and $0.23 per pound, while world market prices ranged between $0.04 per pound and $0.13 per pound (Figure 9.3). Real Caribbean raw sugar prices and U.S. raw sugar import prices have long-term decreasing trends.

The volatility of world sugar prices may be due to the asymmetrical supply response to price changes stemming from high fixed costs of sugar production. An increase in sugar production in response to rising sugar prices requires significant investments in processing facilities. In addition, it takes time for new production capacity to become available. Once the facilities are in place, they tend to be used at full capacity to spread out the fixed costs.

TABLE 9.1. World Sugar Supply and Utilization, 1994 to 1998 Average (1,000 Meteric Tons, Raw Value)

| Country | Crop* | Production | Consumption | Net Exports | Ending Stocks | Per Capita Consumption (kg) |
|---|---|---|---|---|---|---|
| Algeria | B | 10 | 917 | -902 | 96 | 31.44 |
| Australia | C | 5,252 | 884 | 4,293 | 221 | 48.34 |
| Brazil | C | 13,256 | 8,180 | 5,080 | 679 | 51.81 |
| Canada | B | 134 | 1,243 | -1,114 | 160 | 41.50 |
| China | B/C | 7,177 | 8,209 | -1,327 | 2,560 | 6.65 |
| Cuba | C | 3,970 | 646 | 3,300 | 304 | 58.89 |
| Egypt | B/C | 1,120 | 1,735 | -665 | 320 | 28.64 |
| European Union (12) | B | 17,562 | 14,006 | 3,721 | 2,395 | 38.47 |
| Former Soviet Union | B | 5,708 | 9,755 | -3,795 | 1,714 | 33.37 |
| India | C | 15,037 | 14,808 | -242 | 6,012 | 15.53 |
| Indonesia | C | 2,226 | 2,955 | -815 | 537 | 15.01 |
| Japan | B/C | 815 | 2,489 | -1,662 | 135 | 19.79 |
| Mexico | C | 4,576 | 4,238 | 421 | 630 | 43.88 |
| South Africa | C | 1,958 | 1,399 | 552 | 366 | 33.27 |
| South Korea | — | 0 | 1,104 | -1,113 | 134 | 24.25 |
| Thailand | C | 5,176 | 1,517 | 3,673 | 575 | 25.23 |
| United States | B/C | 6,897 | 8,690 | -1,744 | 1,268 | 32.60 |
| Rest of the World | B/C | 28,950 | 34,452 | -7,662 | 6,242 | 18.44 |
| World Total | | 119,825 | 117,228 | 34,888 | 24,346 | 20.03 |

*B = sugarbeet, C = Sugarcane.
*Source:* USDA-ERS, 1996.

157

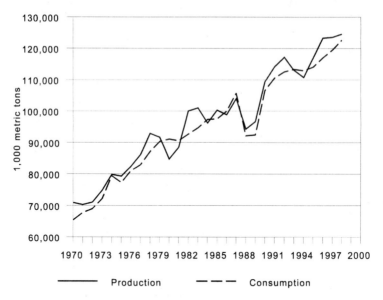

FIGURE 9.1. World sugar production and consumption, raw sugar equivalent. (*Source:* USDA-ERS, various issues.)

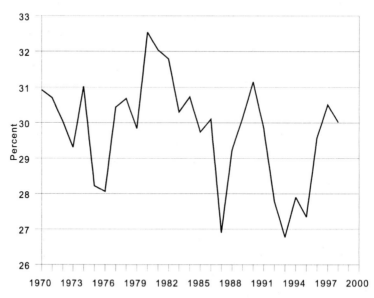

FIGURE 9.2. World sugar exports to production ratio. (*Source:* USDA-ERS, various issues.)

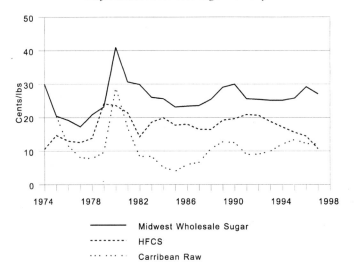

FIGURE 9.3. U.S. sugar and HFCS price. (*Source:* USDA-ERS, various issues.)

Thus, when prices fall, production remains at full capacity. Sugar production is relatively price inelastic in the short run, implying that relatively small changes in supply can have significant price effects with inelastic demand for sugar. The United States produces both beet and cane sugar. Cane sugar is produced mainly in Florida, Louisiana, Texas, and Hawaii, while beet sugar is largely produced in the Great Lakes, Upper Midwest, Great Plains, and far western states. Total sugar production increased by approximately 20 percent from 6.1 million tons in 1985-1986 to 8.2 million tons in 1998-1999 (USDA-ERS, various issues). However, beet sugar production increased faster than cane sugar production (Figure 9.4). Domestic consumption of sugar also increased slightly from 8.9 million tons in 1991-1992 to 9.8 million tons in 1997-1998. The balance was imported from more than forty countries. High-fructose corn syrup (HFCS) production has not affected sugar production in the United States, but it has affected U.S. sugar imports. U.S. sugar imports have been reduced as production of HFCS has increased. Production of HFCS has stabilized since 1995.

## THE U.S. SUGAR PROGRAM AND POLICIES

The U.S. sugar program was established by the Food and Agricultural Act of 1981. Several modifications have been made by the Food Security

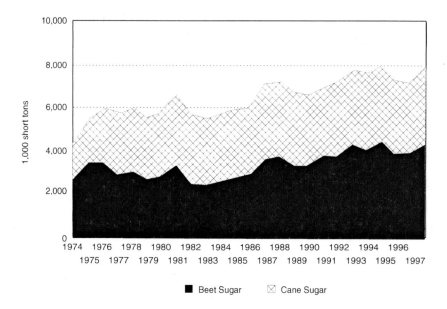

FIGURE 9.4. U.S. production of beet and cane sugar, 1974 to 1998. (*Source:* USDA-ERS, various issues.)

Act of 1985, the Food, Agriculture, Conservation, and Trade Act of 1990, and the Federal Agriculture Improvement and Reform (FAIR) Act of 1996.

The core policy tools in the program are the loan program and import restrictions (Lord, 1996). The main purpose of the loan program is to maintain a minimum market price to U.S. producers. Processors have used sugar produced from cane and beets as collateral for loans from the U.S. Department of Agriculture (USDA). Loans can be taken for up to nine months. Processors pay growers for delivered beets and cane, typically about 60 percent of the loan. The final payments are made, and the loan is repaid after the sugar has been sold. The program permits processors to store the sugar rather than sell it for lower than desired prices.

Under the FAIR Act, the sugar loan rate is set at 18 cents per pound for raw cane sugar and 22.9 cents per pound for refined beet sugar. Loans under the FAIR Act become recourse loans if the tariff rate quota (TRQ) is at 1.5 million or below, regardless of the price. When the TRQ is set above 1.5 million tons, the loans become nonrecourse. Under the nonrecourse loan, a processor forfeits collateral (sugar) to the CCC if market prices fall below the loan rates. The processor must pay a penalty of 1 cent per pound in the case of raw cane sugar. For refined beet sugar, the penalty for forfeiting sugar is

1.072 cents per pound. This reduces the effective level of price support by about 1 cent per pound.

Under the FAIR Act, the Secretary of Agriculture could reduce the U.S. sugar loan rates if major sugar producing and exporting countries reduce their export and domestic subsidies for sugar more than already agreed upon in the URA. The new rates must be at least as high as the level of support in other countries.

Processors who obtain a nonrecourse loan must pay farmers an amount for their sugar beets and sugarcane that is proportional to the loan value of sugar. The USDA is authorized to establish minimum sugar beet and sugarcane prices that processors must pay to growers. This is the same as the previous legislation.

The marketing assessment fee was raised by 25 percent in the FAIR Act. Beginning with fiscal year 1997, sellers of domestic sugar must pay an assessment of 0.2475 cents per pound for raw sugar and 0.2654 for refined sugar. However, the FAIR Act did not change the Harmonized Tariff Schedule of the United States established under the URA on agriculture. This implies that sugar imports are subject to two-tier tariff schedules under TRQ. The 1985 Farm Bill included a provision mandating the President to use all available authority to operate the sugar program established under Section 206 of the Agriculture Act of 1949 at no cost to the federal government. However, Section 206 of the 1949 Act was repealed by Section 701 of the 1996 Act, implying that the no-cost provision is no longer effective in the current sugar program.

The URA on agriculture made minor adjustments for sugar trade. U.S. import quotas on sugar were converted into TRQ, implying that a specified amount of sugar can be imported at the lower of two alternative duty rates. The amount of raw cane sugar subject to the lower duty rate must be no less than 1,117,195 metric tons in a fiscal year (Lord, 1996). The minimum low-duty imports of refined sugar is 22,000 mt. The minimum low-duty imports for raw and refined sugar add up to 1.256 million short tons raw value of sugar a year. The high duty (about 17.62 cents per pound) is imposed on the amount of sugar imported over the import quota. The first-tier duty ranges from 0 to 0.625 cents per pound.

The second-tier duty for raw cane sugar was reduced from 17.62 cents per pound in 1995 to 15.82 cents per pound in 2000 under the URA, while the duty for refined sugar was reduced from 18.6 cents per pound in 1995 to 16.21 cents per pound in 2000. The quota was to remain at the same level for the 1995-2000 period.

The sugar quota has been allocated among over forty quota-holding countries, allowing imports of specific quantities of sugar at first-tier duty rates. The quota allocation is based on historical exports to the United States for the 1975-1985 period.

## *SUGAR AND SWEETENER TRADE UNDER NAFTA*

The NAFTA, modified by an Executive Agreement between the U.S. and Mexican governments, provides increased duty-free access during a fifteen-year transition period to the U.S. sugar market for Mexican "net surplus sugar production" beginning January 1994. Upon completion of this transition period, the U.S. and Mexican sugar markets will be merged into a common market.

Between years one and fifteen, Mexico's allowable duty-free exports to the United States, and U.S. duty-free exports to Mexico, will be the greater of (1) 7,258 mt; (2) the quantity currently allocated by the United States under the sugar program to "other specified countries and areas"; or (3) the quantity allowed under the definition of "net surplus producer" (FAS, 1999). For exports to exceed the current quota level of 7,258 mt, Mexico must become a net surplus producer (production exceeding consumption). During years one through six of the agreement, Mexico's duty-free exports to the United States will be up to 25,000 mt. In year seven, Mexico's exports will increase up to 250,000 mt (FAS, 1999).

Under the U.S. TRQ, the initial 16 cent second-tier tariff rate imposed on Mexican imports will be reduced by 15 percent during years one through six. Mexico has agreed to align its tariff regime with that of the United States, implementing a TRQ with rates equal to those of the United States, by year seven of the agreement (FAS, 1999). Mexico will also adopt the U.S. second-tier tariff as a common border protection to non-NAFTA sugar by year seven of the agreement. During years seven through fifteen, the remaining U.S. and Mexican tariffs on bilateral sugar trade will be reduced to zero (Rosson et al., 1996).

In another key component of the agreement, rules of origin require that for sugar to qualify for preferential tariff treatment, it must be produced in the exporting country. The refining of raw sugar does not demonstrate origin. However, each country will allow duty-free access for raw sugar imported from the other country if it is refined in the importing country and re-exported to the producing country (FAS, 1999). Initial over-quota duties of $0.16 per pound on Mexican imports will be reduced and eventually eliminated (Rosson et al., 1996).

High-fructose corn syrup, a product that is a close substitute for sugar in uses such as soft drinks, plays an integral part in the agreement. The Executive Agreement specifies that consumption of HFCS will be used in the determination of net surplus producer status. To achieve net surplus producer status and increased duty-free access to the United States market, Mexico's production of sugar must exceed its combined consumption of sugar and HFCS (American Sugarbeet Grower's Association, 1993). Both U.S. and Mexican duties on HFCS are to be phased out over ten years. This should allow the United States to export additional HFCS to Mexico as per capita incomes in Mexico increase and import demand for sweeteners expands. It appears unlikely that Mexican capacity exists to keep pace with projected HFCS demand without additional investment in infrastructure and increased corn production (Rosson et al., 1996).

As originally negotiated prior to modifications through the Executive Agreement, Mexican access in year seven would have increased to 150,000 mt, with 10 percent increases annually over the remainder of the fifteen-year transition. In addition, the NAFTA would have granted Mexico unlimited access for its exportable surplus sugar in years seven through fifteen whenever Mexico reached net exporter status during two consecutive years (Haley and Suarez, 1999).

However, the Executive Agreement eliminates the two-year unlimited access clause. As a result, the 250,000 mt access conceded in year seven is an absolute ceiling. Beginning in year seven, and for the remainder of the transition period, Mexico will be allowed to ship its net production surplus to the United States duty-free, up to a maximum of 250,000 mt. United States duty-free access to the Mexican market will, in turn, be determined by the United States net production surplus, also with a cap of 250,000 mt. The calculation of net production surplus for both countries will be carried out annually. For the purposes of this calculation, consumption of HFCS is included with consumption of sugar for both countries. More specific details related to this issue are presented in Table 9.2 (Haley and Suarez, 1999).

The "Rules of Origin" prevent transshipment of sugar from third countries. Implementation and continuation of the common external tariff discourages Mexico's substitution of imported sugar for its domestic needs to export Mexican produced sugar to the United States. Unprocessed cane or beets may be imported for processing, but they must be reexported to the original exporting country.

### The Production and Marketing Environment

Increased capacity in the Mexican sugar industry is at the center of much dispute regarding the Executive Agreement. During the four years immediately following the implementation of NAFTA, Mexican production increased by 1.7 million mt raw value (MTRV) to a record of nearly 5.5 MTRV in 1998. These levels were projected to remain high with production of 5.04 and 5.15 MTRV for marketing years 1999 and 2000, respectively (Haley and Suarez, 1999).

This increase in Mexican sugar production can be attributed, in part, to an increase in the amount of land devoted to sugarcane production combined with several technological and producer incentive measures that have been

TABLE 9.2. Duty-Free Sugar Access Provisions of NAFTA

| NAFTA Sugar Provisions with Side Letter | |
| --- | --- |
| *Mexican Access to U.S. Market* | *Provisions* |
| Years 1-6 (1994-1999) | |
| Mexico not surplus producer[1] | Greater of 7,258 mt or "other country" share of import quota |
| Mexico surplus producer[1] | 25,000 MT |
| Years 7-14 (2000-2007) | |
| Mexico not surplus producer[1] | Greater of 7,258 mt or "other country" share of import quota |
| Mexico surplus producer[1] | 250,000 MT |
| **NAFTA Sugar Provisions Without Side Letter** | |
| *Mexican Access to U.S. Market* | *Provisions* |
| Years 1-6 (1994-1999) | |
| Mexico not surplus producer[2] | Greater of 7,258 mt or "other country" share of import quota. |
| Mexico surplus producer[2] | 25,000 MT[3] |
| Years 7-14 (2000-2007) | |
| Mexico not surplus producer[2] | Greater of 7,258 mt or "other country" share of import quota |
| Mexico surplus producer[2] | Initially 150,000 mt, increasing 10% per year[3] |

[1]Surplus sugar production is calculated as sugar production minus sugar and HFCS consumption.
[2]Surplus sugar production is calculated as sugar production minus sugar consumption.
[3]Maximums can be exceeded if Mexico has achieved net production surplus status for two consecutive marketing years.

*Source:* U.S. Department of Agriculture, Economic Research Service, Markets and Trade Division 1999. *Sugar and Sweetener Situation and Outlook Yearbook,* May 1999, SSS-225, Washington, DC.

implemented. Sugarcane area fell to less than 482,000 hectares in 1992, approximately 18 percent lower than 1987 levels. However, by 1997, a return to 1987 harvested area levels was accompanied by sugar production 22 percent higher than 1987 levels. This can be attributed to new technologies responsible for increased sugar recovery rates, combined with an expansion of the effective milling season from 130 to 175 days (Haley and Suarez, 1999).

Mexican government provided additional enhancements to the infrastructure to enhance its sugar production. These include the provision of several forms of support that enable the Mexican sugar industry to maintain both high domestic prices and high production levels. Among these, a public development bank for the sugar industry, Financiera Nacional Azucarera SA (FINASA), supports the industry by providing over $US1.3 billion of financing to the Mexican sugar sector (Haley and Suarez, 1999).

The Mexican government also controls the quantity of sugar marketed domestically, establishing the amount of sugar that can be exported or must be held in stock. Exportable quantities are divided among sugar companies, with a penalty system used to discourage the domestic sales of targeted exports. In addition, the government provides domestic stockholding subsidies to keep sugar out of the domestic market. At the other end of the spectrum, the government supports the sugar sector through sugar import control. However, under NAFTA Mexico is required to adopt a TRQ system by the year 2000 with third-country rates harmonized to the tariff levels maintained by the United States.

Although increased efficiency in the Mexican sugar industry has created the potential for increased exports to the United States, expanded U.S. HFCS production capability has compounded the problem. The U.S. HFCS industry is hopeful that NAFTA provisions will provide another market for its production. Due to increased capacity, this industry has been plagued in recent years with excess production. Estimates show that HFCS annual production capacity has grown by 3.5 million tons between 1994 and 1997 (Haley and Suarez, 1999).

Although HFCS consumption has increased by more than 13 percent during this same time period, the increases have not kept pace with production capacity. Prices have adjusted accordingly. The ratio of the HFCS-42 spot price to the beet sugar wholesale price fell below 0.60 in the fourth quarter of 1995, dropped to 0.40 for 1997 and 1998, then increased to 0.42 in early 1999. The Bureau of Labor Statistics producer price index for the HFCS industry declined from 117.6 in the final quarter of 1995 to an average of 77.6 in 1998. Given this pressure on prices, the industry was faced with a difficult adjustment process; many small firms left the sector, with others seeking arrangements with larger companies (Haley and Suarez, 1999).

The prospect of increased HFCS exports to Mexico was welcomed by the U.S. industry. Given that HFCS-55 is used primarily in soft drinks and that annual sugar use by the Mexican soft drink industry was approximately 1.4 million mt in the late 1990s, the potential of a nearby market for HFCS excess capacity was welcomed by the U.S. industry. This can be seen in the data as HFCS-55 syrup and solids exports to Mexico rose over a three-year period from 52,000 mt to over 207,000 mt in 1998 (Haley and Suarez, 1999).

## FOREIGN SUGAR POLICIES AND PRACTICES

Sugar policies and practices used by major sugar producing and consuming countries are presented in Table 9.3. The basic tools of the EU's sugar policies include: (1) import restrictions with limited free access for certain suppliers; (2) internal support prices that ensure returns to producers for fixed quantities of production and permit the maintenance of refining capacity; (3) export subsidies for a quantity of domestically produced sugar (Borremans, 1999).

EU member states allocate an "A quota" and a "B quota" to each sugar-producing operation, each isoglucose-producing operation, and each insulin syrup-producing operation established in their territory. Quota levels have been in place since the accession of Austria, Sweden, and Finland to the EU. The total EU sugar production quotas for A and B sugar were 11.98 million and 2.61 million, respectively. Any sugar produced by any member of the EU in excess of its yearly quota is considered "C sugar." A and B sugar production are used for domestic consumption and as subsidized exports, while C sugar must be exported into the world market without a subsidy or carried over into the next marketing year. In general, EU's target price for white sugar is about ECU 30 cents per pound, and its intervention price is ECU 28.72. The export subsidy was ECU 20.0 cents per pound for the 1995 to 1998 period. The EU's internal support is about 30 percent higher than that in the United States.

Since marketing year 1995, subsidized exports of sugar to third world countries have been limited under the Uruguay Round commitments of the EU (Table 9.4). However, the EU did not make an export subsidy commitment on its subsidized exports of a quantity of sugar equal to its preferential imports under the Lome Convention (Borremans, 1999; Steel, 1999). Thus, the cost and volume of those export subsidies, averaging 1.6 million mt in the period 1986-1990, are not included in the table. South Africa has both

TABLE 9.3. Policies and Practices Affecting Sugar Trade

| Countries | Practice/Policy |
|---|---|
| EU, South Africa, Mexico | Internal support, export subsidies |
| Australia, Brazil, China, India | State Trading Enterprises (STEs) |
| Developing Countries | High tariffs, lower labor costs and standards, weak environmental standards |
| Non-WTO Members | Independence from WTO rules on market access, internal support, and export subsidies |

*Source:* GAO, 1999.

TABLE 9.4. EU Export Subsidy Limits, 1995-1996 to 2000-2001

| Year | Volume (1,000 tons) | Budget (million ECU) | Budget/Ton (ECU) |
|------|---------------------|----------------------|-------------------|
| 1995/96 | 1,566.6 | 733.1 | 21.3 |
| 1996/97 | 1,499.2 | 686.3 | 20.4 |
| 1997/98 | 1,442.7 | 639.5 | 20.1 |
| 1998/99 | 1,386.3 | 592.7 | 19.4 |
| 1999/00 | 1,329.9 | 545.9 | 18.6 |
| 2000/01 | 1,273.5 | 499.1 | 17.8 |

*Source:* European Communities (1995). Schedule CXL: Part IV, Agricultural Products. Brussels, Belgium.

internal price supports and export subsidies. South Africa agreed to reduce its quantity of subsidized exports by 200,000 tons to 702,208 tons by the year 2000 under the URA (Steel, 1999). Mexico also has subsidized exports and is subsidizing raw sugar storage (Steel, 1999).

Australia's sugar exports are handled by the Queensland Sugar Corporation (QSC), a statutory authority established under the Sugar Industry Act of 1991 (Boston Consulting Group, 1996). The QSC is responsible for the domestic marketing and export of 100 percent of the raw sugar produced in the state of Queensland, which produces 95 percent of sugar produced in Australia. The QSC supports domestic producers through buyer-seller arrangements, marketing quotas, dual pricing arrangements, and other quasi-government mechanisms that isolate domestic producers from foreign competition. State trading enterprises (STEs) were not included in the URA. Other countries, including Brazil, China, and India, handled their sugar trade through STEs similar to the QSC.

Developing countries are major producers and exporters of sugar, representing about 70 percent of sugar produced in the world. Import protection and lower government standards stimulate continued dependence on sugar as a mainstay of their agricultural sectors (Steel, 1999). Their cost advantage cannot be met by developed countries, which are forced to match their lower prices on the world market.

## MAJOR ISSUES IN THE U.S. SUGAR INDUSTRY

Several major issues will potentially influence the U.S. sugar industry. Among these issues are Mexico's exportable surplus and its impacts on U.S.

domestic prices; HFCS trade under NAFTA; various issues related to WTO negotiations such as internal support and export subsidies, state trading, biotechnology, and environment. All of these factors have the capability to affect U.S. sugar programs and policies.

### Mexico's Exportable Surplus and Its Impacts on U.S. Domestic Prices

Mexican sugar exports to the United States were provisional upon several conditions under the original text of the agreement. During the fifteen-year transition period, Mexican exports were to be capped at no more than Mexico's projected net production surplus of sugar, calculated as sugar production less domestic sugar consumption. Mexico's duty-free access was limited to 25,000 MTRV for the first six years of the agreement. Following this, the maximum duty-free access quantity was to become 150,000 MTRV in year seven, and the maximum duty-free quantity was to increase by 10 percent in each subsequent year. An important point to note is that these maximums could be exceeded if Mexico achieved net production surplus status for two consecutive marketing years (Haley and Suarez, 1999).

Key NAFTA sugar provisions were changed under the side-letter agreement. The sugar provisions in the NAFTA agreement modified by the side letter also link duty-free sugar access to a "net surplus" formula. However, under the amended agreement the net surplus is calculated as the sum of sugar and HFCS consumption minus the production of sugar. For the first six years, through September 30, 2000, duty-free access was limited to the amount of the net surplus but not more than 25,000 MTRV. After this time, duty-free sugar access is limited to the amount of net surplus, but not more than 250,000 MTRV from 2001 to 2007, regardless of Mexico's net surplus producer status. The NAFTA duty on sugar trade above the duty-free levels declines gradually to zero on January 1, 2008 (FAS, 1999). A comparison of both maximum access levels (with and without the side letter) is presented in Figure 9.5.

The validity of the side letter is under dispute by the two countries. Mexico asserts that its version does not include HFCS consumption in the formula defining net surplus-producer status. In addition, Mexico maintains that the side-letter does not limit exports to 250,000 mt per year during 2001-2007. Based on Mexico's interpretation of the NAFTA agreement, the conditions have already been met to permit them to export total net surplus production to the United States on a duty-free basis (Haley and Suarez, 1999).

In early 1998 the Mexican Secretariat of Commerce and Industrial Development (SECOFI) requested consultations with the United States regarding the validity of the NAFTA side letter. When no agreement was

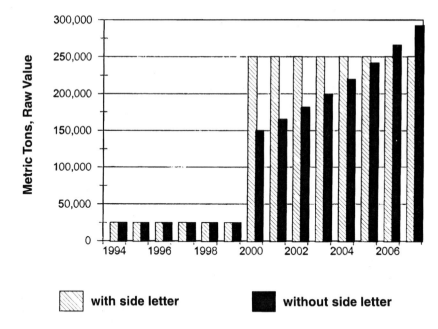

FIGURE 9.5. Duty-free Mexican sugar exportable to the United States under various policy regimes. (*Source:* FAS Online, 1998.)

achieved in November 1998, Mexico formally requested a NAFTA commission to settle the issue. In this process, the commission would consider several options for resolution, none of which, however, are binding unless both parties agree. If the commission cannot resolve the dispute within thirty days after it has convened, or some other time agreed upon by both parties, either of the parties may request an arbitration panel to resolve the issue (Haley and Suarez, 1999). Mexico has been a net exporter of sugar since 1994 and, as a result, NAFTA has allowed some duty-free access to the higher-priced United States market. While Mexico's sugar production has been increasing throughout the 1990s, consumption has been declining. Since 1997-1998 when net exports were estimated at 650,000 MTRV, Mexico has qualified as a net surplus producer and qualified each year for NAFTA duty-free exports up to 25,000 MTRV (Haley and Suarez, 1999).

Figure 9.6 presents comparisons of net surplus producer calculations and maximum access data under the alternative interpretations of the Agreement for years six and seven. It is interesting to note that in both years Mexico achieves a positive net surplus producer status, regardless of whether HFCS is included in the calculation. In addition, net surplus production is

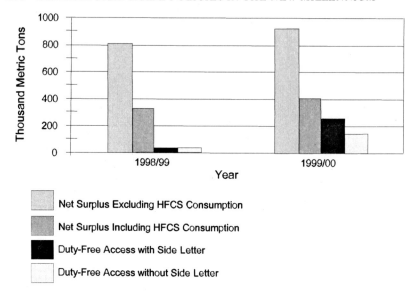

FIGURE 9.6. Calculation of Mexican net surplus production and Mexican sugar exportable duty-free to the United States under alternative NAFTA interpretation, 1999/2000. (*Source:* FAS Online, 1998; USDA, 1999; Haley, 2000.)

well over the duty-free limit, with or without the side letter. Given this information, it becomes clear that the side-letter provision limiting duty-free access to 250,000 MTRV regardless of net surplus production status will benefit U.S. sugar producers during the transition period (USDA, 1999).

If Mexico exports a maximum of 250,000 mt annually to the United States beginning in October 2000 under the terms of the side agreement to NAFTA, the domestic wholesale price of sugar is expected to decrease about 9.6 percent, about 2 cents from the wholesale price of 21 cents per pound. As a result, domestic production of sugar is expected to be about 3.2 percent lower, while consumption is expected to increase about 4 percent.

### HFCS Trade Under NAFTA

Events associated with sugar-sweetener trade between the United States and Mexico have raised concern that the increased use of lower-priced HFCS will displace domestically produced Mexican sugar. Since HFCS costs 10 to 20 percent less than sugar, switching from sugar to HFCS could result in significant cost savings to the agribusiness industry. Approximately one-third of Mexico's total caloric sweetener use of 4.4 million mt is for processes that could utilize HFCS. However, many industrial consumers of sugar are closely associated with sugar producers and have less incentive to switch to HFCS (Salsgiver, 1997). When the United States increased its use of HFCS in the 1970s and 1980s, the move was accompanied by the reduction of imports through the adoption of a tariff-rate quota system. Given the decline in domestic sugar consumption, Mexico's switch to HFCS is apt to result in increased sugar exports. Mexico's sugar industry will most likely attempt to control supply in the higher-priced domestic market, while exporting its surplus at the lower world price. Since the U.S. price is supported at a level that is significantly higher than the world price, the United States is an attractive export market for Mexican surplus production (Salsgiver, 1997).

### Issues Related to WTO Negotiations

Issues related to the U.S. sugar industry for the upcoming WTO agricultural trade negotiations include further reduction in internal supports and export subsidies, state trading enterprises, agricultural biotechnology, and the interface between trade and the environment. These issues are not unique to the U.S. sugar industry, but are fairly common for most agricultural commodities produced in the United States. Issues more directly related to the U.S. sugar industry are expected changes in U.S. sugar programs and policies, mainly loan rates and TRQ.

### Internal Support and Export Subsidies

Although WTO members have made commitments to reduce internal supports and export subsidies, levels of these subsidies differ among countries. For instance, the EU's internal supports for sugar beet growers are about 30 percent higher than those in the United States (Table 9.5). Although the EU will reduce its subsidies on the basis of the committed schedule, the EU's export subsidies will remain at about 18 cents per pound in 2000-2001 and subsidized export will remain at 1.3 million tons. These subsidies have stimulated sugar production in the region and lowered sugar prices in the world market.

### State Trading

Many countries, including Australia, Brazil, and China, use STEs for sugar trade. An example is the QSC in Australia, which handles 100 percent of sugar exports by that country (Boston Consulting Group, 1996). It practices a price discrimination and receives various subsidies from the government.

TABLE 9.5. U.S.-EU Sugar Policy Comparison

| Item | United States | European Union |
|---|---|---|
| Trade Status | Net importer | World's largest exporter |
| Producer Support Price (refined sugar) | 22.90¢/lb | 30-31¢/lb[1] |
| Future Support Price | Effective 6% reduction, 1996-2002 | Frozen through 2001 |
| Retail Price[2] (refined sugar) | 41¢/lb | 61¢/lb |
| Producer Tax on All Sugar Marketed | $41 million/yr[3] | No |
| Export Subsidies | No | $592.7/yr |
| Production or Marketing Controls on Sugar | No | Yes |
| Production or Import Controls on Corn Sweeteners | No | Yes |
| Storage Payments to Producers | No | Yes |
| National Aids to Producers[4] | No | Yes |
| Refiner Subsidies | No | Yes |
| Subsidy for Nonfood Uses of Sugar | No | Yes |

[1]Weighted average of "A," "B," and "C" quotas; dollar value rises with exchange rates.
[2]LMC International, World Retail Sugar Price Survey, June 1997.
[3]Projected revenues of $288 million during 1996/97-2002/03 for federal deficit reduction.
[4]Italy and Spain pay their producers additional subsidies.

*Source:* Landell Mills Commodities International.

State trading will likely be an important issue in the upcoming WTO negotiations, primarily because STEs have the capacity to distort trade flows (Ingco and Ng, 1998). While the agenda of the upcoming WTO negotiations is uncertain with respect to STEs, it is clear that restrictions on STE operations are needed to promote fair trade. STEs should be transparent in terms of their operation and marketing practices; should be subsidy-neutral, implying that STEs should not circumvent domestic and export subsidies; and finally, should be restricted in their ability to exercise market power through price discrimination.

## Biotechnology

Agricultural biotechnology has significant potential for consumers and producers. Genetically modified organisms (GMOs) are a leading edge of this technology; examples include sugar beets, corn, and soybeans that are insect resistant and herbicide tolerant. The GMOs also may increase sugar content in beets. However, GMO beets have not yet been produced in the United States mainly because of expected import restrictions on pulp produced from GMO sugar beets in major foreign markets, including the EU and Japan. The U.S. sugar beet industry exported 555,000 tons of beet pulp at $124 per ton mainly to the EU and Japan in 1998 (USDA, 1998). Differences in GMO regulations across countries pose potential barriers to exports. Clearly there is a need for harmonization of existing regulations among countries, or negotiation of an international standard (Normile and Simone, 2000).

## Environment

Developing countries, including Brazil, Mexico, and India, have much lower environmental standards in producing sugar cane than the United States. As a result, their production costs are lower, causing lower sugar prices in the world market. In addition, trade agreements could exert pressure to reduce national environmental standards to the lowest common denominator and worsen pollution by stimulating economic activity of dirty industries (*The Economist*, 1993). The WTO should regulate environmental standards to level the global playing field for sugar trade.

## U.S. Sugar Programs and Policies

Upcoming WTO agricultural trade negotiations may require TRQs to be converted to a tariff system. If the United States converts TRQs to tariffs and reduces the tariffs gradually, U.S. imports of sugar would be substantially higher. As a result, the domestic sugar price may fall and become more volatile. Even if the United States is able to maintain its TRQ on sugar, the United States might have to raise its quota over the given time period and lower its second-tier duty, implying that more sugar would be imported into the United States. In addition, the United States would import more sugar from Mexico under NAFTA. The increased sugar imports may result in depressed sugar prices in the United States. If the United States maintains the nonrecourse loan program, producers' minimum prices would be the loan rate set by the U.S. government. The difference between the market price and loan rate is the per-unit government expense under the program.

## *CONCLUDING REMARKS*

The U.S. sugar industry has been protected by the U.S. sugar program in the 1996 FAIR Act and the TRQ under the URA. As a result, the domestic sugar price is about 24 cents, while the world sugar price is 8.5 cents per pound. A concern is what major issues the U.S. sugar industry will face in the near future and expected changes in the U.S. sugar programs in the upcoming WTO negotiations on agriculture.

The United States may not be able to maintain the TRQ on sugar. The new WTO negotiations may require member countries to convert TRQ to a tariff system and to reduce the tariff rates over the given period. Even if the United States is able to maintain its TRQ on sugar, the United States would have to raise its quota on sugar and lower its second-tier duties over the given period. In addition, Mexico could have a potential to export sugar to the United States under NAFTA. The United States would import much more and, consequently, the domestic price would fall substantially. If the domestic sugar price is lower than loan rates, producers would get a price equal to the loan rate under the current sugar program. The price difference would be the government subsidy per unit of sugar. However, producers could get the market price, which could be lower than loan rates, if the United States eliminates the sugar program.

Disputes make liberalized sweetener trade between Mexico and the United States uncertain in the near future. However, falling world sugar prices also have the potential to increase the amount of Mexican sugar entering the United States through high-tier quotas. NAFTA established a declining tariff schedule for high-tier raw and refined sugar imported into the United States from Mexico. During the NAFTA adjustment period through 2008, the maximum world price at which it becomes profitable to ship Mexican sugar into the U.S. market increases annually.

Other potential changes in market access through venues such as the WTO will force the modification of the domestic U.S. sugar policy. The upcoming WTO negotiations must address the following issues if they are to create a fair playing field in sugar trade: further reductions in subsidies; restriction of activities of state trading enterprises; standardizing regulations on biotechnology; and improving environmental standards in developing countries.

## REFERENCES

American Sugarbeet Grower's Association. (1993). How the NAFTA Applies to Sugar, Document on Internet: <http://membersaol.com/_ht_a/asga/nafta.htm>, November 8.

Borremans, D. (1999). "Sugar Market Country Report-EU," US Mission to the EU, Office of Agricultural Affair. Web site: <http://www.sugarinfo.co.uk/ sugar_report_eu.htm>.

Boston Consulting Group. (1996). Report to the Sugar Industry Review Working Party: Analysis and Identification of Possible Option. Boston, MA.

*The Economist.* (1993). "Trade and the Environment: The Greening of Protectionism." February 27, p. 28.

Foreign Agricultural Service Online. (1998). NAFTA Agriculture Fact Sheet: Sugar, United States Department of Agriculture, FAS, Document on Internet: <http://www.fas.usda.gov/itp/policy/nafta/sugar.html>.

Foreign Agriculture Service. (1999). World Sugar Situation, FAS online Web site: <http://www.fas.usda.gov/htp/sugar/2000/november/sugarsit.htm>.

GAO. (1999). Sugar Program: Changing the Method for Setting Import Quotas Could Reduce Cost to Users. GAO-RCED-99-209. Washington, DC.

Haley, S. (2000). Mexican Estimates of High Fructose Corn Syrup Consumption. Personal Correspondence with Steve Haley, Agricultural Economist, Specialty Crops Branch, Market and Trade Economics Division, Economic Research Service, USDA. January 18.

Haley, S. and N. Suarez. (1999). "U.S.-Mexico Sweetener Trade Mired in Dispute," *Agricultural Outlook,* United States Department of Agriculture, Economic Research Service, Washington, DC, September.

Ingco, M. and F. Ng. (1998). Distortionary Effects of State Trading in Agriculture: Issues for the Next Round of Multilateral Trade Negotiations, the World Bank, Washington, DC.

Landell Mills Commodities. (1997). *U.S. and EU Sugar Policy Comparison.* Landell Mills, NY.

Lord, R. (1996). "Sugar." Chapter 2, *Provisions of the Federal Agriculture Improvement and Reform Act of 1996, AIB-729,* ERS-USDA. Washington, DC, December.

Normile, M.A. and M. Simone. (2000). *Agriculture in the Uruguay Round,* Economic Research Service, USDA, URAA Issues Series, <http://www.ers.usda.gov/briefing/WTO/uraa.htm>.

Rosson, C.P. III, G.A. Benson, K.S. Moulton, and L.D. Sanders. (1996). *The North American Free Trade Agreement and United States Agriculture,* Southern Agriculture in a World Economy, Leaflet No. 9, Texas Agricultural Extension Service, College Station, Texas.

Salsgiver, J. (1997). HFCS Trade Dispute with Mexico, *Sugar and Sweetener Situation and Outlook,* SSS-221, Economic Research Service, USDA, Washington, DC, September.

Steel, P.M. (1999). Comments from the U.S. Department of Agriculture on GAO Report on Sugar Program, Changing the Method for Setting Import Quotas Could Reduce Cost to Users, GAO/RCED-99-209, Washington, DC.

U.S. Department of Agriculture Economic Research Service (USDA-ERS). (various issues). *Sugar and Sweetener: Situation and Outlook Report.* Washington, DC.

U.S. Department of Agriculture Economic Research Service (USDA-ERS). (1996). *PS&D View* (Computer Files). Washington, DC.

World Trade Organization (WTO). (1998). *Trading into the Future,* Second edition. Geneva: WTO.

Chapter 10

# Major Issues for the U.S. Wheat Industry: The Implications of China's Entry into the WTO

Won W. Koo

## *INTRODUCTION*

Wheat is produced widely across the world. The total world wheat production has increased slightly, from 531 million tons in 1986-1987 to 582 million tons in 1996-1997. China was the largest producer of wheat in 1996 (110 million tons), followed by the European Union (EU) (94 million tons), and the United States (62 million tons). Other major wheat-producing countries are the former Soviet Union (FSU), Canada, Australia, Turkey, India, and Argentina. These nine countries produce about 74 percent of wheat produced in the world. Because of the concentration of wheat production in a few countries, a large volume of wheat is traded in the world market. The total quantity of wheat traded in the world market was 95.4 million tons in 1996, which is about 17 percent of wheat produced in that year. Major exporting countries are the United States, Canada, the EU, Australia, and Argentina.

The world wheat market has changed dramatically in the past decade. Farm support policies in exporting and importing countries have encouraged production, resulting in large stock buildups. Countries used quotas, variable levies, tariffs, and other forms of import restrictions to protect domestic producers. As world trade decreased during the early 1980s due to a depressed world economy, major exporting countries expanded the use of export subsidies or export promotion programs to maintain their grain market share.

The Uruguay Round (UR) of GATT negotiations, which was effective in 1995, has affected trade flows of wheat from exporting countries to importing countries. In addition, recent financial crises in several Asian countries, including South Korea, Thailand, Indonesia, and Taiwan, also have affected the world wheat market. Import demand for wheat from those countries was reduced substantially, resulting in depressed wheat prices in the world market. The average export price of wheat at the Gulf ports decreased from $5.02 per bushel in 1996-1997 to $3.89 per bushel in 1997-1998.

The objective of this chapter is to assess the outlook of U.S. and world wheat industries under the UR Agreement. Special attention is given to evaluating the impacts of China's WTO entry on trade flows of wheat from exporting countries to major importing countries and its implications for the U.S. wheat industry.

## WORLD WHEAT INDUSTRY

World wheat trade is dominated by a few exporting countries: the United States, Canada, Argentina, Australia, and the EU. These countries have handled over 80 percent of wheat traded in the world market. While competition among exporting countries is strong, the world wheat market is not perfectly competitive. Australia and Canada use wheat boards to market their grain, while the United States and the EU rely on export subsidies to increase their market share. In addition, all major wheat exporters use credit guarantees and loan term preferential trade agreements to promote their exports.

### Wheat Classes

Most wheat varieties grown today belong to the broad category of common or bread wheat, which accounts for approximately 95 percent of world wheat production. The remaining 5 percent of world wheat production is durum wheat, used to produce pasta and couscous. Common wheat is further divided into hard and soft wheat. Wheat varieties are highly differentiated in terms of their agronomic and end-use attributes. Based on criteria such as kernel hardness, color, growth habit, and protein content, wheat is divided into several classes. Color and hardness refer to physical properties of the wheat kernel. Based on the color of the outer layer of the kernel, common wheat varieties are described as white, amber, red, or dark, while the hardness of the kernel is used to characterize them as hard or soft.

Growth habitat is an important agronomic feature of wheat varieties. Winter wheat is planted in late summer or fall and requires a period of cold winter temperatures for heading to occur. After using fall moisture for germination, the plants remain in a vegetative phase or dormancy during the winter and resume growth in early spring. In contrast to winter wheat, spring wheat changes from vegetative growth to reproductive growth without exposure to cold temperatures. In temperate climates, spring wheat is sown in spring. Since yields for winter wheat tend to be higher than for spring wheat, spring wheat is produced primarily in regions where winter wheat production is infeasible where frozen soil kills the wheat plants or where winters are too warm. Countries with mild winters, such as Argentina, Australia, and Brazil, produce spring wheat, but plant them in the fall rather than in the spring.

## Wheat Production

China is the largest wheat-producing country in the world. Average production was 112 million metric tons annually for the 1995-1999 period, which is about 20 percent of the world production (Table 10.1). The second largest wheat-producing region is the EU (94 million metric tons), followed by the FSU (66 million metric tons). U.S. wheat production was 64 million metric tons annually for the 1995-1999 period. Australia and Argentina produced 21 and 13 million metric tons, respectively, for the same time period.

Because of differences in soil types and climates, wheat produced in one country generally differs from that produced in other countries. The United States produces hard, soft, and durum wheats. Hard wheat produced in the United States is further divided into hard red winter (HRW) and hard red spring (HRS) wheat and soft wheat into soft red winter (SRW) and white wheat. SRW wheat is produced in the Corn Belt and Southern states. HRS and durum wheat are grown in the Northern Plains, mainly North Dakota, which produces about 80 percent of durum wheat and 60 percent of HRS wheat produced in the United States. HRW wheat is grown primarily in the central plains, particularly Kansas and Oklahoma. White wheat, a type of soft wheat, is grown in the Pacific Northwest, Michigan, and New York. Average U.S. wheat production for the 1995-1999 period was 61.3 million tons of common wheat and 2.8 million tons in durum wheat (Table 10.1).

The majority of Canadian wheat is produced in Saskatchewan, southwestern Manitoba, and southeastern Alberta. Canada produces primarily HRS wheat (Canadian Western Red Spring) and durum wheat. Average Canadian wheat production for the 1995-1997 period included 22.6 million tons of HRS and 3.9 million tons of durum wheat.

The EU produced an annual average of 85.6 million tons of a soft wheat and 8.2 million tons of durum wheat during the 1995-1999 period. France accounted for 40 percent of soft wheat production in the EU in 1995. Germany and the United Kingdom are also major producers. The majority of durum is produced in Italy, Greece, and France. Italy accounted for nearly 60 percent of EU durum production in 1995, followed by Greece (22 percent) and France (13 percent).

Australia primarily produces a winter wheat which is similar to HRW in terms of quality and characteristics (Ortmann, Rask, and Stulp, 1989). Australian average wheat production amounted to 20.7 million tons for the

TABLE 10.1.  World Wheat Supply and Utilization, 1995 to 1999 Average (in 1,000 mt)

| Country | Production | Consumption | Net Exports[a] | Ending Stocks | Per Capita Consumption (kg) |
|---|---|---|---|---|---|
| **Common Wheat** | | | | | |
| United States | 61,287 | 31,547 | 29,433 | 12,579 | 80.90 |
| Canada | 22,661 | 7,050 | 13,852 | 6,396 | 75.49 |
| European Union | 85,652 | 73,862 | 11,949 | 14,918 | 95.19 |
| Australia | 20,741 | 4,636 | 16,313 | 1,874 | 148.79 |
| Argentina | 12,960 | 4,519 | 8,335 | 454 | 127.36 |
| Algeria | 486 | 3,100 | (2,650) | 327 | 113.27 |
| Brazil | 2,295 | 8,444 | (6,115) | 612 | 49.70 |
| China | 112,135 | 114,943 | (3,343) | 26,880 | 92.41 |
| Egypt | 5,792 | 12,562 | (6,851) | 1,067 | 203.41 |
| Japan | 533 | 6,245 | (6,053) | 944 | 49.68 |
| S. Korea | 8 | 3,790 | (3,814) | 790 | 82.58 |
| Mexico | 3,283 | 5,064 | (1,858) | 522 | 10.39 |
| Morocco | 3,164 | 5,154 | (2,171) | 1,822 | 182.15 |
| FSU | 65,813 | 68,950 | (2,880) | 10,299 | 232.36 |
| Tunisia | 89 | 878 | (825) | 206 | 94.26 |
| Taiwan | 1 | 996 | (992) | 261 | 45.85 |
| Venezuela | 0 | 863 | (864) | 54 | 38.18 |
| Rest of World | 161,055 | 203,181 | (40,980) | 37,835 | 34.89 |
| World | 557,955 | 555,784 | 115,636 | 117,850 | 95.43 |
| **Durum Wheat** | | | | | |
| United States | 2,975 | 2,345 | 403 | 1,170 | 8.25 |
| Canada | 3,891 | 1,030 | 3,037 | 1,657 | 5.23 |
| European Union | 8,208 | 7,774 | 486 | 2,259 | 20.77 |
| Algeria | 990 | 2,578 | (1,650) | 590 | 94.41 |
| Tunisia | 891 | 1,143 | (356) | 467 | 122.57 |
| Venezuela | 0 | 327 | (327) | 21 | 14.51 |
| Rest of World | 5,670 | 715 | (1,454) | 1,332 | 0.12 |
| World | 32,625 | 31,911 | 6,177 | 7,496 | 3.80 |

[a]Imports in parentheses
*Source:* USDA, PS&D View, 2002

1995-1999 period. Wheat production is concentrated in the eastern Australian states of New South Wales and Victoria.

Argentina produces a wheat with characteristics of both soft and hard wheat (Harwood and Bailey, 1990). Argentina's average wheat production amounted to 12.9 million tons for the 1995-1999 period.

## Wheat Consumption and Imports

Different wheat classes have their preferred uses. Hard wheat flour has excellent bread-baking properties; soft wheat flour is well-suited for cakes, cookies, and Asian noodles; and durum wheat is used for pasta products and couscous. However, since different types of wheat can be blended to produce flours with certain characteristics, some substitution among wheat classes is possible in flour milling.

Although wheat is used primarily for human consumption, it is also an excellent feed grain for poultry and livestock. Feed use of wheat tends to be highly variable and depends on the quality of wheat crop and on the price relationship between wheat and other feed grains. Generally, only lower-quality wheat is used for feed, and differences among wheat classes are not important for feeding purposes. Wheat is a differentiated product only for human consumption.

Average per capita wheat consumption is the largest in the FSU (232 kg) for the 1995-1999 period, followed by Egypt (203 kg) (Table 10.1). China's per capita wheat consumption is 92 kg for the 1995-1999 period, which is smaller than the world average per capita consumption of 95.4 kg. China, however, is the largest wheat-consuming country in the world, with annual wheat consumption of approximately 115 million metric tons for the 1995-1999 period.

Major importing countries or regions include Algeria, Brazil, China, Egypt, Japan, Mexico, Morocco, South Korea, Taiwan, Tunisia, and Venezuela. Most of these importing countries use various types of barriers to restrict inflow of wheat to their countries. The largest wheat-importing country is Egypt (6.8 million metric tons), followed by Japan (6.1 million metric tons), and South Korea (3.8 million metric tons) for the 1995-1999 period. China was a major wheat importer; however, its wheat imports have been highly volatile, depending upon its domestic wheat production and import policies. China recently reduced wheat imports substantially, from 12.6 million metric tons in 1995 to 1.2 million metric tons in 1999.

## Wheat Exports

The major wheat exporting countries, the United States, Canada, the EU, Australia, and Argentina, supply approximately 80 percent of the wheat traded in the world market. The United States is the largest exporter, followed by Canada and the EU (Table 10.1). The United States leads in exports of HRW and SRW wheats; an annual average of 30 million metric tons were exported in 1995-99, of which nearly 16 million tons were HRW and SRW; another 7.8 million tons were HRS. The United States competes with the EU for market share of SRW wheat exports. Major U.S. markets for SRW wheat include China, West Asia, and the North African markets. EU markets for SRW wheat include the FSU, China, West Asia, and North African markets.

Canada is the leader in exports of HRS and durum wheat. The United States also exports HRS and durum wheat, and competes with Canada. The EU competes with the United States and Canada for market share of durum wheat exports. Major U.S. markets for HRS wheat include Southeast Asia and East Asia, including Japan and South Korea. Major Canadian markets for HRS wheat include China, the FSU, and the East Asian markets. The United States, Canada, and the EU intensely compete for the North African durum markets.

Australia and Argentina compete with the United States in exporting HRW wheat. Major U.S. markets for HRW wheat include the FSU, China, and East Asia. Argentina exports HRW wheat mainly to South America and West Asia. Australia's major markets are the North African countries, China, the FSU, and West Asia.

## TRADE AGREEMENTS AND POLICIES

The major exporting countries use several export promotion policies, including export subsidies, credit arrangements, and long-term agreements, to protect or enhance their positions in the world market.

### Export Subsidies

The EU and the United States are the primary users of direct export subsidies. The EU subsidy is equal to the difference between the EU market price and the world price. The EU's threshold and intervention prices keep domestic market prices well above the world price.

The EU uses two methods to establish export restitutions. First, refund tenders cover the majority of EU exports in which traders apply for refunds on specific quantities exported to specific markets. The exporter receives an export certificate, indicating the refund and a time period within which the certificate is valid. The second method is the "ordinary restitution," which is published regularly. These refunds are designed for particular destinations and often are used for stable import markets. Restitutions are the same for every origin of wheat in the EU, but may differ depending on destination. Subsidies to certain regions, such as Africa, Switzerland, or Scandinavia, are fixed by the commission and published in the EC journal. However, fixed export subsidies apply only to a small proportion of the total EU exports. Most grain is exported by weekly tenders. The commission publishes the quantities to be exported and grain trading firms are invited to submit bids. Recently, restitutions have been increasing since 1997 and averaged at $36/metric ton for wheat and $56/metric ton for barley.

The United States instituted export subsidies under the Export Enhancement Program (EEP) in May 1985 to regain the lost market shares and compete with EU subsidies. The EEP uses a competitive bid process under which the U.S. Department of Agriculture targets a country for a specific quantity of a commodity. U.S. exporters then compete for sales to the targeted market, and bonuses are awarded to the exporters whose sales price and bonus bid fall within an acceptable range. The bonus was calculated by taking the difference between the U.S. market price and world price. The exporters complete the sale, present proof of delivery, and receive a cash subsidy in the form of certificates. The exporter may sell these certificates or exchange them for CCC stocks. Sales targeted to the FSU and China accounted for half of EEP wheat sales and those to North Africa and Middle Eastern countries accounted for one-third of EEP wheat sales during the 1986-1995 period. On average, about 26 percent of wheat exported was under the EEP for the 1986-1999 period. The quantity of wheat traded under the EEP reduced substantially in 1995 (2 percent of the total U.S. wheat exports).

The UR Agreement restricted export subsidies in terms of the quantity of grain subsidized and total expenditure under the subsidy program. The subsidized quantity under the program should be reduced by 21 percent of the average quantity for the 1986-1990 period by year 2000, and total expenditure under the program should be reduced by 36 percent of the average expenditure for the 1986-1990 period. The EEP has not been used for wheat exports since 1996 in the United States.

The Canadian rail subsidy under the Western Grain Transportation Act (WGTA) provided direct government payments to Canadian railroads for shipments of specified commodities, including wheat (U.S. International Trade Commission, 1990). Rail shipments subject to this subsidy include those from any point west of Thunder Bay, Ontario, or Armstrong, Ontario, to Thunder Bay or Armstrong for export or domestic use, and any port in British Columbia for exports (except to the United States).

The WGTA rail subsidy was estimated at $21.31 per metric ton, which was equivalent to 70 percent of the estimated freight rate of $30.31 per metric ton in 1989-1990 (U.S. International Trade Commission, 1990). However, the Canadian government eliminated the subsidy as of August 1, 1995.

New regulations for western grain transportation are contained in the 1996 Canada Transportation Act (CTA), which came into effect on July 1, 1996. A cap on freight rates still exists for western grain in the CAP.

### Export Credit

The United States, the Canadian Wheat Board (CWB), the Australian Wheat Board (AWB), and Argentina offer export credit to their buyers. The EU does not offer credit assistance as a group; however, some member countries do. France, for example, guarantees repayment through the Coface Group (COFACE) and certain commercial banks.

The credit terms and conditions vary widely across countries. Argentina's credit has primarily been granted to other Latin American countries, including Peru and Cuba, and has not exceeded twelve months. Australia extends credit through the AWB for up to three years. Egypt and Iraq are regular recipients. The Canadian government guarantees loan repayment on credit extended which does not exceed three years. Brazil is the largest credit buyer of wheat, followed by Iraq, Egypt, and Algeria. France's COFACE provides short-term credit and guarantees 85 percent of the credit if the purchaser is a private buyer and 90 percent if the purchaser is a foreign government. Medium- and long-term credit financing is provided through the Banque Francaise du Commerce Exterieur.

The U.S. Department of Agriculture operates two credit programs, GSM-102 and GSM-103. The GSM-102 program guarantees repayment of private credit extended to importers in specified countries for up to three years. The GSM-103 program covers private credit extended for between three and ten years. Wheat exports under this program range between 5 percent of the total U.S. wheat export in 1995 and 26 percent in 1991. From 1996 to 2000, the largest users of the U.S. credit guarantee programs for wheat are Algeria, Brazil, Egypt, Jordan, Korea, Mexico, Pakistan, Sri Lanka, Tunisia, and Turkey.

### Long-Term Agreements

Long-term agreements (LTAs) are advantageous to both exporting and importing countries. Exporting countries use LTAs to maintain export shares, attain new markets, and stabilize exports from year to year. Importers use LTAs to ensure reliable supplies. LTAs include provisions for an upper and lower bound on purchases and, in some cases, financing arrangements; they may involve shipments over two or more seasons. Historically, 75 percent of the FSU's wheat imports were through LTAs. Canada and

Australia have an advantage in negotiating LTAs because their grain boards can guarantee these trade commitments. Actual LTA shipments account for a small share of U.S. wheat exports (Harwood and Bailey, 1990).

### *Trade Barriers and Internal Supports*

Importing countries have used various types of trade barriers to protect their domestic producers. Under the WTO, created under UR agreements, nontariff trade barriers were to be converted into tariffs and reduced by 36 percent of the average level for the 1986-1988 period by the year 2000 in developed countries and 24 percent in developing countries by 2004. In 1995, the minimum access was to 3 percent of base level consumption for the 1986-1988 period, which would increase to 5 percent by 2000 in developed countries and by 2004 in developing countries.

In addition, both exporting and importing countries use various types of internal support to enhance their competitiveness in producing grain in the world market. Under the UR Agreement, the internal support in terms of aggregate measures of support (AMS) should be reduced by 20 percent of the average support for the 1986-1988 period by 2000 in developed countries and by 13 percent by 2004 in developing countries.

### *North American Free Trade Agreement*

The Canada-U.S. Free Trade Agreement (CUSTA) was signed in 1988 and implemented in 1989. The objective was to create a U.S.-Canada free trade area so that trade between the two countries would be uninhibited by border measures. The agreement called for conversion of nontariff border measures to tariffs, with all tariffs to be phased out over a ten-year period. The agreement was expanded to the North American Free Trade Agreement (NAFTA) by including Mexico in 1994.

There are no trade barriers in trading wheat between the United States and Canada under CUSTA. NAFTA reduces trade barriers among the United States, Canada, and Mexico. The United States is phasing out a 0.77 cents per kilogram tariff on durum wheat from Mexico over ten years; for other wheat, tariffs are being reduced to zero over five years. Mexico agreed to convert its import license for wheat imported from the United States and Canada to tariffs. U.S. wheat exports to Mexico were to be subject to an initial 15 percent tariff to be reduced in equal installments over a ten-year period. In addition, Canada agreed to eliminate its import license requirement for wheat imported from Mexico.

## MAJOR ISSUES IN THE 2000 ROUND
## OF WTO NEGOTIATIONS

Issues related to the U.S. wheat industry for the 2000 Round of WTO agricultural trade negotiations included further reduction in internal supports and export subsidies, state trading enterprises, China's WTO entry, and agricultural biotechnology. These issues are not unique to the U.S. wheat industry but are fairly common for most agricultural commodities produced in the United States.

### Internal Support and Export Subsidies

Although WTO members have made commitments to reduce internal supports and export subsidies, levels of these subsidies differ among countries. For instance, internal supports in terms of Producer Subsidy Equivalents (PSEs) for wheat growers were $84/metric ton in the EU, $52/metric ton in the United States, and $11/metric ton in Canada in 1998. Although the countries have reduced their subsidies on the basis of the committed schedule, the subsidies are still high and have stimulated wheat production, resulting in decreased wheat prices in the world market.

The EU provides export restitutions for exports of agricultural commodities. In 1999, average export restitution was $36/metric ton for wheat and $56/metric ton for barley. However, other wheat-exporting countries including the United States do not have export subsidy programs. The United States has not used export subsidies since 1995, even though the UR Agreement allows the subsidies on the basis of its scheduled commitments.

### State Trading

Many countries, including Canada and Australia, use STEs for wheat trade. STEs practice price discrimination and receive various subsidies from the government.

State trading is an important issue in the current Round of WTO negotiations, primarily because STEs have the capacity to distort trade flows (Ingco and Ng, 1998). Although the agenda of the WTO negotiations is uncertain with respect to STEs, it is clear that restrictions on STE operations will be needed to promote fair trade.

## Biotechnology

Agricultural biotechnology has significant potential for consumers and producers. Genetically modified organisms (GMOs) are a leading edge of this technology; examples of GMOs include corn and soybeans that are insect resistant and herbicide tolerant. GMO wheat has favorable characteristics for producers. However, GMO wheat has not yet been produced in the United States, mainly because of expected import restrictions on GMO wheat in most importing countries, including the EU and Japan. Differences in GMO regulations across countries pose potential barriers to exports. Clearly there is a need for harmonization of existing regulations among countries or negotiation of an international standard (Normile and Simone, 2000).

## China's Entry into the WTO

The United States and China signed a comprehensive bilateral trade agreement on November 15, 1999. The agreement is based on China's desire and commitment to participate in the global trade communities and was an effort to become a member of the World Trade Organization (WTO) (Economic Research Service, 2000). Under this agreement, China is expected to adopt a TRQ in which the import tariff is 1 percent for import quantities smaller than 7.3 million metric tons (Table 10.2). The import quota will rise to 9.5 million metric tons in 2005. Imports above the quota levels will face a higher duty of 76 percent. This tariff will be reduced to 65 percent by 2004. Since China is the largest wheat-producing and consuming country in the world, the impacts of the bilateral agreement on U.S. agriculture could be significant.

TABLE 10.2. China's TRQ Under the U.S.-China Bilateral Trade Agreement

| | Quota (1,000 mt) | | Private Share (%) | Average Imports (1997-1999) (1,000 mt) |
|---|---|---|---|---|
| | 2000 | 2004 | | |
| Wheat | 7,300 | 9,636 | 10 | 2,000 |
| Corn | 4,500 | 7,200 | 25(40)[a] | 250 |
| Rice | 2,660 | 5,320 | — | 250 |
| Short/medium grain | 1,330 | 2,660 | 50 | — |
| Long grain | 1,330 | 2,660 | 10 | — |

[a]40% is private share of the total import quota in 2005.
*Source:* U.S. Trade Representative, 2000.

## THE IMPACTS OF CHINA'S WTO ENTRY ON THE WORLD AND U.S. WHEAT INDUSTRIES

The impacts of China's entry into the WTO on the world wheat industry are a concern. A large body of literature has focused on the potential impacts of China's agriculture on the world agricultural industry (Brown, 1995; Koo, Lou, and Johnson, 1996) and China's agricultural development (Mao and Koo, 1996; Lin and Koo, 1992). However, a very limited number of studies (Economic Research Service, 2000) has analyzed the impacts of China's accession into the WTO on the world or U.S. wheat industries.

### Global Simulation Model

The Global Wheat Policy Simulation Model, which is operational at the Northern Plains Trade Research Center, was used to analyze the impacts of China's trade liberalization policy on the U.S. wheat industry. The model is an econometric dynamic partial equilibrium model that differentiates wheat into common and durum wheat. The model distinguishes between hard red winter, hard red spring, soft red winter, and white wheat.

The model contains five exporting countries and regions (Argentina, Australia, Canada, the United States, and the EU) and thirteen importing countries and regions (Algeria, Brazil, China, Egypt, the FSU, Japan, Mexico, Morocco, South Korea, Taiwan, Tunisia, Venezuela, and a Rest of the World Region). The model simulates production, consumption, stocks, exports, and trade flows for wheat classes over a ten-year period. The model is solved for a set of equilibrium wheat prices in which demand for each wheat class equals supply for every year.

### Model Structure and Development

Area and yield equations determine the supply of wheat. Since wheat is divided into two classes (common and durum), two separate supply equations are estimated in countries where both wheat classes are produced (Canada, the EU, and the United States).

Wheat area depends upon expected prices of wheat and alternative crops. As a proxy for price expectations, lagged prices are used in the area equation. In addition to commodity prices, the lagged area variable is included to capture dynamics associated with producers' planting decisions. Area harvested is a function of lagged area, prices of wheat and alternative crops, and government policies as follows:

$$a_{i,t}^n = f\left(a_{i,t-1}^n, p_{i,t-1}^n, p_{c,t-1}^n, g_t^n\right) \tag{10.1}$$

where $a_{i,t-1}$ is the wheat area harvested, $p_{i,t-1}$ is the world market price or domestic price of wheat, $p_{c,t-1}$ is the prices of alternative crops, $g$ is policy parameters, $i$ represents index for wheat type ($i = 1$ for common wheat and $i = 2$ for durum wheat), $c$ represents index for competing crops with wheat, and $n$ represents index for country.

Since the wheat classes are generally competing directly for land, area of each type is a function of prices of both wheat types and competing crops. Competing crops are barley, corn, and soybeans in common wheat-producing regions and barley in durum wheat-producing regions.

Assuming that wheat yields depend upon production practices and advances in technology, the total quantity of wheat produced ($qp$) is the product of the area harvested and yield per hectare:

$$qp_{i,t}^n = a_{i,t}^n \cdot y_{i,t}^n \tag{10.2}$$

Per capita wheat consumption is a function of the price of wheat, income, and a time trend representing changes in consumers' tastes and preferences:

$$fd_{i,t}^n = f\left(p_{i,t}^n, cy_{i,t}^n, t\right) \tag{10.3}$$

where $fd^n$ is per capita demand for wheat, $p^n$ is the domestic price of wheat, $cy^n$ is per capita disposable income, and $t$ is a trend.

Total consumption of wheat is calculated by multiplying the per capita consumption by population in the country as

$$qd_{i,t}^n = fd_{i,t}^n * pop_t^n \tag{10.4}$$

where $qd^n$ is the total demand for wheat and $pop^n$ represents population in country $n$.

Carry-out stocks in country $n$ ($qs^n$) are a precaution against unexpected shortfalls in production. These stocks, therefore, are likely related to the level of domestic production. However, since the opportunity cost of holding stocks depends on the price of wheat, the stocks should respond to price changes as

$$qs_{i,t}^n = f\left(qs_{i,t-1}^n, qp_{i,t}^n, p_{i,t}^n\right). \tag{10.5}$$

Net exports in country $n$ ($qx^n$) are the difference between domestic supply (domestic production plus carry-in stocks) and demand (domestic consumption plus carry-out stocks):

$$qx_{i,t}^n = qs_{i,t-1}^n + qp_{i,t-1}^n - qd_{i,t}^n - qs_{i,t}^n \qquad (10.6)$$

If net exports ($qx^n$) in a country are positive, the country is an exporting country. On the other hand, if net exports ($qx^n$) in a country are negative, the country is an importing country.

A market equilibrium condition is expressed as:

$$\sum_{n=1}^{n} qx_{i,t}^n = 0 \qquad (10.7)$$

The equilibrium condition is solved to determine market-clearing prices of wheat classes. The equilibrium world prices of wheat type $i$ ($pm^w$) obtained from Equation 10.7 are converted into domestic prices ($pm^n$) using the official exchange rates ($er^n$) as follows:

$$pm_{i,t}^n = pm_{i,t}^w * er_t^n \qquad (10.8)$$

A price equation, which was estimated by regressing the domestic price of wheat against the world price of wheat ($pm^w$), is used to convert world price into domestic price. To simulate trade liberalization scenarios in China, the domestic price of wheat is enforced to be equal to the world price ($pm^w$), plus handling charges in the home country and average shipping charge from the major exporting countries.

### Assumptions and Data Collection

The baseline simulation is grounded on a series of assumptions about general economy, agricultural policies, and technological changes in exporting and importing countries for the simulation period (2000-2005). Macro assumptions are based on forecasts prepared by the WEFA group and Project Link. Some of the macro variables are GDP growth rates, interest rates, exchange rates, and inflation rates in the countries. It is generally assumed that current agricultural policy will continue in all countries in the baseline simulation. Average weather conditions and historical rates of technological change also are assumed in this simulation. The price of wheat in individual countries and the world market is endogenous, while the prices of other crops are exogenous. Thus, the baseline simulation is based on the forecast world prices of other crops that have substitute and complementary relationships with wheat. The forecast prices were obtained from the Food and Agricultural Policy Research Institute (FAPRI) baseline solution and the *2000 USDA Agricultural Outlook*.

## *Alternative Scenarios*

Two scenarios were developed to estimate the world wheat industry's response to China's inclusion into the WTO. The Base Scenario assumes that China's import tariffs remain at the current levels. Scenario 1 assumes normal wheat imports based on China's trade liberalization policy under the TRQ in which the tariffs on wheat would be reduced to 1 percent if China's wheat imports are less than 7.3 million metric tons in 2000 and 9.5 million metric tons in 2005. In Scenario 2, it is assumed that China imports wheat up to the maximum level of the TRQ (7.3 million metric tons in 2000, increasing to 9.5 million metric tons in 2005).

## *RESULTS*

### *Effects on World and U.S. Domestic Prices*

Under the Base Scenario, the U.S. wheat price is estimated to increase 6.8 percent from $3.49 per bushel in 2000 to $3.73 per bushel in 2005, and world prices are expected to increase 7.6 percent (Table 10.3). This is because of strong demand for wheat in major importing countries, including South Korea and Japan. If China reduces its import barriers based on the bilateral trade agreement with the United States, China will increase its wheat imports from major exporting countries, which will cause increases in prices of wheat in both the world and U.S. markets. The U.S. wheat price is estimated to increase 10.6 percent to $3.86 per bushel in 2005 under Scenario 1, compared to the wheat price in 2000. Wheat price increases further in the U.S. and world markets under Scenario 2. Increases in price in 2005 due to the bilateral agreement are 12.3 percent in the U.S. market and 13.5 percent in the world market under Scenario 2, compared to the wheat price in 2000.

TABLE 10.3. Wheat Prices in the U.S. and World Markets ($/bushel)

| | | Base | | Scenario 1 | | Scenario 2 | |
|---|---|---|---|---|---|---|---|
| | | 2000 | 2005 | 2005 | Change (%) | 2005 | Change (%) |
| U.S. | $/bushel | 3.49 | 3.73 | 3.86 | 10.6 | 3.92 | 12.3 |
| World | $/mt | 3.30 | 3.55 | 3.65 | 10.2 | 3.74 | 13.5 |

### Changes in China's Wheat Industry

In the Base Scenario, Chinese domestic wheat production is expected to increase 1.6 percent for the 2000-2005 period (Table 10.4). However, wheat production would decrease under the trade liberalization scenarios, mainly because the trade liberalization policy would lower the domestic price of wheat in China. Wheat production is expected to decrease 2 percent from 120.7 million metric tons to 118.3 million metric tons for the 2000-2005 period under Scenario 1 and 6.61 percent to 112.8 million metric tons in 2005 under Scenario 2. Small reductions in wheat production in China under the trade liberalization scenarios are mainly because Chinese farmers, in general, are less sensitive to prices. The average farm is small; thus, a large portion of production is used on the farm.

Domestic consumption of wheat in China is expected to increase under both the base and trade liberalization scenarios. Domestic consumption of wheat in the trade liberalization scenarios, however, is larger than the Base Scenario because lower prices under the trade liberalization scenarios tend to stimulate wheat consumption. But increases in wheat consumption are not significant; 3.4 percent in the Base Scenario for the 2000-2005 period, 3.6 percent in Scenario 1, and 3.8 percent in Scenario 2 (Table 10.4). The main reason is that Chinese consumers are not sensitive to domestic prices of wheat. In China, COFCO (the government import agency) imports wheat from foreign markets and resells wheat to domestic millers. Consumers buy wheat flour or wheat products processed by wheat millers and food processors, which are generally owned by the Chinese government. Thus, prices of consumer goods, such as wheat flour and wheat products, are not directly influenced by changes in the domestic prices of wheat.

TABLE 10.4. China's Wheat Industry Under the Base and Alternative Trade Scenarios

|  | Base | | Scenario 1 | | Scenario 2 | |
|---|---|---|---|---|---|---|
|  | 2000 | 2005 | 2005 | Change (%) | 2005 | Change (%) |
| Carry-in | 26,336 | 25,897 | 26,315 | −1.0 | 26,724 | 1.5 |
| Production | 116,481 | 120,735 | 118,340 | 1.6 | 112,760 | −3.3 |
| Import | 1,262 | 772 | 3,422 | 171.4 | 9,176 | 627.4 |
| Consumption | 117,671 | 121,741 | 121,965 | 3.6 | 122,162 | 3.8 |
| Carry-out | 26,409 | 25,662 | 26,111 | −1.2 | 26,497 | 0.3 |

## China's Wheat Imports

China's wheat imports are projected to be about 772,000 metric tons in 2005 under the Base Scenario, which is much smaller than wheat imports in 2000. This is mainly because of (1) expected increases in wheat production (3.6 percent) in China, and (2) slow increases in consumption of wheat (3.4 percent).

China's wheat imports are expected to increase dramatically under both trade liberalization scenarios. China is expected to import 3.4 million metric tons in 2005 under Scenario 1, which is about four times higher than China's expected wheat imports in 2005 under the Base Scenario (Table 10.4). However, these imports are smaller than the import quota (7.5 million metric tons) established by the bilateral trade negotiation. Under Scenario 2, China is expected to import about 9.2 million metric tons of wheat, which is similar to the import quota in 2005. China's total wheat imports under Scenarios 1 and 2 are about 3.0 and 8.0 percent of the total wheat traded in the world market, respectively.

China imports wheat from four major exporting countries: the United States, Canada, Australia, and Argentina. If China liberalizes its wheat imports in Scenarios 1 and 2, China is expected to increase its wheat imports as mentioned previously, but major importing countries are expected to reduce their wheat imports due mainly to increased wheat prices in the world market. The increase in U.S. exports to China is projected to be 1,562,000 metric tons in 2005 under Scenario 1, while U.S. exports to the rest of the world are expected to decrease by 662,000 metric tons, implying that the net change in U.S. exports is 904,000 metric tons (Table 10.5).

Similarly, an increase in other exporting countries' exports to China is expected to be 1,096,000 metric tons in Scenario 1, while their exports to the rest of the world are predicted to decrease by 474,000 metric tons. Thus, the net change in wheat exports by major exporting countries is 622,000 metric tons. Export shares in the Chinese market are 59.2 percent for the United States and 40.8 percent for the rest of the world in Scenario 1.

U.S. exports to China are expected to increase by 5,856,000 metric tons under Scenario 2, while U.S. exports to the rest of the world should decrease by 1,844,000 metric tons. In this scenario, the net change in U.S. exports is 4,012,000 metric tons. The impacts of China's trade liberalization policies on other exporting countries are much smaller than for the United States. An increase in major exporting countries' exports to China is expected to be 2,548,000 metric tons, which is about 30.3 percent of the total wheat imported by China in Scenario 2. The net change in the countries' wheat exports is predicted to be 1,746,000 metric tons. U.S. export shares in China are much larger under the trade liberalization scenarios than the Base Scenario because the United States could be more competitive by exporting

TABLE 10.5. Changes in 2005 Wheat Exports in Base and Alternative Scenarios

|  | Base | Scenario 1 | Scenario 2 |
| --- | --- | --- | --- |
| *Export quantity (thousand metric tons)* | | | |
| United States | | | |
|   Exports to China | 0 | 1,562 | 5,856 |
|   Exports to ROW | 0 | −662 | −1,844 |
|   Net change | 0 | 904 | 4,012 |
| Major exporting countries | | | |
|   Exports to China | 0 | 1,096 | 2,548 |
|   Exports to ROW | 0 | −474 | −802 |
|   Net change | 0 | 622 | 1,746 |
| *Export Values (million U.S. dollars)* | | | |
| United States | | | |
|   Exports to China | 0 | 221 | 842 |
|   Exports to ROW | 0 | −94 | −265 |
|   Net change | 0 | 127 | 577 |
| Major exporting countries | | | |
|   Exports to China | 0 | 147 | 349 |
|   Exports to ROW | 0 | −63 | −110 |
|   Net change | 0 | 84 | 239 |

wheat to China through Pacific Northwest (PNW) ports under the bilateral trade agreement. China has banned wheat imports from PNW ports because of *Tilletia controversa* Kuhn (TCK) smut. However, China agreed to eliminate the import ban on wheat from PNW ports under the new bilateral trade agreement. This implies that China would increase wheat imports through PNW ports under the trade liberalization scenarios.

The total value of world wheat exports to China in 2005 is expected to increase by $147 million under Scenario 1 and $349 million under Scenario 2. However, net increases in the total export value are smaller than those to China, mainly because exports to the rest of the world were reduced under the scenarios. The total value of U.S. wheat exports to China ranges between $221 million under Scenario 1 to $842 million under Scenario 2, while net changes in U.S. exports in the scenarios are much smaller; $127 million under Scenario 1 and $349 million under Scenario 2.

## CONCLUSION

The purpose of this study is to assess the outlook of U.S. and world wheat industries under the UR Agreement. Special attention is paid to the impacts of China's WTO entry on trade flows of wheat from exporting countries to major importing countries and its implications for the U.S. wheat industry.

Issues related to the U.S. wheat industry for the 2000 Round of WTO agricultural trade negotiations include further reduction in internal supports and export subsidies, state trading enterprises, China's WTO entry, and agricultural biotechnology. These issues are not unique to the U.S. wheat industry but are fairly common for most agricultural commodities produced in the United States.

The impacts of China's trade liberalization policy based on the proposed TRQ on the U.S. and world wheat industries are not significant. The trade liberalization policy would raise the world price of wheat about 2 to 5 percent and the U.S. domestic price about 3 to 6 percent. The trade liberalization policy would lower the Chinese domestic price of wheat. However, the lower prices would not affect domestic consumption and production in China, mainly because Chinese producers and consumers are not very sensitive to changes in wheat prices.

## REFERENCES

Brown, Lester R. (1995). *Who Will Feed China?* The Worldwatch Environmental Alert Series, W. W. Norton and Company, New York.

Economic Research Service. (2000). *China's WTO Accession Would Boost U.S. Ag Exports and Farm Income.* Agricultural Outlook, U.S. Department of Agriculture, Washington, DC.

Harwood, Joy L. and Kenneth W. Bailey. (1990). *The World Wheat Market: Government Intervention and Multilateral Policy Reform.* Staff Report No. AGES9007. USDA, CED, ERS, Washington, DC.

Ingco, Merlinda and Francis Ng. (1998). *Distortionary Effects of State Trading in Agriculture: Issues for the Next Round of Multilateral Trade Negotiations,* The World Bank Development Research Group, Policy Research Working Paper 1915, Washington, DC, April 1998.

Koo, Won W., Jianqiang Lou, and Roger G. Johnson. (1996). *Increases in Demand for Food in China and Implications for World Agricultural Trade.* Agricultural Economics Report No. 351, Department of Agricultural Economics, North Dakota State University, Fargo.

Lin, Jinding and Won W. Koo. (1992). "An Inter-Sectoral Prospective on Relationship Between the Agricultural and Industrial Sectors in Chinese Economic Development," *Issues in Agricultural Development,* ed. Margot Bellamy and Bruce

Greenshields, IAAE Occasional Paper No. 6, International Association of Agricultural Economics.

Mao, Weining and Won W. Koo. (1996). *Productivity Growth, Technology Progress, and Efficiency Changes in Chinese Agricultural Production from 1984 to 1993.* Agricultural Economics Report No. 362, Department of Agricultural Economics, North Dakota State University, Fargo.

Normile, Mary Anne and Mark Simone (2000). *Agriculture in the Uruguay Round,* Economic Research Service, USDA, URAA Issues Series, <http://www.ers.usda.gov/briefing/WTO/uraa.htm> Updated December 29, 2000.

Ortmann, Gerald, F. Norman Rask, and Valter J. Stulp. (1989). "Comparative costs in Corn, Wheat, and Soybeans Among Major Exporting Countries." Research Bulletin. Agricultural Research and Development Center, Wooster, Ohio, August.

USDA (2002). *Production, Supply, and Distribution Database,* Economic Research Service, Data on Diskette, <http://www.ers.usda.gov/data/sdp/view.asp?f=international/93002/>.

U.S. International Trade Commission. (1990). *Durum Wheat: Conditions of Competition Between the U.S. and Canadian Industries.* Washington, DC.

U.S. Trade Representative. (2000). *U.S.-China Bilateral WTO Agreement.* Washington, DC.

Chapter 11

# New WTO Agricultural Trade Negotiations: Issues for the U.S. Coarse Grain Market

Linwood Hoffman
Erik Dohlman

## *INTRODUCTION*

New WTO agricultural trade negotiations aimed at continuing the process of multilateral agricultural trade reform initiated by the Uruguay Round are currently underway.[1] As the leading producer and exporter of coarse grains, the United States has a large stake in the outcome of these negotiations. In recent years, exports have accounted for about one-fifth of U.S. coarse grain disappearance, and these exports comprise about 11 percent of total U.S. agricultural export earnings.

Although trade impediments in the world coarse grain market are typically lower than those on other commodities, such as wheat or rice, tariffs, tariff rate quotas (TRQs), export subsidies, and domestic support programs are still a source of trade distortions. Of all the coarse grains, trade distortions appear to be most significant in the barley sector. For example, export subsidies are mostly used for barley, with the European Union (EU) accounting for the largest share of these expenditures. Among coarse grains, expenditures on WTO-limited domestic support programs have also been used mostly for barley, particularly in the EU and Japan. Import tariffs on barley are low for many countries, but for some countries they are higher for processed barley malt—a situation known as tariff escalation.

In addition to trade distortions in the barley market, potential barriers to trade in the corn sector remain a concern. High "bound" (maximum allowable) tariff rates on corn, for example, allow some countries the discretion to considerably raise current applied tariff rates. South Korea has a WTO over-quota tariff binding on corn in excess of 300 percent, while the EU and Japan have a corn over-quota tariff of slightly more than 100 percent AVE (ad

A more detailed version of this chapter can be found at <http://www.ers.usda.gov/briefing/WTO/commodities.htm>.

valorem equivalent). The administration of import licenses for TRQs, the impact of state trading enterprise (STE) activities on coarse grain trade, export credit guarantees, market access for products of new technologies such as genetically modified corn, and export taxes have also emerged as issues. In addition, further liberalization in the world meat market could expand market opportunities for coarse grain producers.

This chapter examines the features of trade in the global and U.S. coarse grain market, discusses major accomplishments of the Uruguay Round, and examines issues relevant to the coarse grain sector that may be addressed in the new agricultural trade negotiations.

## GLOBAL COARSE GRAIN TRADE

Coarse grains, feed wheat, and nongrain feedstuffs, such as tapioca, cassava, citrus pulp, and other by-products are primary sources of energy for livestock, poultry, and hogs. Oilseed meals and other sources of protein are typically used to complement these energy sources. Of global coarse grain supplies, about two-thirds are used as animal feed, with the remainder going to seed, industrial, and food uses. Just over 10 percent of global coarse grain production is traded, and most is destined for feed use. Trade consists primarily of corn (about three-quarters of global trade) and barley (two-tenths), followed by sorghum, oats, and rye. Income growth and corresponding changes in per capita meat consumption are therefore key factors driving consumption and trade patterns for coarse grains, but trade and domestic policies also play an important role.

Between 1971-1972 and 2000-2001, global coarse grain trade rose from 49.3 to 100.4 million metric tons, an annual compound growth rate of 2.5 percent.[2] Coarse grain trade grew at an 8.7 percent annual rate during the 1970s, experienced negative growth in the 1980s, and rose only 0.5 percent annually in the 1990s. Although the volume of world coarse grain exports has been relatively stable since the early 1980s, the import destinations have changed dramatically, with Western Europe and transition economies experiencing large declines and growth coming mainly from East Asia, Latin America, North Africa, and the Middle East.

Nearly three-fourths of global coarse grain is produced in the United States, China, EU, Brazil, India, Canada, Mexico, Argentina, and Romania. The United States, Argentina, EU, China, Australia, and Canada account for more than nine-tenths of global coarse grain exports. Japan, South Korea, Mexico, Saudi Arabia, Taiwan, Egypt, EU, United States, and China account for nearly two-thirds of coarse grain imports.

The United States is the world's largest exporter of corn and sorghum, a minor barley exporter, and the largest importer of oats. In (fiscal) 1999, the

United States exported nearly 58 million metric tons of coarse grains, valued at $5.6 billion. These exports represented 11 percent of total U.S. agricultural exports by value. More than seven-tenths of U.S. coarse grain exports go to Japan, South Korea, Taiwan, and Mexico, while the remainder is dispersed among many countries.

Trade and domestic agricultural policies have had a major influence on the volume and direction of coarse grain trade in recent decades. For example, to encourage self-sufficiency, the EU's Common Agricultural Policy (CAP) established common external import barriers while at the same time removing internal barriers to trade. As a result, coarse grain imports by the EU declined significantly between 1981 and 1986 and became minor thereafter. In the former Soviet Union (FSU), artificially low food prices under the Communist system led to increased coarse grain imports for feed use between 1972 and 1992, but after the breakup of the Soviet Union, grain imports declined rapidly. The North American Free Trade Agreement (NAFTA), on the other hand, has facilitated trade between the United States, Canada, and Mexico and contributed to recent growth in coarse grain trade between these countries. In other areas, such as Asia, Africa, and the Middle East, income growth and increased demand for meat and meat products have played a large role in stimulating greater coarse grain imports. World credit constraints have also influenced coarse grain trade over this time period.

In addition to policies and other developments that directly affect coarse grain trade, liberalization of the world meat (beef, pork, and poultry) market could provide indirect benefits to coarse grain producers. Since 1989, Japan's meat market liberalization has brought significant increases in its meat imports, for example. Although Japan has slightly reduced coarse grain imports, more U.S. corn is being exported in the form of higher value-added beef, pork, and poultry products. Since 1985, total U.S. exports of these products have risen from less than 1 billion pounds (4 percent of world exports) to over 9 billion pounds (19 percent of global trade) in 1999 (USDA, 2000a).

## ACCOMPLISHMENTS OF THE URUGUAY ROUND AND ISSUES FOR NEW AGRICULTURAL NEGOTIATIONS

The Uruguay Round (UR), concluded in 1994, represented the first comprehensive multilateral effort to address agricultural trade issues. Under the Uruguay Round Agreement on Agriculture (URAA), WTO members committed to eliminate nontariff trade barriers, cut tariff levels on all agricultural products, lower the volume of and expenditures on subsidized exports, and reduce aggregate spending on certain trade-distorting domestic support programs for agriculture. A new process for settling trade disputes was also established, and the agreement contained a temporary "peace" provision designed to protect certain subsidy policies from some WTO challenges (WTO, 1995).

Furthermore, under URAA Article 10.2, member countries agreed to work toward the development of internationally agreed disciplines on export credits, export credit guarantees, or insurance programs (WTO, 1995). The WTO's "Understanding on Article XVII" also established a working definition of state trading enterprises (STEs) and created stronger notification requirements for STEs.

The agriculture agreement also recognized the need for special and differential treatment for developing countries, granting them additional time to meet obligations and requiring smaller subsidy and tariff reductions than developed countries. Other special provisions were made for least developed countries and countries that rely on food imports.

Finally, the Uruguay Round established a new agreement on the use of sanitary and phytosanitary (SPS) measures. The SPS agreement requires that regulations for food safety and animal and plant health be based on science and that such regulations should not be arbitrary or discriminate between countries with similar conditions.

Although the URAA was an important step toward identifying and disciplining trade distortions in agriculture, the impact of the agreement on global coarse grain markets is difficult to separate from other market-related events that occurred during the implementation period—such as the Asian financial crisis and increased exports by ( then non-WTO member) China. Average global coarse grain exports did rise slightly after the enactment of the URAA (from 92 million metric tons in 1990-1994 to 92.5 million tons during 1995-1999), but at a slower rate of growth than for global production. U.S. market share rose by 2 percentage points during this period.

In the new agricultural negotiations, issues important to the U.S. coarse grain industry include those raised in the Uruguay Round, such as increased market access and continued reduction in trade-distorting domestic support and export subsidies. Other issues that could be addressed by the WTO, or in other negotiating arenas, include disciplines on the operational activities of state trading enterprises, the treatment of export credits and credit guarantees as an export subsidy, trade in biotech grains, a "zero for zero"[3] proposal for barley and malt trade, and a curb on export taxes.

The following sections discuss URAA accomplishments and key issues pertinent to coarse grains in the new agricultural trade negotiations. In each section, accomplishments are presented first, followed by a discussion of the issues. Further elaboration on many of these subjects can be found in (USDA, 1998) and the ERS WTO Web site <http://www.ers.usda.gov/briefing/wto/>.

## CONTINUING ISSUES

### Market Access

The URAA required WTO members to reduce and bind existing tariffs and to convert all nontariff agricultural trade barriers—such as quotas or discretionary import licenses—into tariffs, a process referred to as "tariffication." In some cases, the new bound (maximum) tariff levels were still prohibitively high, so a TRQ system was created to maintain pre-URAA import levels and/or assure minimum access import opportunities. The TRQ is a two-tiered tariff system, in which in-quota import quantities are subject to lower, generally nonprohibitive tariffs, but with higher tariffs on over-quota quantities. All bound tariffs (except in-quota rates) are to be reduced by an average of 36 percent (24 percent for developing countries) from "base" period (usually 1986-1988) levels to the final bound level by the end of the implementation period, with a minimum tariff cut of 15 percent per product (10 percent for developing countries).

The majority of global coarse grain imports are concentrated among a small number of countries that currently maintain very low applied (actual) tariff rates, or have TRQs with low in-quota tariff rates. Four countries—Japan, Egypt, Malaysia, and Taiwan—account for 40 percent of world corn imports and have applied tariffs on corn of 0 to1 percent. Five other countries—Colombia, EU, Mexico, South Korea, and Venezuela—account for 27 percent of corn imports and have TRQs, but often do not apply the over-quota rate, thus permitting imports equal to or well above quota levels.

### Tariffs

Countries that currently import a large amount of coarse grains generally have low applied tariff rates and/or large quotas. However, in many countries, including some that currently allow low tariff access, maximum allowable (bound) rates on over-quota imports were set at very high levels due to exaggerated estimates of the tariff equivalent of nontariff barriers during the base period, a process referred to as "dirty tariffication." The base period was also a time of high protection for agriculture generally, and many less-developed countries claimed base period tariffs that were much higher than actual rates during that time. Consequently, the actual reduction of base level tariffs to final bound rates was fairly modest. Finger estimated that tariffs affecting less than 15 percent of world agricultural trade will be reduced from base period levels by the end of the URAA implementation period (USDA, 1998).

Because applied tariffs are often much lower than bound rates, exporters face uncertainties due to the possibility of sudden tariff increases based on changing domestic supply and demand conditions or policy objectives of importing countries. Brazil, Peru, Romania, and Turkey, for example, have bound tariffs on corn ranging from 55 to 240 percent, but much lower applied tariffs ranging from 11 to 50 percent. In the past two years, Turkey has raised and lowered its applied tariff on corn several times, most recently from 30 to 50 percent in July 2000 (USDA, various issues b). Very low tariffs on unprocessed coarse grain imports are also accompanied in many cases by higher tariffs on processed products (tariff escalation) in order to protect domestic processors. Japan, Turkey, and non-WTO members Russia and Saudi Arabia impose higher applied tariffs on barley malt than barley, for example.

Trade patterns in coarse grains and products may not reflect nations' comparative advantage due to other market access policies, such as regional trading arrangements giving preferential access to a group member or differential tariffs for substitute commodities. Brazil's Southern Cone Common Market (MERCOSUR) partners (Argentina, Uruguay, and Paraguay) can export corn to Brazil duty-free, whereas non-MERCOSUR imports are subject to an 11 percent tariff.

### Tariff Rate Quotas

By replacing nontariff barriers with TRQs, the URAA objective was to increase the transparency of protection in agriculture, ensure that historical trade levels were maintained, and to expand trade through minimum access commitments. The URAA required minimum access opportunities in cases where imports had been less than 5 percent of domestic consumption during the base period (1986-1988), and required that minimum access quotas rise to that percentage of consumption by the end of the implementation period. In cases where imports exceeded 5 percent of consumption, countries had to maintain existing access opportunities.

Although several of the larger U.S. corn importers have TRQs, very few were required to increase their quota level over the commitment period because they already met or exceeded the minimum access requirements. Nevertheless, the URAA was expected to create new access commitments for about 1 million metric tons in coarse grains by 2004, largely due to commitments by Japan, South Korea, South Africa, and the Philippines. For example, the Philippines was required to replace a ban on corn imports with a TRQ having a final quota level of 217,000 metric tons.[4] Philippine corn imports subsequently rose from zero during 1991-1993 to 500,000 tons in 1999, well above its quota level. Recently, India established a TRQ for corn with a quota of 350,000 tons, an in-quota tariff of 15 percent and an over-quota tariff of 50 percent. India imported virtually no corn from 1989 to

1997, but imported over 200,000 tons of corn last year, including its first commercial imports (85,000 tons) from the United States.

Despite some increased access opportunities created by the establishment of TRQs, a number of complications related to the administration of TRQs have emerged. One issue is the practice of allocating the quota to suppliers based on the historical distribution of trade. This practice can perpetuate past patterns of trade into the future despite changing market conditions. Some countries have also assigned import rights to STEs or producer associations. These organizations may lack the incentive to increase market access, resulting in quota "underfill," or may bias the quota distribution to favored suppliers. In Venezuela, for example, there have been some claims that licenses for in-quota imports have not been issued with the same degree of transparency as in the past (USDA, various issues b).

Another potential hindrance to trade are the special emergency measures (agricultural "safeguards") that WTO members can use to protect domestic producers from a sudden drop in prices or a surge in imports. Some countries have used these provisions to restrict imports of sensitive products during the URAA implementation period, but they have been used only sparingly on coarse grains and products, and their use is subject to strictly defined conditions. Special safeguards are only permitted on products converted from nontariff restrictions to TRQs and only on imports exceeding minimum access quota levels. The country imposing the safeguard also has to reserve the right to do so in its schedule of commitments on agriculture. About twenty countries have reserved this right for coarse grains.

Although the URAA provided for some increased market access, there is potential for additional and more stable access if high bound tariff rates are reduced, tariff quotas are increased, or if out-of-quota duties are substantially reduced. This is especially true for some of the projected growth markets such as Latin America and South East Asia. The current U.S. objective on market access is to reduce tariff levels and tariff escalation among countries and to ensure market access opportunities for all products in all markets. Continued reduction of bound tariffs seems possible because of the often high tariff levels established through the URAA negotiations. Additional U.S. objectives include expanding quota levels in countries with TRQs, improving TRQ administration, and eliminating the transitional special agricultural safeguard as defined in Article 5 of the Agreement on Agriculture (WTO, 2000).

## Export Subsidies

The URAA began the process of reducing the use of export subsidies in agricultural trade by prohibiting their use on agricultural products unless the specific commodity is listed under the WTO member's schedule of export subsidy reduction commitments. Of the 140 member countries, twenty-five countries that had export subsidies in the base period agreed to reduce the volume and value of subsidized exports on specific products by a set percentage over a period of time. The remaining countries and commodities are bound at zero subsidies.

Of the twenty-five countries making commitments, twelve notified the WTO of commitments to reduce export subsidies on coarse grains. The EU, a major global exporter of barley, oats, and rye, is the largest subsidizer of global coarse grains, accounting for 96 to 100 percent of all subsidized coarse grain exports between 1995 and 1997. The EU's export subsidies for coarse grains totaled $397 million for 1995, $493 million for 1996, and $310 million for 1997. South Africa's coarse grain export subsidies amounted to $11.5 million in 1995, but it eliminated export subsidies in 1997. Coarse grain subsidies by Hungary totaled $4.9 million in 1995 and $3.4 million in 1997, and Slovakia had expenditures of $0.7 million in 1997.

The United States used the Export Enhancement Program (EEP) to subsidize coarse grain (mostly barley) and barley malt exports between 1985-1986 and 1994-1995. Coarse grain subsidies under EEP peaked in FY 1987 when about 3.5 million tons (with a bonus value of $143 million) of coarse grain exports were assisted. However, since 1994-1995 EEP has not been used for coarse grains, with the exception of a $1.2 million barley export subsidy made in 1997. All countries have remained within their commitment levels other than Hungary in 1995, when its actual subsidy was 282 percent of its commitment level. However, Hungary received a waiver that year under WTO Article XXVIII consultations.

As the largest user of coarse grain export subsidies, the EU could face the greatest challenge conforming to URAA subsidy limitations, particularly those on the volume (rather than value) of subsidized exports. Although the EU has not exceeded its commitment level on export subsidy expenditures for coarse grains, it has rolled over unused volumes from previous years and applied them to subsequent years. In 1998, the EU subsidized 14.8 million tons of coarse grain exports, about 2.8 million tons above its original 1998 commitment level. This rollover will no longer be possible in 2000-2001, when the commitment level is 9.97 million metric tons. However, the recent depreciation of the EU currency and its Agenda 2000 reforms—which reduced the intervention (support) prices for grains—have lowered the per unit subsidy needed to match world prices (Leetmaa and Bernstein, 1999). During 2000 the EU exported barley without subsidies because of a depreciating currency. Whether the EU will need to provide subsidies on future coarse grain exports depends largely on exchange rate developments.

A current U.S. objective for the negotiations is the complete elimination of export subsidies, including both budgetary outlays and quantity commitments (WTO, 2000). Some countries are opposed to such an approach, while others have focused on ways to prevent members from circumventing commitments through state trading enterprises or the use of subsidized export credits.

### Domestic Support

Domestic policies that support prices or subsidize production may encourage excess production and distort trade flows by causing a decline in imports in some markets and/or increasing the use of export subsidies. The URAA distinguished between policies that are considered production and trade distorting (amber box) and nondistorting (green box), and required WTO member countries to annually report and reduce amber box support provided to domestic agricultural producers. The total value of support related to policies in the amber box is referred to as the aggregate measurement of support (AMS). Countries agreed to keep their AMS from exceeding limits specified by the URAA for 1995-2000. These limits decrease from 97 percent of the 1986-1988 base support level in 1995 to 80 percent of the base level in 2000 (87 percent of the base level for developing countries by 2004).

Amber box policies subject to reduction include price supports, marketing loans, direct payments based on current production or price levels, input subsidies, and certain subsidized loan programs. If support for a specific crop is equal to or less than 5 percent of its production value (10 percent for developing countries), it is not counted toward the AMS limits. This de minimis exemption provides some flexibility to a country in the design of its domestic support policies for specific commodities. But much more flexibility for commodity support is provided by the use of the aggregate support measure concept, since the reduction commitments do not apply to specific commodities, only to the total value of support for a country.

Direct producer payments under certain production-limiting programs (referred to as blue box policies) are exempt from reduction (not included in the current AMS) as long as they satisfy specific criteria. Specifically, the program must be production limiting, with payments based on fixed area and yield, or on 85 percent or less of the base level of production.

Support from policies with minimal impacts on trade or production (green box policies) is also excluded from the AMS. Examples of these policies include public stockholding, natural disaster relief, marketing and promotion, inspection, extension services, pest and disease control, and research. They also include producer payments that are minimally distorting to production, such as certain forms of decoupled income support not tied to production like the United States' production flexibility payments.

Currently, thirty-one WTO member countries have AMS reduction commitments (WTO, 1999). In 1997—the most recent year with comparable data—the EU, Japan, and the United States accounted for about 90 percent of total AMS notifications. However, the coarse grain contribution to the AMS for each of these countries was relatively low. For example, in 1997 the EU's coarse grain (mostly barley and corn) nonexempt amber box payments totaled $4.6 billion, or 5.2 percent of its commitment level. Barley was the only coarse grain receiving amber box support in Japan, and represented just 0.5 percent of its $39.7 billion AMS commitment level that year. Although U.S. coarse grain (mostly corn) amber box expenditures totaled $155 million in 1997, they were excluded from the AMS calculation due to the de minimis exemption.

Of these three countries, the EU and Japan have come closest to their AMS limits. In 1997, AMS outlays equaled about 66 percent of URAA limits for Japan, about 63 percent for the EU, and about 29 percent for the United States. Other smaller countries such as South Korea, Norway, Switzerland, Iceland, Israel, and Slovenia have come closer to their AMS limits during this time period.

Countries have stayed within their AMS commitment limits for two main reasons. First, reductions are being made from a base period (1986-1988) which was characterized by high levels of domestic support. Second, several countries such as the EU, United States, Japan, South Korea, Switzerland, Norway, Iceland, Israel, and Slovenia have "reinstrumented" policies to meet these commitment levels. Commodity prices were also relatively high during 1995-1997, so fewer domestic subsidies were needed during that period. When AMS levels are reported for more recent years, some countries are likely to be much closer to their commitment ceilings.

In the EU, policy changes since the base period have resulted in an increase of compensatory payments, which are classified as exempt blue box payments, since support is tied to production limitations based on a fixed area and yields. These payments for EU corn and other cereals averaged about $13 billion between 1995 and 1997. In contrast, the intervention market price support provided for EU coarse grains is counted as amber box payments and averaged $4.6 billion during 1995-1997. In coming years, the EU's Agenda 2000 reforms are expected to reduce the level of the EU's amber box price supports and increase the level of blue box income supports, continuing a policy direction which began with the MacSharry CAP reforms in 1992 (Leetmaa and Bernstein, 1999). This assumes the blue box will continue to be available to the EU, which relied on it heavily. One of the issues for the current round of negotiations is whether the blue box will be continued.

The major domestic support policies affecting the U.S. coarse grain sector are provided for in the Federal Agriculture Improvement and Reform (FAIR) Act of 1996, which replaced deficiency payments for feed grains (classified as blue box due to production limitations) with potential marketing loan gains or loan deficiency payments (classified as amber box). Marketing loan benefits were not made during 1996 due to relatively strong prices, but feed grains did receive amber box interest subsidies of $29.6 million related to participation in the Commodity Credit Corporation's (CCC) commodity loan program. In 1997, $52 million in interest subsidies and $103 million in marketing loan gains (mostly to corn producers) were made, but payments in both years were excluded from the AMS because of the de minimis exemption. Crop and revenue insurance premiums subsidized by the government are considered amber box non–commodity-specific support policies, but these subsidies have also been excluded from the AMS because of the de minimis exemption. Feed grain production flexibility contract payments, which totaled about $3 billion in 1999, are classified as green box since payments are not tied to current production or price.

Some countries have begun to shift away from amber box policies and toward more green box policies, which presumably reduces production and trade distortions. Negotiations appear likely to focus on continued reduction of the AMS. The current U.S. objective on agricultural domestic support is to substantially reduce trade-distorting domestic support in a manner that corrects the disproportionate levels of support used by some WTO members. Specifically, the United States proposes that each country with current AMS commitments further reduce their AMS from final bound levels to a new level equal to a fixed percentage of the member's value of total agricultural production in a fixed base period, and that the fixed percentage be the same for all members. The U.S. proposal also suggests simplifying the domestic support disciplines into two categories: exempt support, with minimal trade distorting effects on production, and nonexempt support subject to a reduction commitment (WTO, 2000).

## OTHER ISSUES

### State Trading Enterprises

STE activities affect trade by influencing domestic and international prices (Ackerman and Dixit, 1999). A particular concern with STEs is that their lack of price transparency can be used to mask export subsidies or import barriers, and can therefore be used to circumvent URAA commitments on market access and export subsidies. STEs may also benefit from advantages unavailable to commercial firms that compete against them. Several factors influence the tariff/subsidy equivalents associated with STE practices, including its degree of control over the domestic market, its policy objectives, the extent of its international market power, and its range of authorities and government support. State trading is also an issue that pertains to countries seeking WTO accession, such as China, Taiwan, Russia, and Vietnam, which use STEs.

In 1995 and 1996, more than thirty WTO countries reported nearly 100 STEs involved in their agriculture sector. STEs are generally not as significant for coarse grains as they are for wheat, rice, and sugar, but state traders are important in the world barley market (Ackerman and Dixit, 1999). The Canadian Wheat Board and Australia's marketing board, for instance, maintain about 14 and 23 percent of global barley export market shares, respectively. Although Australia plans to eliminate its marketing board for barley exports in 2000, smaller exporters such as Turkey and Russia also use STEs to exercise some control over their exports.

On the import side, Saudi Arabia's STE, the Grain Silos and Flour Milling Organization (GSFMO), accounted for an average of about one-third of world barley imports during 1996-1999. China and Japan, which also regulate imports with STEs, held world barley import shares of 11 and 9 percent, respectively. Saudi Arabia permitted private traders to import barley for the first time in 1998, and Japan followed suit for feed barley in 1999.

Corn trade has also been affected by some STEs. South Africa, for example, had a maize marketing board, but it was terminated in 1997. China's National Cereals, Oils and Foodstuffs Import and Export Corporation (COFCO) controls imports and exports of corn, but as part of its WTO accession agreement with the United States, China has agreed to initially reserve 25 percent of TRQ imports for the private sector, rising to 40 percent by the end of the implementation period. China also agreed to allow out-of-quota trade through entities other than STEs.

Discussions on STEs are likely to revolve around strengthening WTO rules governing these enterprises and imposing additional disciplines on their exclusive authorities and the policies they implement. The current U.S. proposal regarding state trading enterprises calls for termination of exclusive import and export rights possessed by an STE, elimination of government funds or guarantees to support single-desk exporters, and promotion of increased transparency in the STE's operations.

### Export Credit Guarantees and Export Credits

A potential issue related to the new negotiations in agriculture is the ongoing discussion on export credit guarantees and export credits being conducted in the Organization for Economic Cooperation and Development (OECD). Some U.S. competitors have argued that export credits and credit guarantees should be treated as an export subsidy. While negotiations to discipline government-sponsored credit and credit guarantee programs are ongoing in the OECD, some WTO members are calling for additional negotiations to take place in the WTO.

Export credit programs are used extensively by the United States and other countries. In fiscal year 1999, about $3 billion in U.S. export sales were conducted under these types of programs. U.S. credit guarantees are used for a portion of U.S. coarse grain exports, typically around 10 percent but as high as 20 percent in some years.

Under the URAA, Article 10.2, member countries agreed to work on disciplines for the use of government-sponsored export credit programs in agriculture (USDA, 1999). The United States prefers that negotiations continue in this forum since much of the work has already been completed. The United States also takes the position that all WTO members offering such credits and export credit guarantees be included in any agreement.

### Country Accession to WTO

Improved/increased market access or subsidy reduction may be the primary interest of many WTO members, but they will also have an interest in the implications of potential entrants into the WTO. Although most of the world's major trading partners are members of the WTO, several key countries, including China, Taiwan, Russia, and Vietnam, were not members until recently. A U.S.-China agreement signed on November 15, 1999, and the subsequent decision by the United States to grant permanent normal trading relations (PNTR) to China in October 2000 represented a major step toward China joining the World Trade Organization (Colby, Price, and Tuan, 2000). China is the third largest exporter of corn with a global export market share of about 7 percent, and is a consistent importer of barley and rye. However, China is expected to increase its net imports of coarse grains with this agreement, which could provide an opportunity for U.S. corn exporters.

As of 2000, China had an applied tariff of 114 percent on all corn imports, and these imports were controlled by its state trading agency, COFCO. Terms of the bilateral agreement included the establishment of a TRQ for corn, with an initial quota of 4.5 million tons, rising to 7.2 million tons within five years.[5] Within-quota imports were subject to a low duty (1 percent), while over-quota duties were high—77 percent at the beginning of implementation, declining to 65 percent within five years. Non-state trade companies were initially allocated 25 percent of the quota, rising gradually to 40 percent within five years. With China's accession to the WTO, its net corn imports are projected to increase by an average annual $497 million beyond USDA's baseline estimates for the period 2000 to 2009 (Colby, Price, and Tuan, 2000).

### Trade in Genetically Engineered Commodities

Foreign regulations and labeling initiatives governing products from genetically engineered organisms have created concerns among U.S. coarse
grain producers and companies about the loss of corn exports and potential
loss of exports of processed by-products such as corn gluten feed and meal
(Lin, Chambers, and Harwood, 2000).[6] U.S. corn exports to the EU, for example, have recently declined about $200 million per year because of an EU
moratorium on import approvals for new corn varieties being grown in the
United States. The share of U.S. corn exports destined for the EU subsequently declined from over 3 million tons in fiscal year (FY) 1994-1995 to
less than 150 thousand tons in FY 1999-2000.

Export sales of U.S. corn by-products to the EU, which were valued at
$403 million in FY 1998-1999, could also potentially be affected. For ex-
ample, concern over the loss of corn by-product exports to the EU caused
several actions on the part of U.S. processors in 1999. In April 1999, A.E.
Staley and Archer Daniels Midland (ADM) stated that they would not accept biotech corn varieties for processing if they were not approved by the
EU. In the summer of 1999, ADM advised producers to segregate biotech
crops from nonbiotech crops, although this decision was reversed in early
February 2000 because of an apparent weak demand for the higher priced
nonbiotech grain.

The EU has also recently adopted labeling regulations for foods containing biotech ingredients and is currently drafting feed labeling regulations.
Japan has finalized its biotech labeling regulation and was to begin its implementation in April 2001. Korea, Australia, and New Zealand, among
other countries, are proposing mandatory labeling policies for some
bioengineered foods as well. The current U.S. proposal on this issue calls
for a focus on disciplines to ensure that processes regarding trade in products developed through new technologies are transparent, predictable, and
timely.

### Zero for Zero

The International Barley and Malt Coalition for Free Trade, an industry
association that includes barley producer organizations, malting companies,
and malting industry representatives from Canada and the United States, is
proposing an accelerated zero-for-zero trade liberalization for barley and
malt.

A zero-for-zero strategy was successfully used in the UR to bring about
complete elimination of tariffs on selected industrial goods and some members explored this approach for the global oilseed market, but no agreement
could be reached. By allowing market forces to determine production and

trade flows, a zero-for-zero agreement could increase the market share of competitive barley producers, mainly because the EU would have to eliminate high levels of domestic support and export subsidies for these products.

### Export Taxes

Discussions during the new agricultural trade negotiations could include consideration of a ban on export taxes, which have been used by some countries to generate government revenue and redistribute income. Export taxes restrict the availability of a commodity on the global market and consequently tend to raise global prices.

Argentina once taxed the export of corn and sorghum to subsidize its manufacturing sector. Although this policy has been abandoned, it tended to discourage domestic production of corn or sorghum and reallocate resources into the manufacturing sector. Another example is the EU's temporary tax on barley exports during its 1995-1996 marketing year. By taxing exports, the EU's goal was to control tight internal supplies, ease prices for domestic grain users, and encourage the rebuilding of government-owned intervention stocks. The U.S. goal for the new agricultural trade negotiations is to prohibit the use of export taxes, including differential export taxes, for competitive advantage or supply management purposes (WTO, 2000).

### CONCLUSION

Although the URAA was an important step toward identifying and disciplining trade distortions in agriculture, the agreement's impact on the global coarse grain market so far appears to be limited. Since the agreement, global coarse grain trade has risen slightly but is well below growth in global production.

In the new WTO agricultural negotiations, improved market access and more market-oriented trade depend largely on progress in issues addressed during the Uruguay Round. These include increased market access by further lowering bound tariff rates, expanding quotas for products with TRQs and improving TRQ administration, continued reduction in export subsidies, and domestic support. Additional issues could include a zero-for-zero proposal for barley and malt, tighter disciplines on STEs, and improved market access opportunities for genetically engineered coarse grains and products. Market opportunities for coarse grain producers could also be enhanced by further liberalization of the global meat market. Progress on these issues could expand the market opportunities for the U.S. coarse grain sector.

## NOTES

1. WTO members have submitted proposals setting out negotiating objectives. These proposals were to be received by the end of December 2000, with some flexibility for those who cannot meet that date or want to make additions. Negotiating sessions for this first phase were held in June, September, and November 2000. Based on the negotiating positions submitted in 2000, a commitment was made to further comprehensive negotiations at the Doha Ministerial in November 2001. A date of January 2005 was set for completing the majority of the negotiations.

2. Excludes intra-EU trade.

3. Under this industry association proposal, trade-distorting policies such as export subsidies, import restrictions, and export taxes would be eliminated. In addition, domestic support programs would continue to be decoupled from current production and prices, and state trading enterprises would be required to eliminate exclusive trading rights and operate on market principles.

4. Japan committed to increase a zero-duty quota for industrial use corn by 450,000 metric tons and increase a TRQ for barley by 51,000 tons; South Korea committed to establishing a TRQ for barley and barley products other than malting barley and barley malt, which will grow to 23,582 tons; and South Africa established a TRQ for corn of 260,000 tons.

5. The TRQ is not a minimum purchase requirement, but the agreement does require China to establish access opportunities for the full quota amount. The agreement also introduces private trade and increased transparency of the import process to maximize the likelihood that quotas will fill.

6. About 25 percent of U.S. corn acreage was planted to genetically engineered varieties in 2000 (USDA, 2000b).

## REFERENCES

Ackerman, Karen and Praveen Dixit. (1999). *An Introduction to State Trading in Agriculture.* Agricultural Economic Report, No. 783. U.S. Department of Agriculture, Economic Research Service, Washington, DC, October.

Colby, Hunter, J. Michael Price, and Francis C. Tuan. (2000). "China's WTO Accession Would Boost U.S. Ag Exports and Farm Income." *Agricultural Outlook,* AGO-269, Washington, DC, March. pp. 11-16.

Leetmaa, Susan and Jason Bernstein. (1999). "An Analysis of Agenda 2000." *The European Union's Common Agricultural Policy: Pressures for Change.* International Agriculture and Trade Reports, Situation and Outlook Series, U.S. Department of Agriculture, Economic Research Service. WRS-99-2, Washington, DC, October. pp. 14-20.

Lin, William, William Chambers, and Joy Harwood. (2000). "Biotechnology: U.S. Grain Handlers Look Ahead." *Agricultural Outlook.* U.S. Department of Agriculture, Economic Research Service. AGO-270, Washington, DC, April. pp. 29-34.

U.S. Department of Agriculture (Various issues b). Foreign Agricultural Service, Global Agriculture Information Network (GAIN) Reports.. <http://www.fas.usda.gov/scriptsw/attacherep/default.asp>.

U.S. Department of Agriculture. Economic Research Service. (1998). *Agriculture in the WTO*. International Agriculture and Trade Reports, Situation and Outlook Series, WRS-98-4. December.

U.S. Department of Agriculture. Foreign Agriculture Service. (1999). *Backgrounder: International Negotiations on Export Credit Programs*. November. <http://www.fas.usda.gov/info/factsheets/ec-backgrounder.html>.

U.S. Department of Agriculture. Economic Research Service. (2000a). *PS&D View*. January. <http://www.ers.usda.gov/data/psd/>.

U.S. Department of Agriculture. National Agricultural Statistics Service. (2000b). *Acreage*. Agricultural Statistics Board. June. p. 28.

World Trade Organization. (1995). *The Results of the Uruguay Round of Multilateral Trade Negotiations*. The Legal Texts. Geneva, Switzerland.

World Trade Organization. Secretariat. (1999). *Domestic Support-Background Paper by the Secretariat: Revision*. September.

World Trade Organization. Committee on Agriculture. (2000). *Proposal for Comprehensive Long-Term Agricultural Trade Reform: Submission from the United States* (G/AG/NG/W/15). June.

Chapter 12

# The World Trade Organization and Southern Agriculture: The Cotton Perspective

Darren Hudson

Major strides toward trade liberalization have occurred as a result of the implementation of the Uruguay Round Agreement (URA) and the World Trade Organization (WTO). However, further advances are proving more difficult. Agriculture continues to be a problem area for WTO negotiations. Resistance to further reductions in export subsidies and domestic support has generated considerable debate. In addition, the growing proliferation of regional trade agreements such as the proposed Free Trade Area of the Americas (FTAA) may serve to complicate the issues surrounding the WTO negotiations.

The implications of further WTO negotiations for some Southern agricultural commodities could be important. Cotton is a commodity that is important both to the southern United States and to many countries around the world. Cotton (and textiles) may be prominent in WTO negotiations because of their importance to many developing countries. However, most developing countries do not currently subsidize exports of cotton, thus limiting the potential influence of any regulations that the WTO may administer. This would seem to suggest that, at least for cotton, future WTO negotiations will have little impact on trade. However, a significant event is the accession of China into the WTO. The impact of that change depends on the rules under which China was admitted.

The implications of the ongoing WTO negotiations are the subject of a continuing dialogue. Commodity-specific impacts are both relevant and a source of concern for economists, policymakers, and farm leaders. The purposes of this chapter are to: (1) describe the position of the United States in world cotton production and trade, (2) explore the ramifications of potential changes in the WTO, and (3) examine the potential indirect impacts of other WTO policy changes.

*215*

## *THE CURRENT COTTON SITUATION*

The United States is both a large producer of cotton, consistently ranking third or fourth in the world in terms of total output, and the largest exporter of cotton (USDA, various issues). U.S. exports have accounted for an average 23 percent of world exports from 1970 to 1997 (Figure 12.1), and about 33 percent of U.S. production over the same period. Given the relative proportion of the world market, U.S. cotton exports are influenced by both domestic and international policies that influence trade.

### *Domestic Policy Affecting Trade*

The United States has utilized an export subsidy for cotton since 1985 (see Hudson and Ethridge, 2000a). Although there is little direct evidence of the positive impacts of this policy on U.S. cotton exports, anecdotal evidence suggests that the export subsidy has increased exports (a recent preliminary analysis by Hishamunda et al. [2000] suggests that the export subsidy has increased U.S. exports, increasing U.S. cotton price by about 7 percent). Figure 12.1 shows that exports after 1985 increased over the levels prior to 1985.

U.S. export subsidy levels are currently below required levels agreed upon in the URA. Thus, unless lower levels are negotiated in the new round of WTO talks, additional reductions in export subsidization are not likely to

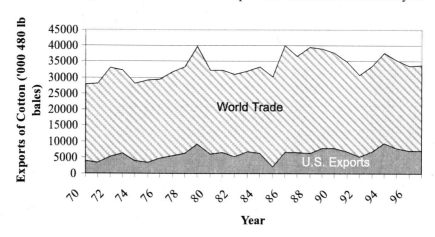

FIGURE 12.1. U.S. and world cotton exports, 1970-1997. (*Source:* USDA, various issues.)

be required. The primary constraint to export subsidization in cotton has been domestic budget concerns. That is, the appropriated funds for cotton export subsidies for the 1996-2002 (under the Federal Agricultural Improvement and Reform [FAIR] Act) period were exhausted in 1999, leading to some concerns about the budgetary cost of the program. However, in 1999, the U.S. Congress appropriated money for continuation of the export subsidies through 2002.

Other domestic policies indirectly affect trade as well. The FAIR Act was designed, at least in part, to limit government involvement in agricultural production by decoupling payments from acreage decisions. Thus, if acreage/production distortions are minimized, trade distortions may be minimized as well. This appears to be the maintained hypothesis espoused by U.S. trade negotiators in negotiating for reduced domestic subsidization. However, other government programs such as subsidized crop insurance, of which cotton is a major beneficiary, may also distort acreage decisions and trade. To the extent that these programs are addressed in the WTO negotiations, they may ultimately affect the U.S. cotton industry.

### The Foreign Cotton Market

Besides the United States, major cotton producers include China, Uzbekistan, Pakistan, India, Turkey, and Australia. With the exception of China and Australia, all these major producers have practiced or continue to practice negative protection through either direct taxation or export restraints (see, e.g., Hudson and Ethridge, 1998, 1999, 2000b; Isengildina, Herndon, and Cleveland, 1998; and Hudson, 1997). These policies are aimed at indirectly subsidizing domestic textile industries by manipulating cotton production sectors. For example, Hudson and Ethridge (2000b) found that a cotton export tax in Pakistan significantly lowered Pakistani cotton exports while increasing Pakistani textile exports, thus altering global cotton and textile trade flows.

China remains an enigma for the cotton market. In some years, China is a net importer of cotton, while it exports significant quantities of cotton in other years. China has increased its man-made fiber production capacity of late to reduce reliance on cotton fiber for textile products. This shift, at least in part, is a result of refocusing agricultural production efforts to food and feed grains to support a growing population. Nevertheless, whether China is a net exporter or importer of cotton will continue to have important implications on global trade flows and world cotton prices.

## *THE WTO AND COTTON*

The fact that the majority of the major cotton producing countries are considered less developed countries (LDCs) and that these countries practice either no or negative protection for their cotton production sectors somewhat complicates the outlook for world cotton trade. First, LDCs receive special considerations under the WTO such as longer tariff phase-out periods and smaller subsidy reduction requirements. More importantly, however, the WTO attempts to reduce export subsidies, but does little to regulate a country from restraining its own exports. In this context, the URA and WTO have done little to *directly* alter world cotton trade, nor will further reductions in export subsidies as a result of further WTO negotiations likely directly alter cotton trade. Thus, the narrow commodity focus suggests that the WTO negotiations will have little impact on cotton. However, expanding the analysis to include textiles changes the results.

### *Multifiber Arrangements*

Several textile policies embodied within the URA have had indirect impacts on global cotton trade, and thus, Southern agriculture. Among these has been the phase-out of the Multifiber Arrangements (MFAs). The MFAs are bilateral trade agreements between textile producing/consuming countries that limit the flow of textile products from textile producers to consumers (Meyer et al., 2000). For example, the United States may have an agreement with India that allows the importation of a given quantity of Indian textiles in exchange for India importing some quantity of U.S. textiles. Under the URA, these MFAs were to be phased out over a ten-year period (Varangis and Thigpen, 1995).

The elimination of the MFAs has the potential to redirect where textile products are produced globally, thus changing global trade flows of the primary input, cotton. That is, as restrictions on exports of textile products from textile-producing countries are lifted, textile output and exports by these countries is expected to increase. This is likely to displace some textile production (or at least accelerate the displacement already occurring) in developed countries. As the location of textile production shifts, the demand for cotton will shift to those areas as well, thus altering global cotton trade flows.

As Meyer et al. (2000) point out, this is especially true for labor-intensive portions of the textile production process such as cutting, sewing, and assembly of garments. The United States and other developed countries currently maintain a competitive advantage in capital-intensive processes such as spinning and weaving where capital and technology are easily substitutable for labor. The United States has continually increased exports of cotton-containing textile products (Figure 12.2), due mainly to increased exports of semiprocessed goods such as yarns and fabrics. These products are exported primarily to textile producing countries, which further process those yarns and fabrics into finished consumer goods and reexport them to the United States.

### Net Trade Balance

The United States is traditionally considered a large net exporter of cotton, which is true when only raw (unprocessed) cotton is considered. However, when one considers the cotton content of textile exports and imports as well, a substantially different picture is revealed. Since 1970, the ratio of total cotton (including the cotton content of textiles) exports to total cotton imports has steadily declined (Figure 12.3). In fact, the United States is now a net importer of cotton on this basis. This is not to say that cotton or textile exports have declined. Rather, the increase in imports has outstripped increases in exports, resulting in a declining trade balance. Potentially more important is the fact that these textile imports are higher valued, finished goods as compared to the unprocessed or semiprocessed exports, suggesting a more rapid decline in the net balance of value-added.

Part of this shift is likely a result of macroeconomic variables such as wage rate differentials and the relative abundance of labor in these textile-producing countries, giving them a comparative advantage in labor-inten-

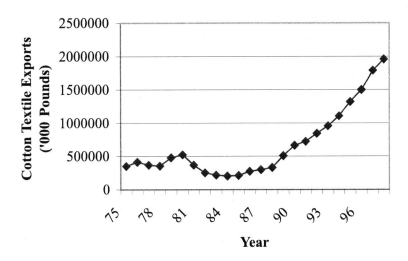

FIGURE 12.2. Cotton equivalent of U.S. textile exports, 1975-1998. (*Source: USDA, various issues.*)

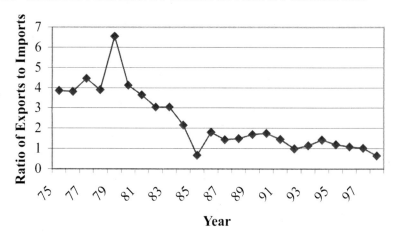

FIGURE 12.3. Ratio of U.S. total cotton exports (including the cotton fiber equivalent of textile and apparel exports) to U.S. total cotton imports, 1975-1998. (*Source:* USDA, various issues. *Note:* A value of 1 indicates that exports are equal to imports.)

sive processes. If this is the case, however, the phase-out of the MFAs can only serve to hasten the movement of textile production from developed to developing countries (Meyer et al., 2000). That is, the phase-out of the MFAs has allowed developing, textile-producing countries to more fully exploit their comparative advantage in these labor-intensive production processes.

### Economic Arguments

The real question becomes "Is all this shifting of textile production and changes in global cotton trade flows hurting or helping U.S. cotton farmers and the U.S. cotton industry?" The answer to that question is neither simple nor clear. There is a body of literature dealing with the concept of "intra-industry" trade (IIT) (Dixon and Stiglitz, 1977; Krugman, 1981; Ruffin, 1999). That is, traditional international trade theory focuses on "inter-industry" trade, or the trading of corn for automobiles for example, on the basis of comparative advantage under the assumption of perfect competition. The intra-industry argument stems from an assumption of imperfect competition and suggests that countries may specialize in a differentiated product within the same product grouping. For example, one country may specialize in the production of sport utility vehicles while another specializes in luxury sedans (Ruffin, 1999).

The central point of the IIT argument is that it is entirely possible for two countries to produce similar products, trade those products, and maintain (and in some cases, enhance) the gains from trade. Thus, it would seem plausible that U.S. (and other developed country) specialization in the production of yarns and fabrics (capital-intensive products) and developing country specialization in apparel and finished goods (labor-intensive products) can enhance the overall welfare of both groups.

The IIT argument, however, is not directly on point in this case. Textile production occurs in relatively discrete stages, each of which can be carried out in various locations around the world. The preponderance of world trade is between these stages of the value chain. That is, yarn is exported to weavers, fabrics are exported to assemblers, and final products are exported to consuming countries. This fact would seem to limit any argument on the basis of intra-industry trade.

The counterargument, grounded in the more traditional Stolper-Samuelson (SS) theory, would suggest that these shifts in textile production are a result of comparative advantage. That is, the labor-intensive portion of textile production has migrated to areas abundant in labor, while capital-intensive processes have migrated to areas abundant in capital. In this sense, elimination of the MFAs is simply allowing a more efficient allocation of resources, resulting in increased global welfare.

There are, however, some potential negative consequences to this trade liberalization. First, the SS theory shows that importation of labor-intensive goods necessarily decreases the welfare of domestic workers employed in that industry. This hypothesis appears to be borne out by the fact that garment manufacturing is disappearing from the United States, thus resulting in lost employment. However, this lost welfare is, at least in part, mitigated by the fact that the jobs that have been created in the U.S. textile complex as a result of technological advances are skilled, higher-wage jobs than their garment-producing counterparts. Thus, the welfare loss appears to be concentrated on unskilled labor.

Second, the negative trade balance in cotton suggests that, on balance, the United States is importing foreign-grown cotton (at least a composite of labor and cotton). Thus, the relocation of textile processing to other parts of the world appears to have a negative, albeit small, impact on the cotton production sector. In theory, it should not matter to the cotton producer whether all of the cotton is consumed domestically or exported, as long as the producer receives the world price. Pragmatically, the existence of a strong domestic textile industry has provided a stable source of demand for U.S. cotton and more effectively buffered the U.S. cotton producer from world demand and supply shocks.

## *CONCLUSION*

The available evidence to date suggests that the accession of China into the WTO will be the main trade factor influencing the U.S. cotton industry. This conclusion is based on the fact that the majority of global cotton production is produced under the conditions of no or negative protection. Thus, further reductions in domestic or export subsidization are not likely to result in any significant changes in global cotton production or trade.

However, the URA and the WTO have had an indirect impact on global cotton trade through the phase-out of the MFAs. The structural adjustments that have been taking place in textile production (Meyer et al., 2000) would suggest an increase in global welfare when viewed from the perspective of the Stolper-Samuelson theory. The ultimate effect on the U.S. cotton producer is unclear, but appears to be only marginal.

More profoundly, the URA and the phase-out of the MFAs appear to be detrimental to unskilled labor employed in garment and other labor-intensive textiles in the United States. Skilled labor appears to benefit from this change through the creation of higher-wage jobs as compared to garment manufacturing. The phase-out of the MFAs also appears to benefit the owners of capital used in the capital-intensive textile processes.

## REFERENCES

Dixon, A. and J. Stiglitz. (1977). "Monopolistic Competition and Optimum Product Diversity." *American Economic Review,* 67: 297-308.

Hishamunda, N., P. Duffy, C. Jolly, and J. Raymond. (2000). "Effect of Decreasing Government Involvement in the U.S. Cotton Industry." *Beltwide Cotton Conferences, Proceedings,* Cotton Economics and Marketing Conference, National Cotton Council, Memphis, TN, pp. 298-300.

Hudson, D. (1997). "The Turkish Cotton Industry: Structure and Operation." *Beltwide Cotton Conferences, Proceedings,* Cotton Economics and Marketing Conference, National Cotton Council, Memphis, TN, pp. 285-287.

Hudson, D. and D. Ethridge. (1998). "The Pakistani Cotton Industry: Impacts of Policy Changes." *Beltwide Cotton Conferences, Proceedings,* Cotton Economics and Marketing Conference, National Cotton Council, Memphis, TN, pp. 294-297.

Hudson, D. and D. Ethridge. (1999). "Export Taxes and Sectoral Economic Growth: Evidence from Cotton and Yarn Markets in Pakistan." *Agricultural Economics,* 20: 263-276.

Hudson, D. and D. Ethridge. (2000a). "Cotton and Fibers." Chapter 8 In D. Colyer, W. Amponsah, S. Fletcher, and L. Kennedy (Eds.), *Competition in Agriculture: The U.S. in the World Market,* Binghamton, NY: The Haworth Press, Inc.

Hudson, D. and D. Ethridge. (2000b). "Income Distributional Impacts of Trade Policies: A Case in Pakistan." *Journal of Agricultural and Applied Economics,* 32: 49-61.

Isengildina, O., C. Herndon, and O. Cleveland. (1998). "Cotton Industry in Uzbekistan: Structure and Current Developments." *Beltwide Cotton Conferences, Proceedings,* Cotton Economics and Marketing Conference, National Cotton Council, Memphis, TN, pp. 297-301.

Krugman, P. (1981). "Intraindustry Specialization and the Gains from Trade." *Journal of Political Economy,* 89: 959-973.

Meyer, L., S. MacDonald, A. Somwaru, and X. Diao. (2000). "Effects of the MFA Phase-Out on the U.S. Cotton and Textile Markets." *Beltwide Cotton Conferences, Proceedings,* Cotton Economics and Marketing Conference, National Cotton Council, Memphis, TN, pp. 385-390.

Ruffin, R. (1999). "The Nature and Significance of Intra-Industry Trade." *Economic and Financial Review,* Federal Reserve Bank of Dallas, Fourth Quarter, pp. 2-9.

U.S. Department of Agriculture (USDA). (various issues). *Cotton and Wool Situation and Outlook Yearbook,* U.S. Department of Agriculture, Economic Research Service, Washington, DC.

Varangis, P. and E. Thigpen. (1995). "The Impacts of the Uruguay Round Agreement on Cotton, Textiles, and Clothing." *Beltwide Cotton Conferences, Proceedings,* Cotton Economics and Marketing Conference, National Cotton Council, Memphis, TN, pp. 370-373.

*PART III:*
*MULTILATERAL TRADE*
*NEGOTIATIONS:*
*ISSUES AND CONCERNS*

# Chapter 13

# The Impacts of Export Subsidy Reduction Commitments in the Agreement on Agriculture and International Trade: A General Assessment

Lilian Ruiz
Harry de Gorter

## *INTRODUCTION*

Export subsidy reduction is one of the three pillars of commitments made in the Agreement on Agriculture in the World Trade Organization (WTO).[1] Discipline on export competition measures in agriculture is deemed important, because it was included for the first time in the Uruguay Round of negotiations. The purpose of this chapter is to assess the extent to which export subsidy commitments have been effective, and what modifications are needed to improve their performance. To do so, an overview of policy developments is given first, and then we focus on the economics of the bindings imposed on export subsidies.

Several commentators argue that export subsidies are the most distorting of agricultural policy interventions, and so should be outlawed (e.g., Josling, 1998; McCalla, 1999). However, an export subsidy has the same trade-distorting effects as an import barrier for the same market supply/demand conditions. The source of this confusion may be that competing export subsidy programs by two or more countries is self-defeating. However, this is true for import restrictions as well, which lower world prices and require other countries to increase tariffs to maintain the same level of producer support. Indeed, import barriers by some countries increase the need for export subsidies by others and vice versa. Hence, disciplines on export subsidies should receive the same urgency as that on market access.[2]

Limits on and reductions in the volume and value of export subsidies for twenty-five countries (and the prohibition of new subsidies) was the cornerstone of export subsidy commitments in the Agreement. Additional countries have recently joined the reduction commitments. Each country agreed to reducing the volume of subsidized exports by 21 percent over six years from a 1986-1990 average base period level (14 percent over a ten-year period for developing countries), and reduce the value of export subsidies by 36 percent (24 percent over ten years for developing countries). The Agreement provides flexibility by allowing countries to redistribute the value of subsidies or the volume of subsidized exports between years but the cumulative totals through the year 2000-2001 are not to exceed those that would have resulted from full compliance. Countries were also permitted to aggregate products within a commodity group in their commitments.

WTO notifications have the European Union spending the lion's share (over 80 percent) of total world export subsidies (Table 13.1). The U.S. export subsidy expenditures were less than several countries including South Africa (which eliminated export subsidies in 1997) and Switzerland. Canada eliminated transportation subsidies for exports and so reduced their support in this category dramatically (see column 1 for Baseline and column 2 for 1995-1996 levels). Tables 13.2 and 13.3 present the countries and commodity sectors that have exceeded their value and volume commitments. In the commitments, the largest share of the expenditure allowances goes to wheat, coarse grains, dairy products, and meats. The largest share of the actual expenditures is represented by the export subsidies for meat, dairy, and incorporated products. In terms of volume, wheat and coarse grains used a smaller proportion of the allowances, but the share has been increasing in recent years for dairy, pig meat, and sugar as well.

The effects of value and volume limits vary under different scenarios. We isolate the factors affecting export subsidy expenditures in the intervening years, in order to measure the binding effect of limits on volume versus value limits on export subsidies. We analyze the relative effectiveness of volume versus value constraints under changing market conditions. For example, the volume limit for the scheduled reductions is always binding for a given static scenario, but value can become binding if the ratio of the reductions follows certain proportions, or if changes in baseline conditions occur. The choice of base period and world market developments may be contributing to subsidy reductions rather than genuine efforts by governments to reduce price supports that determine subsidy levels. Export subsidy limits on both expenditures and physical quantities are also negated somewhat by aggregation across products and over time.

TABLE 13.1. Export Subsidy Expenditure Commitments During the Implementation Period (in US$ million)

| | Base | 1995-1996 | 1996-1997 | 1997-1998 | 1998-1999 | 2000 bound |
|---|---|---|---|---|---|---|
| **European Union** | 14,800 | 6,100 | 6,300 | 4,800 | 4,800 | 9,400 |
| **United States** | 930 | 26 | 122 | 113 | 150 | 600 |
| **World Total** | 20,000 | 6,900 | 7,000 | 5,500 | 5,500 | 13,400 |

TABLE 13.2. Countries Exceeding the Value Commitments

| Member | Commodity | 1995-1996 | 1996-1997 | 1997-1998 | 1998-1999 |
|--------|-----------|-----------|-----------|-----------|-----------|
| Colombia | Sugar confection | | 105 | | 95 |
| Cyprus | Cheese | 406 | 100 | 100 | 100 |
| EC-15 | Alcohol | | 90 | | 106 |
| EC-15 | Pig meat | | | | 155 |
| EC-15 | Rice | | 141 | | |
| EC-15 | Sugar | | | 122 | 134 |
| EC-15 | Wine | | 111 | | |
| Hungary | Corn | 282 | | | 413 |
| Hungary | Red pepper meal | 147 | 141 | | |
| Norway | Bovine meat | | | | 106 |
| Norway | Cheese | | | 113 | 117 |
| Norway | Poultry meat | 243 | | | |
| Norway | Sheep meat | | | 112 | |
| South Africa | Cocoa and prep. | 108 | 323 | 99 | |
| South Africa | Tea | 112 | | | |
| South Africa | F&V and nut prep. | | 114 | | |
| South Africa | Waters | | 144 | | |
| South Africa | Wine products | 94 | 179 | 107 | |
| Switzerland | Processed Prods. | | | | 100 |
| Turkey | F&V | | 99 | 99 | 100 |
| Turkey | Eggs/dozen | | | 100 | |
| United States | Other milk prods. | | | 100 | 129 |
| United States | Skim milk powder | | | | 136 |

## THE ECONOMICS OF VOLUME VERSUS VALUE COMMITMENTS ON EXPORT SUBSIDIES

A key policy issue is whether the formula for further cuts in export subsidies should focus on volumes or expenditures. Indeed, the volume exported with subsidies determines part of the value of export subsidies (the other factor is the per-unit subsidy). The effects of both types of bindings are not obvious, even when restrictions in volume and value are both binding in the baseline in a static economic environment (excess supply and demand

TABLE 13.3. Countries Exceeding the Volume Commitments

| Member | Commodity | 1995-1996 | 1996-1997 | 1997-1998 | 1998-1999 |
|---|---|---|---|---|---|
| Colombia | Fruits | | | 115 | 138 |
| Colombia | Processed prod | 129 | 91 | 19 | |
| Colombia | Sugar confection. | 316 | 473 | 332 | 1033 |
| Cyprus | Cheese | 189 | 100 | 99 | 100 |
| EC-15 | Coarse grains | | 90 | | 123 |
| EC-15 | Sugar | | | 118 | 112 |
| EC-15 | Rice | | 144 | 103 | 99 |
| EC-15 | Other milk prod | 98 | 100 | 102 | 91 |
| EC-15 | Cheese | 99 | 99 | | |
| EC-15 | Olive oil | 96 | 104 | | |
| EC-15 | Beef meat | 90 | 110 | 94 | |
| EC-15 | Poultry meat | 96 | 99 | 105 | 99 |
| EC-15 | Wine | | 111 | 115 | 98 |
| EC-15 | F&V, fresh | 99 | 99 | 98 | 93 |
| EC-15 | Eggs | | | 90 | 104 |
| Norway | Sheep meat | | | 142 | 106 |
| Norway | Pig meat | | | 106 | |
| Norway | Cheese | | | 102 | 122 |
| Norway | Poultry meat | 214 | | | |
| Poland | Sugar | | 116 | 149 | 119 |
| Slovak Rep. | Other dairy prods. | | 95 | 99 | 110 |
| South Africa | Beer | 106 | | 105 | |
| South Africa | Other milk prods. | 139 | | | |
| South Africa | Wine products | 103 | 619 | 227 | |
| Switzerland | Cattle for breeding | 111 | | | |
| Turkey | F&V, fresh or proc. | 100 | 97 | 100 | 100 |
| Turkey | Poultry meat | 100 | | | |
| United States | Other milk prods. | | | 100 | 107 |
| United States | Skim milk powder | | | 104 | 154 |
| United States | Cheese | | | 100 | 93 |

curves do not shift over time). In this latter static case, consider a 21 percent reduction in the volume of exports with an export subsidy, which may result in exports being equal to or less than the free trade level $Q_{ft}$ (where the per unit export subsidy will fall to zero). In this case, the volume constraint will be more liberalizing, and any reduction beyond $Q_{ft}$ would not be necessary. Reducing the volume of exports receiving a subsidy results in a lower price gap, which becomes zero at $Q_{ft}$. One needs a less than 100 percent reduction in the volume of subsidized exports to reach free trade, unlike the case for reductions in export expenditures.

The per-unit subsidy and the volume exported with subsidies are limited by the shape and position of excess supply and excess demand curves. If the volume commitment is binding, it is not possible to apply a larger per-unit subsidy in a static scenario, because the per-unit subsidy increases with volume. Therefore, in a static analysis, value bindings would not be necessary if there is a volume constraint.

For the case where only value is constrained, it is not possible to subsidize more exports with a decreasing per-unit subsidy, because the per-unit subsidy increases with exports. Therefore, if value commitment is binding, there is no need for a volume binding. Hence, one needs to analyze the relative effectiveness of each type of binding in a dynamic situation as well.

We develop an analytical framework that allows us to analyze the economics of volume versus value reductions as a static framework, and then under different supply and demand conditions. In the static analysis, we first analyze the case where there is no change in baseline market conditions before reductions take place. Export subsidy reductions are applied only once. The dynamic analysis will consider the scenarios in which market conditions change after the baseline (shifts in excess supply and demand curves). Such an analysis is also relevant for market changes that occur after the reduction commitments are met.

### Static Analysis

Assume in the baseline scenario that a given country applies a per-unit subsidy $s_0$ to a given commodity, resulting in exports $Q_0$. Assume that $s_0$ and $Q_0$ determine the initial value and volume that will be reduced as specified in the commitments.

Consider an inverse excess supply curve

$$P_d = a + bQ \qquad (13.1)$$

where $P_d$ is the domestic price, and an inverse excess demand curve

$$P_w = c + dQ \qquad (13.2)$$

where $P_w$ is the world price, $c > 0$, $d < 0$, and $Q$ is imports (imports = exports). The per-unit export subsidy in the base time $t = 0$ is given by:

$$s_0 = P_d - P_w = (a-c) + (b-d)Q_0 \qquad (13.3)$$

The value of the export subsidy in the baseline is given by

$$V_0 = [(a-c) + (b-d)Q_0]Q_0 \qquad (13.4)$$

Define $\beta$ as the proportion of baseline export subsidy value that is allowed. This is the value of export subsidies remaining after a reduction is accomplished, so that for a 36 percent reduction in the value of the export subsidy, $\beta = 0.64$. Hence, $\beta = 1 - V_{red}$, where $V_{red}$ is the required percentage reduction in the value of export subsidies. For the current reduction commitments, $V_{red} = 0.36$. Define $\alpha$ as the proportion of baseline volume of subsidized exports remaining after the reductions. Hence, $\alpha = 1 - Q_{red}$, where $Q_{red}$ is the percentage reduction in the volume of subsidized exports. For the current commitments, $\alpha = 0.79$, because $Q_{red} = 0.21$.

To compare the effect of a reduction in volume versus value, define $V_\beta$ as the value of export subsidy remaining, after the value reduction, and $V_\alpha$ as the value of export subsidy remaining, after the volume reduction.

$$V_\beta = [(a-c) + (b-d)Q_0]\beta Q_0 \qquad (13.4a)$$

$$V_\alpha = [(a-c) + (b-d)\alpha Q_0]\alpha Q_0 \qquad (13.4b)$$

By equating equations (13.4a) and (13.4b), the conditions under which $\alpha$ and $\beta$ generate volume exported and value of subsidies that are equally binding can be determined. In other words, when $V_\beta = V_\alpha$:

$$V_\beta = (a-c)\beta Q_0 + (b-d)\beta Q_0^2 = (a-c)\alpha Q_0 + (b-d)\alpha^2 Q_0^2 = V_\alpha \quad (13.5)$$

Solving equation (5) for $Q_0$ gives:

$$Q_0 = \frac{(\alpha - \beta)(a-c)}{(\beta - \alpha^2)(b-d)} \quad \text{where } 0 < \alpha < 1, 0 < \beta < 1 \qquad (13.6)$$

Solving equation (3) for $Q_0$, substituting in equation (5) and solving for $\beta$ gives:

$$\beta = \alpha^2 - s_0(c-a)(1-\alpha)\alpha \qquad (13.7)$$

*Proposition 1.* For identical reductions in volume and value ($\alpha = \beta < 1$), the volume reduction is always binding (holds independent of the elasticities of excess supply and excess demand).

*Proof.* By definition, baseline value and volume are equally binding, and $\alpha = \beta = 1$ when no reductions were imposed. By imposing the restriction $\alpha = \beta < 1$ in equation (13.5), we can show that the volume reduction becomes binding, because the right-hand side of equation (5) becomes larger (note that $\alpha^2 < \alpha$ because $0 < \alpha < 1$). $V_\alpha$ is the value of export subsidy after the volume reduction commitment is met, and if it is smaller than $V_\beta$, the volume reductions are more binding for any $\alpha = \beta < 1$.

However, we cannot infer from Proposition 1 that reductions in value have to be larger than reductions in volume in order to be binding at the same $Q_0$. To determine the values of $\alpha$ and $\beta$ where reductions are equally binding, we impose the following restrictions on equation (13.6): $Q_0 > 0$, $c > a$ (excess supply intercept is larger than excess demand intercept) and $d < 0$ (slope of the excess demand curve is negative). Therefore, $(a - c) < 0$, and $(b - d) > 0$. The conditions that satisfy $Q > 0$ on (13.6) are: $\beta > \alpha$ or $\beta < \alpha^2$. These are necessary but not sufficient conditions for volume and value reductions to be equally binding at positive quantities traded. Note that these conditions do not determine the exact relationship between $\alpha$ and $\beta$, but they do rule out some combinations of $\alpha$ and $\beta$ that are not equally binding.

*Corollary.* For the current reduction commitments (21 percent in volume and 36 percent in value[3]), the value reduction will never be binding because:

$\beta = 0.64$ or $0.76$ ($1 - V_{red}$ for developed and developing countries, respectively)

$\alpha = 0.79$ or $0.86$ ($1 - Q_{red}$ for developed and developing countries, respectively)

$\alpha^2 = 0.62$ or $0.74$

These values of $\alpha$ and $\beta$ do not satisfy $\beta > \alpha$ or $\beta < \alpha^2$, and therefore cannot be equally binding. If we substitute these values of $\alpha$ and $\beta$ in equation (13.5), $V_\beta$ becomes larger than $V_\alpha$, showing that $V_\alpha$ is binding. Therefore, the value of export subsidies after the volume reduction is smaller than the volume left from a 36 percent value reduction, and so the volume limit is binding.

*Proposition 2.* A reduction in value is binding only when the ratio of the reduction commitments in value and volume is smaller than the ratio of the initial and reduced per unit export subsidy.

*Proof.* We rewrite equation (13.5) in general form, where $ES(.)$ is the excess supply curve, and $ES = Pd = a + bQ$; $ED(.)$ is the excess demand curve, and $ED = Pw = c + dQ$:

$$V_\beta = \left[ ES(\beta Q_0) - ED(\beta Q_0) \right] \beta Q_0 > \left[ ES(Q_0) - ED(Q_0) \right] \alpha Q_0 = V_\alpha \quad (13.8)$$

where $V_\beta > V_\alpha$ determines that the final value of the subsidy after a $1 - \alpha$ reduction in value will be smaller (and therefore binding) than the final value of the subsidy after a $1 - \beta$ reduction in volume. Solving equation (13.8) for $\beta/\alpha$ gives:

$$\frac{\beta}{\alpha} < \frac{\left[ ES(\alpha Q_0) - ED(\alpha Q_0) \right]}{\left[ ES(Q_0) - ED(Q_0) \right]} \Rightarrow \frac{\beta}{\alpha} < \frac{s}{s_0} \quad (13.9)$$

When $\beta/\alpha < s/s_0$, value is binding; and when $\beta/\alpha > s/s_0$ volume is binding (but still the relative values of $\alpha$ and $\beta$ have to satisfy $\beta > \alpha$ or $\beta < \alpha^2$ for value to be possibly binding). Therefore, at baseline $\alpha = \beta = 1$ and $s = s_0$ (no reductions). In order to get $s_0 < s$ (reduction in per-unit subsidy), $\alpha$ has to be larger than $\beta$, and according to proposition 2, smaller than $\alpha^2$. It confirms the fact that reductions in value $(1 - \alpha)$ have to be larger in order to bind.

This relationship allows us to forecast the effects of bindings on the reduction of per-unit subsidies for a given set of parameters. Therefore, when reductions are negotiated, a country can analyze the final per-unit export subsidy resulting from the proposed $\alpha$ and $\beta$ reductions for each commodity sector, and determine the reduction exports for given supply and demand conditions.

*Proposition 3.* For a larger initial per-unit export subsidy, ceteris paribus, volume reduction commitments are more effective.

*Proof.* In the baseline, both value and volume are binding, no reductions were applied, and $\alpha = \beta = 1$. Then reductions are applied to value and volume. From equation (13.7), we can show that the larger the initial subsidy $s_0$, the larger the reductions in value have to be for each $(1 - \alpha)$ reduction in volume:

$$\beta \leq \alpha^2 - s_0(c - a)(1 - \alpha)\alpha \quad (13.7)$$

For expenditure reductions to be binding, $\beta$ has to be sufficiently small, and therefore reduction in expenditures has to be sufficiently large (because $\beta = 1 - V_{red}$ or the proportion of the value of export subsidies allowed by the reduction in value). We can conclude that the higher the initial subsidy, the more likely that volume reductions will be binding, because $\beta$ is decreasing in $s_0$. Intuitively, we can understand that volume reductions happen in one dimension, while value reductions involve reduction of an area (exports multiplied by the per-unit export subsidy). Therefore, for a larger area (which implies a larger per-unit export subsidy, ceteris paribus) the value reductions would affect volume of exports less dramatically than a reduction in volume itself.

*Proposition 4.* For larger initial quantities exported (ceteris paribus), value reductions are more likely to bind.

*Proof.* Define the equilibrium (free trade) quantity as:

$$Q* = \frac{c - a}{b - d} \qquad (13.10)$$

Solving equation (3) for $Q_0$ gives:

$$Q_0 = \frac{s_0 - c - a}{b - d} \qquad (13.11)$$

where $Q_0 > Q*$. There is a maximum reduction in volume that will make $Q_0 = Q*$, which is the free trade equilibrium and $s_0 = 0$. Let us define $\alpha$ as the minimum $\alpha$ (or the maximum reduction in volume) that will bring exports to the free trade equilibrium such that

$$\hat{\alpha} Q_0 = Q* \Rightarrow \hat{\alpha} = \frac{Q*}{Q_0} \qquad (13.12)$$

The range of volume-reduced $\alpha$ is then in the interval $[\alpha, 1]$ and the range of value-reduced $\beta$ is in the interval $[0,1]$. Figure 13.1 can represent the relationship between the range of reductions in volume and value. It shows the range of percent of volume and value left, $\alpha$ and $\beta$, respectively. The percent value of the export subsidy remaining after the reductions, $\beta$, can be any fraction from 0 to 100 percent of its original value. The percent volume of subsidized exports left $\alpha$, however, does not need to be reduced by 100 percent. The volume of subsidized exports will be reduced up to the free trade equilibrium, where the export subsidy falls to zero. In the diagram, the free trade equilibrium is represented by $\alpha$, where $\alpha_1$ and $\alpha_2$ represent the minimum percent volume remaining (or the maximum reduction in volume) that will make $Q_0 = Q*$ for two different initial quantities $Q_0^1 > Q_0^2$. Equation (13.12) shows that when the initial quantity subsidized ($Q_0$) is larger, the fraction of the original volume that results in free trade ($\alpha$) becomes smaller, which means that a larger reduction in volume is necessary to achieve free trade. Figure 13.1 shows that the same $\beta$ (and therefore the same reduction in value) is equally binding to increasing reductions in volume (decreasing $\alpha$), when $Q_0$ increases.

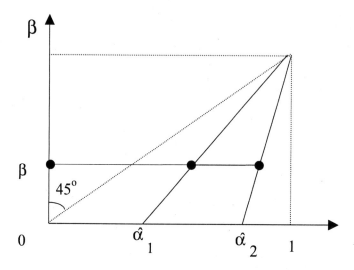

FIGURE 13.1. Volume of exports and the bindings.

In order to determine the relationship between volume reduction and initial quantity, we differentiate $\alpha$ with respect to $Q_0$ in equation (13.12), which gives:

$$\frac{\delta\hat{\alpha}}{\delta Q_0} = -\frac{Q*}{Q_0^2} < 0 \qquad (13.13)$$

For larger initial quantities $Q_0$, ceteris paribus, the reductions in volume to achieve free trade have to be larger (for the same per-unit export subsidy). Also, $\alpha$ has to be smaller for the same reduction in expenditures $\beta$. Therefore, a reduction in volume has to be larger to be as binding for the same reduction in expenditures, for larger $Q_0$. Value commitments are more likely to be binding for larger volumes exported, ceteris paribus.

Note from (13.13) that $\alpha$ is not linear in volume. For larger volumes we need increasing reductions in $\alpha$ to achieve free trade. Also, the area to the right of the diagonal solid lines on the diagram is the area of volume binding. Recalling that $\alpha = 1 - Q_{red}$, and $Q_{red}$ is the percentage reduction in volume of subsidized exports, the smaller the area, the more volume reductions are binding, and that occurs with smaller $Q_0$. The range of reductions for volume to achieve free trade is smaller. Volume is generally more binding.

*Proposition 5.* For more elastic curves, ceteris paribus, value reduction commitments become more effective.

*Proof.* Note that in equation (13.7), $\beta$ is a function of volume restrictions $\alpha$, initial per-unit subsidies $s_0$, and intercepts $a$ and $c$. The value reduction $1 - \beta$ is not a function of elasticities.

However, we can use the findings in Propositions 3 and 4 to set the framework to analyze how bindings are affected by the value of trade elasticities. Proposition 3 proves that a value reduction is more effective for smaller per-unit subsidies: we then keep $a$, $c$, and (implicitly) $Q_0$ constant in (13.7) and decrease $s_0$. This is only achieved by making the excess demand and supply curves more elastic. In Figure 13.2, this corresponds to decreasing the per-unit subsidy from distance $gh = s_0$ to the bold segment $s'$ at $Q_0$. This proves that more elastic curves (smaller per-unit export subsidies, ceteris paribus) make the value reduction more effective.

Proposition 4 states that for larger initial quantities exported (ceteris paribus), value reductions are more likely to bind. So we keep $c$, $a$, and $s_0$ constant, and increase $Q_0$. Again, this is only achieved by making the excess demand and supply curves more elastic. For increasing $Q_0$, the curves become more elastic. Therefore, for more elastic curves value reductions are more likely binding, given that in Proposition 4 we proved that for increasing $Q_0$ expenditure reductions are more likely to bind. In Figure 13.3, this corresponds to increasing initial volume from $Q_0^1$ to $Q_0^2$. This proves that more elastic curves (larger initial volume, ceteris paribus) make the value reduction more effective.

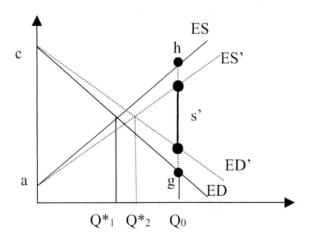

FIGURE 13.2. Proof that more elastic curves (smaller per unit export subsidies) make the value reduction more effective.

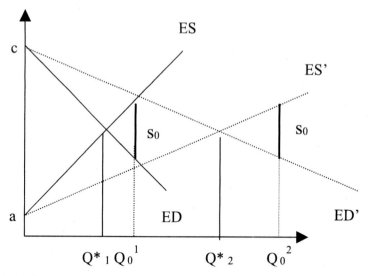

FIGURE 13.3. Proof that more elastic curves (larger initial volume) make the value reduction more effective.

## Dynamic Analysis: Shifts in Free Trade Equilibrium

The analysis in Propositions 1 to 5 is in static framework. But conditions of trade may change, and therefore the analysis needs to be augmented to include a dynamic analysis where the free trade equilibrium shifts during the export subsidy reduction implementation period, or after the reductions have been accomplished. Figure 13.4 depicts two cases: the free trade equilibrium point shifted to the left in panel (A) with positive shifts in excess supply and/or excess demand; or the reverse in panel (B) where the free trade equilibrium shifted to the right. The effectiveness of a volume versus an expenditure constraint depends on the goal of the government. Politicians have the option to maximize any one of at least four potential policy targets to fulfill political demands as well as possible under changing market conditions and WTO export subsidy commitments: the per-unit export subsidy, total export subsidy expenditures, the volume of exports subsidized, or domestic price levels.

If the free trade equilibrium shifts to the left, volume limits become less binding. We can have "water" in the volume limit, if value commitments bind at a lower $Q$. It is more likely that countries that reduced their schedule of exports over the implementation period will have to increase per-unit subsidy on a smaller volume of exports. It is possible, though, that countries

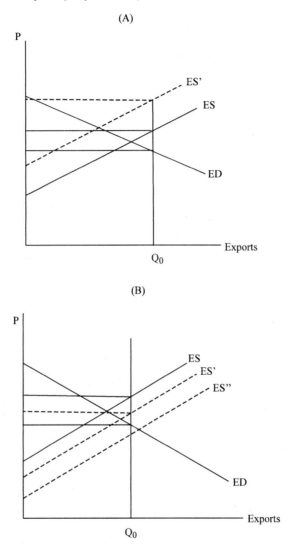

FIGURE 13.4. Shifts in the free trade equilibrium with export volume/value limits.

will not be able to spend all their value allowances, if it means exports above the volume limit. It would be possible to keep the level of domestic prices.

If the trade equilibrium shifted to the left, value limits became less binding, so we can have "water" in the value limit. It is more likely that countries will have to reduce the per-unit subsidy, and that the domestic price will have to fall due to the volume binding. The volume of exports will be maximized if the value limit does not bind at a lower $Q$.

*Proposition 6.* Assume that both volume and value reductions are equally binding at $Q_0$ after the reductions have occurred. When exogenous shifts in *ES* and *ED* occur, volume constraints are always binding in the case of a positive shift, and expenditure constraints are always binding in the case of a negative shift.

*Proof.* Suppose that reductions in value and volume are equally binding at $\alpha Q_0$, such that $V_\beta = V_\alpha$ and $s_0(\alpha) = s_0(\beta)$.

### Outward (Positive) Shift

For volume restriction $\alpha$, the reduced per-unit export subsidy equals:

$$s_\alpha = (a-c) + (b-d)\alpha Q_0 \qquad (13.14)$$

If a positive shift in excess demand occurs, the reduced per-unit subsidy becomes

$$s_{\alpha,k} = (a - kc) + (b-d)\alpha Q_0 \qquad (13.15)$$

where $k > 1$ is an intercept shifter. Clearly $s_{\alpha,k} < s_\alpha$, and because volume is limited by $\alpha Q_0$, the subsidy value is unambiguously reduced after the shift.

For value-binding $\beta$, the reduced export subsidy expenditure equals

$$V_\beta = [(a-c) + (b-d)Q_0]\beta Q_0 \qquad (13.4a)$$

If a positive shift in excess demand occurs, volume is not limited by $\alpha Q_0$, so the new subsidy value becomes

$$V_{\beta,k} = [(a - kc) + (b-d)Q_k]\beta Q_k \qquad (13.16)$$

We can have the same value of export subsidy ($V_\beta = V_{\beta,k}$) with a smaller per-unit export subsidy and a larger quantity. Value of export subsidy does not change for value-binding $\beta$ after the shift.

Therefore, for an outward shift in excess demand, expenditures have to be reduced due to a volume binding, but do not need to change due to the value binding: that means that a volume binding will be more trade liberalizing. The same reasoning is valid for a positive shift in excess supply, where the new supply intercept is $ka$, $0 < k < 1$.

*Inward (Negative) Shift*

If a negative shift in excess demand occurs, the same reasoning is valid for volume and expenditure reductions. In the case of a volume binding, $Q_0$ remains unchanged in equation (13.15), so the value of the subsidy $S_{\alpha,k}$ does not change. In this case, $0 < k < 1$.

In the case of a value binding, in order to make $V_\beta = V_{\beta,k}$ (equations (13.4a) and (13.16), $Q_0$ has to be reduced, and the expenditures are reduced with respect to the initial $Q_0$.

Therefore, in the case of a negative shift in excess demand, expenditures become binding and are more trade liberalizing. The same reasoning is valid for a negative shift in excess supply, where the new supply intercept is $ka$, $k > 1$.

Proposition 6 is only true if volume and expenditures are equally binding in baseline. If value reduction commitments are binding after the reductions, however, a positive shift on excess demand or excess supply can result in either volume or value commitments that are binding, depending on the ratio $\alpha/\beta$.

If volume reduction commitments are binding after the reductions, a negative shift on excess demand or excess supply can result in either volume or value commitments that are binding, depending on the ratio $\alpha/\beta$.

### Level of Agricultural Prices

Given variations in world prices with shifts on the excess demand and supply curves, a binding on volumes exported will become limiting with any changes in domestic or world prices that reduce the per-unit subsidy (an outward shift in excess demand and/or in excess supply). Therefore, in periods of expansion of agricultural production and consumption, volume bindings are essential for keeping export subsidies from escalating. This reasoning may seem counterintuitive that for higher levels of world prices, the need for subsidies is reduced. This is true only with respect to the per-unit export subsidy, with no regard to what happens to total expenditures and volume of subsidized exports. The volume binding is more effective than the value binding with outward shifts in excess supply or excess demand curves.

A binding on the value of subsidy will become limiting with any changes in domestic or world prices that increases the per-unit subsidy (an inward shift in excess demand and/or in excess supply). In those periods of contraction of agricultural production and consumption, a limit on expenditures (value) on export subsidies will help to force consumers and producers to respond to real market conditions instead of consume and produce at levels that are artificially maintained by an increase in subsidies.

## *Fixed Internal Price Supports*

In a scenario where domestic price is fixed (guaranteed), the per-unit export subsidy must fluctuate to accommodate the difference between domestic and world prices (an open-ended export subsidy). However, the value and volume reduction commitments will both limit the level of the price support.

If the value commitment is binding, and therefore the total value of subsidies cannot be increased, domestic price will have to fall if the gap between world and domestic price becomes wider (an outward shift in world demand). But countries can use another type of mechanism. Countries can subsidize only part of the exports by the total amount of the gap and spend all the value allowed, and the rest will be traded at lower world prices. This represents "seasonal" subsidies, where exports receive subsidies during alternative periods, such that the total of the period would not exceed the constraint, but individual periods would. The mechanism allows countries to circumvent volume bindings. In these situations, a limit on subsidy expenditures would be required to ensure trade liberalization. In contrast, for the case of an increase in world demand, the same binding volume can be subsidized by a smaller per-unit export subsidy, and therefore fixed domestic prices will not change.

## *FINAL CONSIDERATIONS AND CONCLUSION*

This study showed that a less than 100 percent reduction in volume of exports subsidized is necessary to achieve free trade. For equal percent reductions in volume and value, volume is always binding. Moreover, for the current reduction commitments, the volume limit is always binding. In order for the value reduction to be binding, it has to be sufficiently large. For a larger per-unit subsidy, ceteris paribus, volume restrictions are relatively more binding. Therefore, volume restrictions are more effective in decreasing distortions in the world price. For larger quantities traded, value restrictions are relatively more binding. Therefore, value restrictions are relatively more effective in decreasing subsidized exports. For more elastic trade curves, value restrictions are relatively more binding, so they are more effective in the case of small countries' exporters.

The dynamic analysis showed that the combination of volume and value bindings is necessary to limit the increase in expenditures or volumes subsidized in situations where excess supply and demand curves shift. In periods of expansion in agricultural production and consumption, volume restrictions keep export subsidies from escalating in terms of volumes subsidized. Conversely, in periods of contraction of agricultural production and consumption, the value binding limits the increase in the per-unit subsidy, thereby limiting the possibility of using variable export subsidies. Therefore, the combination of both bindings is important to limit subsidies that could increase even within the commitments, because they will seldom bind at the same time.

One cannot rank volume versus value limits without taking into consideration the objective of the government. The volume restriction limits increases in the per-unit subsidy. Volume limits allow for higher domestic prices with increases in excess demand, and value bindings allow for higher domestic prices with decreases in excess supply. The volume restriction also limits government expenditures in most cases, so it is more desirable that volume limits bind when governments want to cut back on export subsidy expenditures. In order to maximize exports, more binding volume limits are desirable, because if value is binding, not all the volume allowances will necessarily be exported.

The per-unit export subsidy is not regulated in the discipline, so the reduction commitments do not limit distortions in world price. This is because the same value of subsidy can lead to different per-unit subsidies, and different levels of distortion in world price. The short-term trade distortion from export subsidies might not have decreased as a result of the current rules. As the per-unit subsidy is not regulated, the discipline cannot reduce the level of asymmetry of support, because the bindings do not avoid different rates of subsidization across commodities, countries, and over time. It is possible to circumvent the volume commitments through discretionary subsidies by awarding larger subsidies to a smaller quantity and pooling the revenues to pay a higher price to producers. This is possible because the agreement does not require that subsidies be accounted for as if all the volume exported were subsidized by the same amount.

The current WTO Round should address the issues related to front-loading and banking, and include other modalities of export subsidy in the Agreement.[4] The per-unit subsidy should be regulated, and the notifications should account for the expenditures on export subsidies "as if" all the exports had received the highest per-unit offered in each period, as a means to avoid circumvention of the volume commitments.

## NOTES

1. The other two pillars are commitments on market access and domestic support.

2. Export subsidies are illegal in the GATT, which is an indication of this political bias.

3. For developing countries, the reductions are 14 percent in volume and 24 percent in value.

4. As examples, this is the case of consumer-only financed export subsidies, export credits and insurance, and food aid.

# REFERENCES

Josling, T. 1998. *Agricultural Trade Policy: Completing the Reform.* Washington, DC: Institute for International Economics.

McCalla, A. 1999. Closing comments at the World Bank Conference "Agriculture and the New Trade Agenda from a Developmental Perspective: Interest and Options in the Next WTO Negotiations," Geneva, October 2.

# Chapter 14

# Agricultural Trade Liberalization Beyond Uruguay: U.S. Options and Interests

John Gilbert
Thomas Wahl

## *INTRODUCTION*

Under the Uruguay Round Agreement on Agriculture (URAA), the rules governing international trade in agricultural products changed in a fundamental way. Members of the World Trade Organization (WTO) agreed to a process of tariffication of agricultural nontariff barriers (NTBs), to bind those tariffs, and to subject them to reductions. They also agreed to institute disciplines and reduction commitments on export subsidies and domestic support mechanisms. Significant progress has therefore been made toward subjecting trade in agricultural commodities to the same disciplines as trade in manufactures.

On the other hand, "dirty" tariffication and flexibility in implementing the tariff reduction formula has led to increased tariff dispersion and in some cases to highly distorting "mega-tariffs." Concern also remains over issues such as tariff escalation and the effect of certain features of the URAA, such as tariff rate quotas and the de minimis provisions.

A range of issues therefore must be addressed in the new negotiations, when they are eventually started. All economies, including the United States, are faced with the task of identifying the negotiating options that best serve their interests. This requires an understanding of the relative importance of the various issues that remain on the agenda, knowledge of the tradeoffs involved in adopting certain negotiating positions, and an understanding of the costs and benefits of the various alternatives.

Ideally, evaluation of the various aspects of U.S. interests should be conducted within a single consistent framework. This suggests that computable general equilibrium (CGE) simulation, which has already made substantial contributions to the quantitative analysis of trade reform in a wide variety of contexts, is suitable. In this chapter we use a recursive dynamic CGE model based on the GTAP4 database to analyze some of the important issues for the United States.

The chapter is organized as follows. Under Methodology we outline the salient features of our modeling approach. The next three sections use the model to evaluate various issues that are important in developing the U.S. position in the coming negotiations. In Unfinished Business we consider the potential effect of removal of the outstanding protection remaining after the implementation of the URAA, i.e., what gains remain from tariff, export subsidy, and the domestic support reform. The simulation results help identify the aspects of a new agreement to which the United States should devote most of its energy. China and the WTO discusses the looming China issue, in particular focusing on the consequences of extending of normal trade relations. Finally, Regional Alternatives considers the potential of two alternative liberalization forums (APEC and the Pacific-5). A summary and concluding comments follow.

## METHODOLOGY

Computable general equilibrium or CGE models are in essence numerical models based on general equilibrium theory, which are implemented in the form of a computer program. These models have a number of features that have made them increasingly popular in recent decades. They are multisectoral, and in many cases multiregional, and the behavior of economic agents is modeled explicitly through utility and profit-maximizing assumptions. In addition, they differ from other multisector tools of analysis in that economy-wide constraints are rigorously enforced. Distortions such as trade barriers in an economic system will often have second-best repercussions far beyond the sector in which they occur. Where the distortions are wide-ranging, general equilibrium techniques are effective in capturing the relevant feedback and flow-through effects. The model used in this chapter is recursive dynamic, consisting of an intraperiod and an interperiod component.

### Intraperiod Model

The intraperiod model is a modified version of Rutherford's model (Rutherford, 1998). Because it is of a well-established class, we present only brief details. Perfect competition and full employment prevail. Factors are mobile within each economy, but not internationally. There are limits on labor mobility within economies in the intraperiod model, such that rural (agricultural) unskilled labor in the developing economies and Japan can only move between agricultural production sectors, and urban unskilled labor can only move between industrial production sectors. Factor supply is exogenous within each period. Production is specified using constant elasticities of substitution (CES) functions. All factors enter in the production functions in variable proportions, with a single elasticity for each production sector governing rates of substitution. Intermediate inputs enter in fixed proportions.

International trade in the model is modeled using the Armington approach (CES). The final demand specification assumes government expenditure is fixed, while households maximize a Stone-Geary Linear Expenditure System (LES) utility function. Savings is a fixed proportion of income. We close the model by fixing the current accounts of each economy.

### Interperiod Model

The interperiod linkages of the model generate the growth scenario. Investment in each period augments the capital stock according to a standard capital accumulation equation—which is subject to steady state. Productivity growth is modeled as Hicks-neutral technological change.

We make a departure from the standard specification by not allowing technological progress to continue at a constant rate in all economies. One of the well-known features of historical growth patterns is a slowdown as economies become more developed. We incorporate this stylized fact into our model structure. The paths are determined in a base simulation with no liberalization, and are thereafter fixed for the liberalization scenarios.

Labor growth is where our model differs most from others. We have three categories of labor: skilled, rural unskilled, and urban unskilled. The economy-wide labor endowment in each region grows at a predetermined rate. Unskilled labor grows at the economy-wide rate, but is drained by movements to the skilled category (up-skilling) and augmented by rural-urban migration. We adjust the initial rate of labor growth in developing economies as per capita real incomes approach average developed country levels. Once again, the paths are determined in a base simulation with no liberalization, and are thereafter fixed. Second, incremental growth rates by category respond to changes in wage differentials (see Gilbert, Scollay, and Wahl, 2000, for further details).

### Data and Parameter Values

The source of the input-output and protection data for this model is the Global Trade Analysis Project (GTAP-4) database, described in McDougall, Elbehri, and Truong (1998). The base year for the simulations is thus 1995. The database has been aggregated to the level of fifteen commodities and fifteen regions. The commodities are: paddy rice, wheat, grains, vegetables and fruit, other nongrain crops, livestock, forestry, fisheries, processed rice, meat products, dairy products, other food products, light manufactures, heavy manufactures, and services. The regions are identified in Table 14.1.

Factor returns data in the GTAP database are normalized to unity in the base year. For our purposes the actual (per unit) returns are required for the three categories of labor. We supplement the GTAP data with information

TABLE 14.1. Assumptions Used in the Projections (Annual Percentage Growth Rates)

| Region | Labor[a] | Rural Labor[b] | Skilled Labor[c] | Capital[d] | TFP[e] |
|---|---|---|---|---|---|
| Australia | 0.80 | − 0.68 | 6.65 | 3.00 | 0.30 |
| Canada | 0.60 | − 4.54 | 4.67 | 4.80 | 0.30 |
| China | 0.90 | − 0.39 | 2.58 | 10.70 | 1.60 |
| Europe | 0.00 | − 3.35 | 9.30 | 5.10 | 0.30 |
| Indonesia | 2.10 | − 0.26 | 7.64 | 7.10 | 1.60 |
| Japan | − 0.03 | − 4.20 | 4.73 | 5.90 | 0.30 |
| Malaysia | 2.80 | − 3.77 | 7.30 | 9.20 | 0.70 |
| Mexico | 2.30 | − 1.88 | 2.93 | 1.40 | 0.90 |
| New Zealand | 0.40 | 0.57 | 7.07 | 2.40 | 0.30 |
| Other APEC | 1.00 | − 2.14 | 4.95 | 7.90 | 1.60 |
| Philippines | 2.50 | − 1.23 | 3.22 | 2.20 | 0.50 |
| ROW | 1.70 | − 0.50 | 4.92 | 3.50 | 1.00 |
| South Korea | 1.20 | − 6.05 | 4.94 | 7.80 | 1.60 |
| Thailand | 0.90 | − 0.43 | 6.34 | 6.60 | 1.60 |
| USA | 0.70 | − 2.00 | 4.57 | 3.20 | 0.30 |

[a]World Bank (1999) projections 1997-2010.
[b]Cumulative rates of growth, based on trend in preceding ten-year period (five years in China). Figures from FAOSTAT (http://apps.fao.org/), except Taiwan from Taiwan Agricultural Yearbooks.
[c]Cumulative rates of growth, based on projections of Ahuja and Filmer (1995) and trends from UNESCO (1997).
[d]Growth rate based on projections in Anderson et al. (1997b) and historical trend in preceding ten-year period (Penn World Tables http://pwt.econ.upenn.edu/). Depreciation rates on capital selected to calibrate to this rate in base year, thereafter growth rates endogenous.
[e]Implemented as a Hicks-neutral change across all inputs. Figures based on estimates from Chen (1977), Young (1994), Drysdale and Huang (1997), and World Bank (1997).

on agricultural and nonagricultural labor body counts from the Food and Agricultural Organization Statistical Databases (FAOSTAT) and the Taiwan Agricultural Yearbook, using the skill breakdowns in Liu et al. (1998) to obtain consistent measures of the average unskilled rural wage, unskilled urban wage, and skilled wage for all economies in the model.

Armington and production elasticities are from GTAP-4. The Armington values are doubled at both levels, as in Anderson et al. (1997b). This has the effect of dampening terms-of-trade responses to a given volume of trade change, and expanding the trade response to the removal of distortions. It has been shown to improve the accuracy of backcasting exercises.

The data used in specifying the dynamic path of the model is from a number of sources. Productivity growth rates have been obtained from the existing literature. The primary source for labor force growth projections is the World Bank (1999). Where available, World Bank projections are used for skilled labor growth; otherwise we use historical trends (full details are in Table 14.1). Both capital stock growth and rural labor growth are based on historical rates over the preceding ten-year period. As with other models of this type, the main purpose of the growth path exercise is to establish a baseline to which the experimental simulations can be compared.

### Reference and Alternative Scenarios

Our baseline simulation implements the growth path as described above, and also accounts for the major existing liberalization agreements, the completion of the Uruguay Round, NAFTA, and AFTA (the ASEAN [Association of Southeast Asian Nations] Free Trade Arrangement). For the URAA completion and China's WTO membership, we use post-UR tariff rates contained in the GTAP-3 database as the target, which is reached in linear steps by the end of 2000 for developed economies and 2004 for developing economies. Where GTAP-4 contains a rate that is lower than the target rate, we assume that the economy has already met its UR obligations and no further reductions are made.

The URAA-required export subsidies on agricultural commodities are to be reduced by 24 percent for developing economies and 36 percent for developed economies, along the same time frame as previously stated. Because the GTAP-4 database has 1995 as its base year, some reductions should already have taken place. We assume linear reductions over the entire implementation period, including 1995, and calculate the remaining reductions accordingly.

In the case of production subsidies, the reduction requirements were 13 percent for developing economies and 24 percent for developed economies. Once again, we assume linear reductions over the implementation period, and eliminate reductions that should already be reflected in the base year data contained in GTAP-4. In order to reflect the effects of the de minimis rule, where support is less than 10 percent for developing economies and 5 percent for developed economies, no reductions are made. These levels also form a floor for the reductions.

Because China was not a WTO member until recently, it is not included in the UR liberalization scenarios. China neither makes concessions nor receives the benefits of concessions by other economies. In the simulations that consider China's WTO membership, China is assumed to make similar concessions to those required of developing economies in the UR: 24 percent reductions in agricultural tariffs and export subsidies, and a 13 percent reduction in domestic support (subject to the de minimis rule), implemented by 2004.

The Pacific-5 group brings together the United States, Australia, New Zealand, Singapore, and Chile. These last two economies form the majority of our "Other APEC" category. We consider the removal of agricultural tariffs, export subsidies, and domestic support among these economies on a preferential basis. Linear reductions over the period 2001-2005 are used.

The AFS proposal calls for an extensive and rapid liberalization of food and agricultural products trade and production. We interpret this as the removal of import tariffs, export subsidies, and production subsidies on all agricultural and food products in the APEC region. The original AFS proposal calls for quick action to eliminate distortions before 2006, and we therefore assume that the liberalization is implemented as a set of linear reductions in distortion levels over the five-year period 2001-2005. We interpret "open regionalism" as requiring liberalization on an unconditional most-favored nation (MFN) basis.

## UNFINISHED BUSINESS

While the URAA went a considerable way toward imposing the same disciplines on agricultural trade as exist for most other products, for various reasons it did not go far in terms of actually reducing trade distortions. A significant proportion of the new round will therefore be devoted to negotiating further reductions. Sumner (2000) has argued that the WTO should focus these negotiations on border measures, in particular tariffs. His argument rests on two foundations. The first is a political economy of protection/negotiation argument. It may be easier to obtain agreement on tariffs if economies are free to use whatever domestic support measures they wish. Moreover, since domestic support measures are more transparent than tariffs, we might expect internal pressure to build for their eventual removal. The second argument is that there is simply a "bigger bang" for the negotiating buck.

In Figure 14.1 we present the estimated welfare effects of the complete removal of domestic support, export subsidies, and import protection, from a post-UR dataset. The results support Sumner's hypothesis in the global sense. Estimated global gains from agricultural tariff reform are $75 billion per year, compared with $16 billion in the case of domestic support and a

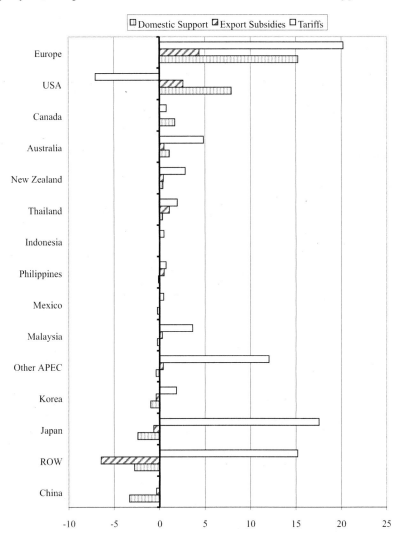

FIGURE 14.1. Post-URAA reform net welfare (US$ billion)

mere $2 billion in the case of export subsidies. These (and all subsequent) figures are the regional equivalent variation, and can thus be interpreted as the increase in regional income that is equivalent to the proposed change at constant 1995 prices. The results suggests that the WTO should indeed be concentrating its efforts on tariff reform.

However, for the United States, it is clear that domestic support is the main issue, followed by export subsidies. This is a reflection of the large domestic support programs in the United States, in particular for grains, which carry with them a substantial welfare cost. In other words, at least in net welfare terms, the United States should devote efforts to cleaning up its own backyard. The result also reflects the fundamental difference between the theory of protection, and the dynamics of trade negotiation. While negotiations have tended to concentrate on issues of market access for exports, the arguments of trade theory emphasize the importance of imports as a source of improvements in allocative efficiency. The United States and other economies should not lose sight of this during the negotiation process. However, to the extent that negotiating concurrent reductions in the high levels of European and Japanese domestic support may help to ease the pressures associated with domestic reform, it will be in U.S. interests to pursue them.

## CHINA AND THE WTO

China withdrew from the precursor to the WTO (GATT) in 1950, and applied to rejoin the organization and its successor beginning in 1986. In late 1999, the United States reached an agreement with China to support WTO membership. The decision to support permanent normal trading relations with China passed through the U.S. Congress in May 2000. A similar agreement was reached with the European Union. China is now a member of the WTO.

Wang (1997) has considered the benefits of China's accession, estimating the benefits to China at approximately $US30.3 billion using the long-run closure that is comparable to the dynamic model used here (see also Anderson et al., 1997a,b). Our own simulation results (Figure 14.2) indicate global gains of $45 billion, of which $33 billion accrues to China—so there is considerable consistency. The estimated net welfare gain to the United States is small—$1.2 billion.

One issue that has not been considered is the question of permanent extension of normal trading relations (NTR) with China, which was approved in 2000. In conformity with WTO rules, China has made it quite clear that it will not reduce its barriers to US imports unless it receives unconditional permanent normal trade relations status (Bergsten, 2000). What are the economic stakes of backtracking on the current U.S. position? The simulation

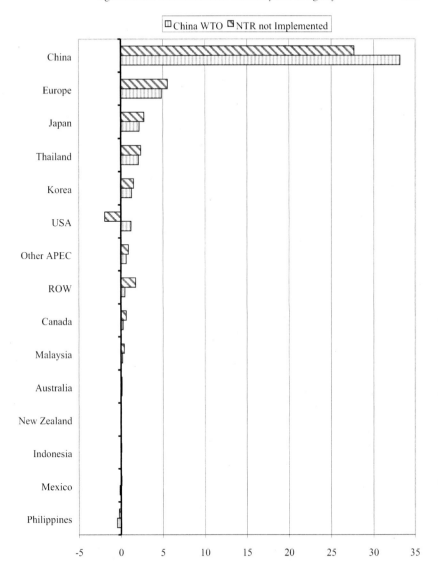

FIGURE 14.2. Chinese WTO entry net welfare (US$ billion)

results indicate an annual welfare cost of $2 billion relative to baseline, or $3.2 billion relative to reaching an agreement on NTR. The United States, in effect, hands the benefits of China's open markets to Europe, Japan, and the ROW (Rest of the World).

Of course, as Bergsten (2000) has noted, the real costs of permanently extending NTR go far beyond the allocative efficiency costs estimated in this type of model. The United States will need to consider whether it is wise to allow other economies to gain an uncontested first foothold in the world's sixth largest and most rapidly growing economy, even if the initial gains are relatively small. The real benefits of China's accession may well be felt further down the road, as China grows, market reforms are set in place, and the Chinese economy integrates more extensively with world markets.

### REGIONAL ALTERNATIVES

In addition to the multilateral negotiations of the WTO, the United States is involved in several regional trading arrangements. A number of these, in particular NAFTA and APEC (Asia-Pacific Economic Cooperation), have been the subject of numerous quantitative modeling exercises (see Scollay and Gilbert, 2000, in the case of APEC). Others, such as the recent Pacific-5 arrangement between the United States, Australia, New Zealand, Singapore, and Chile, have yet to receive any attention. A key question in developing a U.S. position in the WTO negotiations is whether or not these regional forums form a viable alternative to the multilateral process embodied in the WTO.

Figure 14.3 presents the results of simulating agricultural reform in APEC along the lines of the APEC Food System proposal (AFS) and similar reform in the Pacific-5. Clearly, agricultural reform in the Pacific-5 context provides little in the way of benefits to the Pacific-5 economies. Only the U.S. experiences substantial net welfare gains, and again these are largely a reflection of the gains from eliminating the distortions arising from its own domestic support policy. This result should not be too surprising as the Pacific-5 member economies are (with the exception of Singapore) agricultural exporting countries. Thus, while a formal arrangement might be of political significance and may help to boost the negotiating strength of the group, it has negligible welfare consequences.

The same cannot be said of the AFS proposal, which was endorsed at the APEC meetings in Auckland in 1999. There are clearly substantial welfare gains available to APEC members from this proposal. APEC provides an alternative negotiating forum for access to almost all of the United States' key markets. The key question is whether APEC can seriously expect to deliver the comprehensive reform package outlined in the AFS. While the group as a whole has committed to free trade within the region by 2020, debate over agricultural reform has at times threatened to derail the process. On the one hand APEC does not have the Europeans or a large number of least-developed countries (LLDCs) to deal with as the WTO does, which should

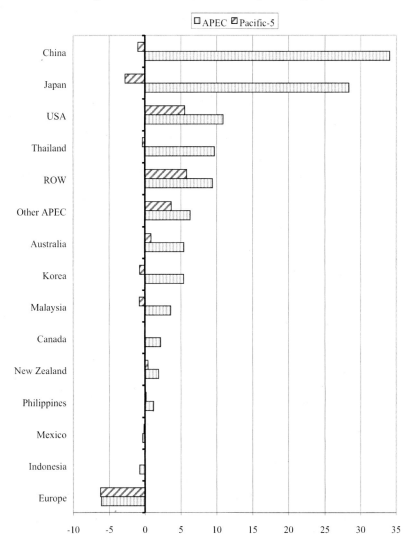

FIGURE 14.3. APEC and Pacific-5 net welfare (US$ billion)

in principle make it easier to reach agreement. On the other hand, it has Japan and South Korea, which vehemently oppose agricultural trade reform in key sectors and have yet to successfully implement any substantial reforms through the process of concerted unilateralism that distinguishes APEC from other regional trading arrangements. In sum, while APEC does bring together a substantial group of potential markets for agricultural products, and protection levels in the region are currently high enough to suggest substantial net gains from reform, a question mark remains over the final outcome.

## *CONCLUDING COMMENTS*

In this chapter we have presented the results of three sets of simulations that address various issues concerning developing a U.S. position in the upcoming multilateral negotiations. Of course, there is much work to be done. This chapter has focused on net welfare effects of agricultural reform, but trade and production effects, and effects of reform on income distribution, will also need to be analyzed.

In interpreting CGE results, it pays to bear in mind the fundamental limitations of the technique, which is designed to paint a broad picture of the consequences of substantial policy changes in an internally consistent framework. The actual numbers should be interpreted as guidelines more than forecasts. However, with the usual caveats in mind, the simulation results presented in this chapter raise some interesting issues. They support the position of Sumner (2000), and other economists, that the real business of the WTO should be the elimination of border measures. However, for the United States, the biggest distortion is its own domestic support. To the extent that global agreement may help to alleviate the pain associated with domestic reform, it may be in U.S. interests to pursue them.

The simulations indicate that permanent extension of normal trading relations was the correct decision for the U.S. economy in net welfare. Although the immediate efficiency costs of opting out of U.S. obligations toward China may not be great, some $3.5 billion per year, the long-term costs of failing to enter the Chinese market, could be substantially higher.

Finally, on the role of regional trading arrangements, there are many questions left unanswered. Hufbauer (1999) has argued for keeping these discussions at a purely technical level, echoing the line that international commerce does not need to be burdened with arrangements that distract from multilateral negotiations. In the case of agriculture in the Pacific-5, this argument holds up well—the only plausible benefit from this forum is greater cohesion among some of the major agricultural exporters in the WTO negotiations. The APEC forum potentially offers greater benefits, because of the diversity of its members, but has yet to develop an effective mechanism for achieving actual reductions. Without this it cannot be viewed as a viable alternative to multilateral negotiations.

# REFERENCES

Ahuja, V. and D. Filmer. (1995). "Educational Attainment in Developing Countries: New Estimates and Projections Disaggregated by Gender." World Bank Policy Research Paper No.1489, World Bank, Washington, DC.

Anderson, K., B. Dimaranan, T. Hertel, and W. Martin. (1997a). "Asia-Pacific Food Markets and Trade in 2005: A Global, Economy-Wide Perspective." *Australian Journal of Agricultural and Resource Economics* 41:19-44.

Anderson, K., B. Dimaranan, T. Hertel, and W. Martin. (1997b). "Economic Growth and Policy Reform in the APEC Region: Trade and Welfare Implications by 2005." *Asia-Pacific Economic Review* 3:1-18.

Bergsten, C.F. (2000). "The Next Trade Policy Battle." International Economic Policy Brief 00-1, Institute for International Economics, Washington, DC.

Chen, E.K.Y. (1977). "Factor Inputs, Total Factor Productivity, and Economic Growth: The Asian Case." *Developing Economies* 15:121-143.

Drysdale, P. and Y. Huang. (1997). "Technological Catch-Up and Economic Growth in East Asia and the Pacific." *Economic Record* 73:201-211.

Gilbert, J., R. Scollay, and T. Wahl. (2000). "The APEC Food System: Implications for Agricultural and Rural Development Policy." *Developing Economies* 38:308-329.

Hufbauer, G.C. (1999). "World Trade After Seattle: Implications for the United States." International Economics Policy Brief 99-10, Institute for International Economics, Washington, DC.

Liu, J., N. van Leuween, T. Vo, R. Tyers, and T. Hertel. (1998). "Disaggregating Labor Payments by Skill Level." In McDougall, R.A., A. Elbehri, and T.P. Truong (Eds.), *Global Trade, Assistance, and Protection: The GTAP 4 Data Base*. West Lafayette: Center for Global Trade Analysis, Purdue University.

McDougall, R.A., A. Elbehri, and T.P. Truong (Eds.) (1998). *Global Trade, Assistance, and Protection: The GTAP 4 Data Base*. West Lafayette: Center for Global Trade Analysis, Purdue University.

Rutherford, T. (1998). "GTAPinGAMS: The Dataset and Static Model." Working Paper, Department of Economics, University of Colorado.

Scollay, R. and J. Gilbert. (2000). "Measuring the Gains from APEC Trade Liberalization: An Overview of CGE Assessments." *World Economy* 22:175-197.

Sumner, D. (2000). "Domestic Support and the WTO Negotiations." Working Paper, Department of Agricultural Economics, University of California at Davis.

UNESCO. (1997). *Statistical Yearbook*. Paris: UNESCO.

Wang, Z. (1997). "China and Taiwan Access to the World Trade Organization: Implications for US Agriculture and Trade." *Agricultural Economics* 17:239-264.

World Bank. (1997). *Global Economic Prospects and the Developing Countries*. Washington, DC: World Bank.

World Bank. (1999). *World Development Indicators.* Washington, DC: World Bank.

Young, A. (1994). "Lessons from the East Asian NICs: A Contrarian View." *European Economic Review* 38:964-973.

Chapter 15

# Regional versus Multilateral Trade Arrangements: Which Way Should the Western Hemisphere Go on Trade?

Karen M. Huff
James Rude

## INTRODUCTION

Countries in the Western Hemisphere have a long history of participating in multilateral trade negotiations, as well as actively pursuing numerous regional trade liberalization efforts. Although the media's focus on trade issues has faded considerably since the failed World Trade Organization (WTO) Ministerial meeting in Seattle, work continues at the WTO on agriculture and services trade liberalization. Additionally, in the Western Hemisphere, trade ministers have been quietly pursuing an ambitious regional trade agenda through the Free Trade Area of the Americas (FTAA). While the timing of a new multilateral trade round remains uncertain, recent accounts (*BRIDGES,* 2001) indicate that the FTAA is proceeding on schedule for completion in 2005.

The purpose of this chapter is to discuss the potential interaction and/or conflict between these two approaches to liberalizing trade, as well as providing an empirical assessment of the impacts of regional and multilateral trade liberalization on the Western Hemisphere. A computable general equilibrium model (CGE) of the global economy is employed to quantify the outcomes of these regional and multilateral trade liberalization scenarios with the main focus of the analysis on the agri-food sector.

First, the formation of a potential FTAA agreement is simulated, followed by a quantitative assessment of the combined impact of both a new WTO agreement and the FTAA. The impact of each scenario is measured in terms of changes in trade flows, output, and welfare. Since the FTAA is a regional trade agreement, the results are also examined for potential trade creation and diversion effects. The CGE approach employed accounts for impacts on regions outside the FTAA such as the European Union.

This chapter is organized as follows. The following two sections provide some details about the FTAA process, and the prospects for a new multilateral trade round. This is followed by sections that describe the research methods and data used in the study, and present the results of the two trade liberalization scenarios. The final section of the chapter discusses the conclusions and policy recommendations of this research.

## THE FREE TRADE AREA OF THE AMERICAS

In April 1998, the heads of state of thirty-four Western Hemisphere countries met at the second Summit of the Americas in Santiago, Chile to launch formal negotiations for a Free Trade Area of the Americas. There are currently nine negotiating groups covering the areas of market access: investment, services, government procurement, dispute settlement, agriculture, intellectual property rights, subsidies, antidumping and countervailing duties, and competition policy. Given that the Western Hemisphere presently encompasses some forty regional and bilateral trade agreements (ERS, 1998), the FTAA process represents an opportunity for the region to consolidate these individual agreements into a single, barrier-free trading area.

Currently, there are four customs unions within the Western Hemisphere: the Andean Pact, the Southern Cone Common Market (MERCOSUR), the Central American Common Market (CACM), and the Caribbean Common Market (CARICOM). There are also a number of regional free trade agreements (FTAs) including the North American Free Trade Agreement or NAFTA (Canada, Mexico, and the United States); the Group of Three (Colombia, Mexico, and Venezuela); and Chilean FTAs with Mexico, Canada, Colombia, and Ecuador. As well, there are a number of other agreements among countries within the region.[1]

Prior to the setback at the third WTO Ministerial meeting in Seattle, FTAA trade ministers met in Toronto, Canada. At that time, they reaffirmed their commitment to the WTO and supported a launch of a new round of multilateral negotiations in Seattle. Indeed, Western Hemisphere trade ministers expressed the view that such impediments to trade as agricultural export subsidies and domestic support expenditures were best addressed in the multilateral setting. At the regional level, the ministers stressed their intention to pursue the establishment of the FTAA by 2005. To that end, the FTAA ministers acknowledged the close relationship between their regional negotiations and the next WTO Round. However, the anticipated synergy between these two negotiating processes is now uncertain due to the failure of the Seattle Ministerial to result in an agreed-upon agenda for a new multilateral trade round. The prospects for such a round of multilateral trade negotiations are addressed in the following section.

## THE PROSPECTS FOR A NEW MULTILATERAL TRADE AGREEMENT

Despite the failure in Seattle of WTO trade ministers to agree on an agenda for a new round of comprehensive multilateral trade negotiations, mandated negotiations on liberalization of agriculture and services trade have proceeded in Geneva. However, optimism for speedy and significant progress in these negotiations is likely unwarranted. Many developing countries remain skeptical of the benefits of multilateral trade liberalization, particularly in the area of agri-food trade, and the major agricultural exporters—the United States, the EU, and the Cairns Group—remain far apart in their goals for agriculture reform. Perhaps the biggest impediment to a new multilateral trade round is the general lack of political will around the globe to champion the case for trade liberalization (Meilke and Huff, 2000). Political scandals in Europe, a controversial U.S. presidential election result, the continued absence of U.S. fast track authority, as well as the division between developed and developing countries over the functioning and leadership of the WTO have all contributed to this problem. With all of these complications, the probability of a new multilateral trade deal being completed by 2005 appears slim. Nevertheless, for the purposes of this chapter, it is instructive to explore the potential consequences of such a multilateral deal completed in conjunction with the formation of the FTAA. The modeling approach used for this purpose is discussed in the next section.

## THE MODELING FRAMEWORK

Modeling multicountry, multisector trade liberalization reforms such as the FTAA requires an accounting of all of the interactions in an economy. A general equilibrium model accounts for these interactions in all product and factor markets. In this study, the standard Global Trade Analysis Project (GTAP) model (Hertel, 1997) is employed to determine the changes in regional patterns of output and trade and the welfare impacts resulting from hemispheric and global trade liberalization.[2] The database is version 4 of GTAP (McDougall, Elbehri, and Truong, 1998) which is based on 1995 input-output and trade data. Regional protection is represented by ad valorem equivalents of tariff and nontariff barriers, domestic support transfers, and export subsidies.

Table 15.1 describes the countries and aggregated regions employed in the model, as well as the commodity aggregation. Regions included in the model are Canada, the United States, Mexico, MERCOSUR, the Andean Pact, Chile, other Latin American and Caribbean countries, the European Union, Oceania (Australia and New Zealand), and the rest of the world. Agriculture is subdivided into seven aggregate commodities along with other primary products, three manufacturing sectors, and a single aggregated service sector. Agri-food commodities include wheat; a grains and oilseeds aggregate (coarse grains, rice, and oilseed crops); fruits and vegetables; an aggregate of other crops (sugar, unprocessed tobacco, and all crops not elsewhere specified); an animal products aggregate (live animals, wool, and hides); meats; and an aggregate of other processed food (dairy, beverages, tobacco products, and all other processed foods).

### Establishing a Baseline

A baseline experiment is run to update the protection data to correspond to the year 2005.[3] The baseline experiment reflects full implementation of existing trade agreements including the Uruguay Round, NAFTA, MERCOSUR, and Andean Pact agreements.

Post-Uruguay Round (i.e., 2005) tariff rates for agriculture import tariffs are taken from Podbury et al. (2000) and the authors' own estimates. Post-Uruguay Round tariff vectors for manufactured goods and services are taken directly from Francois and Strutt (1999) and used as baseline target rates. The levels of domestic support and export subsidization are not adjusted in the reference scenario since their levels in the GTAP database are

TABLE 15.1. Regions and Commodities in the Model

| Regions | Commodities |
|---|---|
| 1. Canada | 1. Wheat |
| 2. United States | 2. Grains and oilseeds |
| 3. Mexico | 3. Fruits and vegetables |
| 4. MERCOSUR | 4. Other crops |
| 5. Andean Pact | 5. Animal products |
| 6. Chile | 6. Meats |
| 7. Other Latin America and Caribbean Countries | 7. Other processed food |
| | 8. Natural resource industries |
| 8. European Union | 9. Textiles and clothing |
| 9. Oceania | 10. Automotive |
| 10. Rest of the World | 11. Other manufactures |
| | 12. Services |

assumed to be in line with Uruguay Round commitments for 2005. Although the Agreement on Textiles and Clothing calls for the elimination of all quantitative restrictions associated with the Multifibre Arrangement by 2005, many of these barriers are expected to be perpetuated beyond this date through the application of contingent protection measures (Reinert, 2000). As such, the tariff equivalents for textiles and clothing quotas are not adjusted in the baseline simulation.

The results of the baseline experiment are not presented in this chapter. The purpose of this experiment is to establish a meaningful starting point (i.e., post-Uruguay Round) for the FTAA and multilateral trade liberalization scenarios studied in detail in this chapter.

### Scenario 1: The FTAA Process

The FTAA scenario models the formation of a free trade area covering the entire Western Hemisphere. Tariffs on all intrahemispheric trade are reduced to zero, but each member's tariffs on third country imports remain unchanged as there is no attempt to harmonize external tariffs. In the area of domestic agricultural support, the FTAA process is assumed to repeat the pattern of reductions obtained in the Uruguay Round. In an attempt to liberalize only the production-distorting level of support, programs assumed to be more than minimally distorting, such as market price supports and input subsidies, are subject to 20 percent reduction. Domestic support levels in the developing countries in the region are minimal, and for the purposes of this study are not subject to reductions. In terms of export competition in the region, only U.S. and Canadian export subsidies on dairy products are reduced via a 36 percent reduction in the export subsidy rate on the Other Processed Food commodity. All reductions are made from the post-Uruguay Round base.

### Scenario 2: Multilateral Liberalization with the FTAA

The second scenario examined in this chapter simulates multilateral trade liberalization through a comprehensive WTO agreement combined with the FTAA liberalization. The new WTO agreement is assumed to result in a set of reforms similar to the Uruguay Round agreement. Agricultural import tariffs are reduced by 36 percent, production-distorting domestic support is reduced by 20 percent, and export subsidy expenditure is reduced by 36 percent from the post-Uruguay Round baseline (Josling, 1998). Industrial tariffs are reduced by 20 percent (Nagarajan, 1999).[4] In addition to these cuts, the FTAA experiment outlined previously is also assumed to occur and is implemented with the multilateral reforms just described.

## RESULTS

Table 15.2 highlights the trade diversion and to a lesser extent, the trade creation effects of a preferential FTAA.[5] Trade diversion occurs if the preferential tariffs cause importers to switch from more efficient suppliers outside the region to less efficient suppliers within the FTAA. Trade creation

TABLE 15.2. Trade Diversion and Creation (Difference from Base in Millions of 1995 US $)

|  | Scenario 1 | Scenario 2 |
|---|---|---|
| *NAFTA* | | |
| Imports from FTAA | 21,280 | 14,383 |
| Imports from elsewhere | −12,164 | − 4,405 |
| Agri-food imports from FTAA | 1,428 | 1,178 |
| Agri-food imports from elsewhere | −1,481 | − 662 |
| *MERCOSUR* | | |
| Imports from FTAA | 2,264 | 311 |
| Imports from elsewhere | 187 | 2,656 |
| Agri-food imports from FTAA | 1,531 | 1,260 |
| Agri-food imports from elsewhere | −147 | 2,040 |
| *Andean Pact* | | |
| Imports from FTAA | 1,700 | 1,342 |
| Imports from elsewhere | 277 | 597 |
| Agri-food imports from FTAA | 646 | 568 |
| Agri-food imports from elsewhere | 26 | 195 |
| *Chile* | | |
| Imports from FTAA | 844 | 738 |
| Imports from elsewhere | − 31 | −2 |
| Agri-food imports from FTAA | 68 | 49 |
| Agri-food imports from elsewhere | − 39 | − 61 |
| *Other Latin America & Caribbean* | | |
| Imports from FTAA | 6,901 | 5,729 |
| Imports from elsewhere | − 2,969 | − 2,296 |
| Agri-food imports from FTAA | 805 | 574 |
| Agri-food imports from elsewhere | −1,002 | − 347 |

occurs when the lowering of trade barriers within the region permits members to sell into previously protected neighboring markets with no effect on imports from nonmembers. The change in the value of imports from the baseline is reported in Table 15.2 for both scenarios for all goods in total, as well as for the agri-food commodities in total.

Looking at the results for the FTAA scenario (Column 2), it is clear that NAFTA, Chile, and the other Latin America and Caribbean (Other LAC) region have all shifted the sourcing of total imports, and agri-food imports toward the FTAA region and away from traditional sources outside of the region. MERCOSUR has also clearly diverted sourcing of agri-food imports from outside the region to FTAA partner sources under the FTAA scenario.

In addition to the strong evidence of trade diversion, some trade creation is also apparent as a result of the creation of the FTAA. In the case of NAFTA, total imports from within the region increase by nearly twice as much as the decline in imports from outside the region. The Other LAC aggregate also experiences some trade creation in total imports, while Chile is the only region to experience trade creation in both total imports and agri-food imports.

When a new WTO agreement is also simulated with the FTAA (Column 3), the degree of trade diversion from the regions outside the Western Hemisphere is reduced. Indeed, in the case of MERCOSUR, the introduction of multilateral liberalization is sufficient to reverse the switch in import sourcing toward the Western Hemisphere that occurred under the FTAA agreement. Some trade-diverting effects of the FTAA are still apparent, however, for the regions of NAFTA, Chile, and Other LAC for both total imports and agri-food imports.

Table 15.3 details the impacts on the agri-food sector of both scenarios. Percentage changes in aggregate agri-food output, exports, and imports are reported. The FTAA results in a very small expansion of agri-food output in all member regions, except Chile,[6] while agri-food output in the regions outside the Western Hemisphere falls slightly.

Under Scenario 1, agri-food exports increase for all regions within the FTAA while they decline for nonmembers. Agri-food imports increase for some members of the FTAA, although they do fall for Other LAC, with NAFTA also experiencing a very slight decline. Imports of agri-food commodities also fall for regions outside of the FTAA.

When a new multilateral trade deal is implemented along with the FTAA (Scenario 2), the agri-food output results for NAFTA, Andean Pact, and Chile change little from the FTAA results. MERCOSUR and Other LAC have much larger increases in output with multilateral trade liberalization, and non-FTAA regions now experience an increase in their agri-food output.

TABLE 15.3. Impacts of the FTAA on the Agri-Food Sector with and Without Multilateral Liberalization (Percentage Change from Base)

| | Scenario 1:<br>FTAA | Scenario 2:<br>New WTO Round with FTAA |
|---|---|---|
| *NAFTA* | | |
| Output | 0.3 | 0.1 |
| Exports | 0.6 | 1.7 |
| Imports | − 0.03 | 0.4 |
| *MERCOSUR* | | |
| Output | 0.7 | 2.4 |
| Exports | 5.7 | 13.9 |
| Imports | 3.8 | 9.0 |
| *Andean Pact* | | |
| Output | 0.5 | 0.5 |
| Exports | 9.6 | 11.3 |
| Imports | 6.6 | 7.5 |
| *Chile* | | |
| Output | − 0.1 | − 0.1 |
| Exports | 2.0 | 2.3 |
| Imports | 0.6 | − 0.2 |
| *Other LAC* | | |
| Output | 1.6 | 3.4 |
| Exports | 3.0 | 7.5 |
| Imports | −1.2 | 1.4 |
| *Non-FTAA* | | |
| Output | − 0.1 | 0.2 |
| Exports | − 0.3 | 5.1 |
| Imports | − 0.3 | 2.8 |

Multilateral liberalization combined with the FTAA improves agri-food export prospects for all of the regions in the model including the non-FTAA region. Imports rise more under multilateral liberalization for MERCOSUR and Andean Pact; while for NAFTA, Other LAC, and the non-FTAA region (all of which experienced decreases under Scenario 1), imports now increase. Only Chile experiences a decline in imports relative to the base, and in contrast to the increase under the FTAA scenario.

Table 15.4 details the welfare impacts of both scenarios, reporting the equivalent variation (EV)[7] measure of welfare change for each region in the model. These welfare changes are driven by the impact of tariff liberalization on regional terms of trade and allocative efficiency.[8] A positive terms of trade effect indicates an improvement in a region's terms of trade. The formation of a regional trade agreement generally improves the terms of trade of its members and lowers them for nonmembers (Burfisher, 1998). Since the GTAP model and data feature many preexisting distortions or taxes, resources are not allocated efficiently and, as such, there are second-best effects associated with any policy change.

The impact of the creation of the FTAA is presented in Column 2. All of the regions in the Western Hemisphere experience a welfare gain with the exception of the Andean Pact, which experiences a small loss due to a deterioration in its terms of trade. For the regions outside of the FTAA, namely the European Union, Oceania, and the Rest of World aggregate, the Western Hemisphere trade agreement results in a welfare loss. Not only do all of these regions experience a deterioration in their terms of trade, but they also experience allocative efficiency losses since they do not liberalize any of their trade and domestic distortions in this scenario.

The third column of Table 15.4 presents the welfare results from the implementation of a new multilateral trade round at the same time as the for-

TABLE 15.4. Welfare Effects of the FTAA with and Without Multilateral Trade Liberalization[a] (Million 1995 US $)

| Region | Scenario 1: FTAA | Scenario 2: New WTO Round with FTAA |
|---|---|---|
| Canada | 144 | − 69 |
| USA | 5,404 | 3,957 |
| Mexico | 341 | 362 |
| MERCOSUR | 1,326 | 2,450 |
| Andean Pact | − 28 | −10 |
| Chile | 35 | 44 |
| Other LAC | 2,012 | 2,143 |
| European Union | − 2,425 | 3,324 |
| Oceania | −85 | − 313 |
| Rest of World | − 4,147 | 9,342 |
| TOTAL | 2,578 | 21,230 |

[a]Welfare effects are represented by the equivalent variation.

mation of the FTAA. In this instance, both Canada and the Andean Pact experience welfare losses. Although both of these regions experience larger efficiency gains as a result of the FTAA liberalization and the additional multilateral liberalization, these are offset by terms of trade losses.[9]

Mexico, MERCOSUR, Chile, and the Other LAC region all experience a larger welfare gain than under the FTAA scenario. The United States still experiences a welfare gain from this scenario, but it is smaller than that enjoyed under the FTAA scenario. Now that they are liberalizing, the EU and Rest of World region both experience a welfare gain from the new multilateral trade round. Although both regions still experience terms of trade losses, these are outweighed by gains in efficiency arising from their policy reforms. Oceania experiences a welfare loss that is larger than that associated with the FTAA alone. This result is driven by the fact that this region had the lowest degree of protection initially and its terms of trade losses are larger than under the FTAA scenario given the EU and Rest of World are also now liberalizing. Overall, global welfare gains are considerably higher under the multilateral liberalization scenario, although most of the additional gains arise in the European Union and Rest of World and not in the Western Hemisphere.

## CONCLUSION AND POLICY RECOMMENDATIONS

One of the main findings of this chapter is hardly surprising—the formation of the FTAA is beneficial to the majority of its members, while it creates welfare losses in regions outside the Western Hemisphere. Given this result, is the FTAA a worthwhile effort for countries such as the United States and Canada to pursue? Some argue that the formation of regional trade agreements such as the FTAA is a good thing in the absence of any potential multilateral trade liberalization, much like the current situation. Any liberalization is considered positive, and a step toward global free trade. As well, regional trade processes such as the FTAA are better able to address issues too complex for global negotiation, such as policy harmonization.

Others disagree with this line of argument since they feel that increased regionalism is detrimental to the multilateral trading system.[10] Certainly given the current trade environment, any resources expended on a regional effort such as the FTAA could hurt efforts underway at the WTO to promote a new, multilateral trade round. Opponents of regional trade liberalization efforts would go so far as to state that prominent members of the Western Hemisphere should be using their international clout to promote multilateral negotiations rather than focusing any efforts on the FTAA process.

The analysis in this chapter suggests that the formation of the FTAA results in a high degree of trade diversion away from lower cost, efficient producers outside of the region toward less efficient producers within the FTAA. As pointed out by Burfisher (1998), the costs of trade diversion are borne partially or in full by those outside of the regional trade agreement when their terms of trade deteriorate. For those interested in strengthening the multilateral trading system, this result for the FTAA is very unfavorable.

While the FTAA has a positive impact on agricultural production in the region, for many of the members, the expansion of this sector is shown to be greater with the inclusion of multilateral reforms. Furthermore, it is probably unrealistic to expect certain agricultural tariffs to be reduced dramatically (Canadian and U.S. dairy tariffs, for example), given the example set by NAFTA in this regard. The multilateral setting would be a much more effective venue for negotiating meaningful reductions in the level of agricultural support given major U.S. reluctance to proceed in this area without obtaining similar concessions from the EU.

The question remains—will the formation of the FTAA be a building block or a stumbling block to a new multilateral trade agreement? To a large extent, this depends upon whether the efforts countries such as the United States and Canada invest in the FTAA process help promote multilateral trade liberalization or draw resources away from the launch of a new trade round. In agri-food trade, the FTAA process relies heavily on negotiations at the WTO to address export subsidization and domestic support practices. However, negotiations on market access within the region do not share the same synergy with the multilateral process and clearly, as the results of this analysis suggest, the formation of such a large preferential trading area will injure the economies of all regions outside the Western Hemisphere.

The results of the second scenario demonstrate that the inclusion of multilateral trade liberalization mitigates some of the negative, trade-diverting impacts of the FTAA. However, the realization of a new multilateral trade round within the next five years will require a great deal of political will to gain momentum. So, if political resources and attention in the Western Hemisphere are focused regionally until 2005, a follow-up to the Uruguay Round could be more than a decade in the waiting. This fact likely placed pressure on all WTO members, particularly those outside the Western Hemisphere, to work hard toward another successful round of trade negotiations.

## NOTES

1. See Stout and Ugaz-Pereda (1998) for more detail on each of these trade arrangements.

2. GTAP is a publicly available model and global trade database that is frequently used to examine trade policy reforms. The model assumes perfectly com-

petitive behavior, constant returns to scale, goods differentiated by country of origin (the Armington assumption), and a static model closure (no capital accumulation or dynamics).

3. The projections do not involve endowment growth, so regional endowments of land, labor, and capital remain at their 1995 levels.

4. Most protection in the service sector comes in the form of non-tariff barriers that are not adequately captured by the GTAP Version 4 database. Given this, and the fact that the focus of this study is on agri-food, no service sector liberalization is undertaken.

5. Although the general equilibrium modeling exercise that generated the output for this chapter includes a more detailed set of results for all of the regions and commodities detailed in Table 15.1, these have been aggregated in some cases to produce the results presented here.

6. Although detailed simulation results are not reported in this chapter, manufacturing and services output expansion exceeds that of the agri-food sectors in all of the FTAA regions under the first scenario.

7. The Hicksian equivalent variation is a money metric measure of welfare change and is defined as the expenditure required at base prices to achieve the post-simulation level of utility.

8. The GTAP model provides for a decomposition of the regional welfare changes into terms of trade effects, and allocative efficiency effects (Huff and Hertel, 1996). These are not reported here, but Huff and Rude (2000) include a summary of these effects along with a detailed discussion by region and scenario.

9. In the case of Canada, the deterioration in terms of trade is driven by the automotive and other manufacturing sectors which face increased competitive pressures in export markets now that FTAA members are opening up their borders to trade from outside the region.

10. For a more comprehensive discussion of the arguments for and against regionalism and its relationship to the multilateral liberalization process, see Bhagwati and Panagariya (1996).

# REFERENCES

Bhagwati, Jagdish and Arvind Panagariya. (1996). "Preferential Trading Areas and Multilateralism: Strangers, Friends or Foes?" In Jagdish Bhagwati and Arvind Panagariya (Eds.), *The Economics of Preferential Trade Agreements* (pp. 1-78). Washington, DC: AEI Press.

*BRIDGES Weekly Trade News Digest.* (2001). 5(2) January 23.

Burfisher, Mary E. (1998). "The Economics of Regional Integration." In Mary E. Burfisher and Elizabeth A. Jones (Eds.), *Regional Trade Agreements and U.S. Agriculture*. Economic Research Service, U.S. Department of Agriculture. Agricultural Economic Report No. 771, Washington, DC.

Economic Research Service (ERS). (1998). *Free Trade in the Americas*. Situation and Outlook Series, WRS-98-1, U.S. Department of Agriculture, Washington, DC.

Francois, Joseph and Anna Strutt. (1999). "Post-Uruguay Round Tariff Vectors for GTAP Version 4." World Bank Mimeo, Washington, DC.

Hertel, Thomas W. (Ed.) (1997). *Global Trade Analysis.* New York, NY: Cambridge University Press.

Huff, Karen M. and Thomas W. Hertel. (1996). "Decomposing Welfare Changes in the GTAP Model." GTAP Technical Paper No. 5, Center for Global Trade Analysis, Purdue University, West Lafayette, IN.

Huff, Karen M. and James Rude. (2000). "Regional versus Multilateral Trade Arrangements: Which Way Should the Western Hemisphere Go on Trade?" Paper presented at the Global Agricultural Trade in the New Millennium Conference in New Orleans, LA, May.

Josling, Timothy (1998). *Agricultural Trade Policy: Completing the Reform.* Washington: Institute for International Economics.

McDougall, Robert A., Aziz Elbehri, and Truong P. Truong (Eds.). (1998). *Global Trade, Assistance, and Protection: The GTAP 4 Database.* Center for Global Trade Analysis, Purdue University, West Lafayette, IN.

Meilke, Karl and Karen M. Huff. (2000). "After Seattle: Where from Here." Paper presented at the Policy Disputes Information Consortium meeting in San Diego, CA, February.

Nagarajan, Nigel. (1999). "The Millennium Round: An Economic Appraisal." Economic Paper No. 139, European Commission, Brussels.

Podbury, Troy, Ivan Roberts, David Vanzetti, and Brian S. Fisher. (2000). "Increasing Benefits to Australia from WTO Agricultural Trade Liberalization." ABARE Conference Paper 2000.1, Canberra, Australia.

Reinert, Kenneth A. (2000). "Give Us Virtue, But Not Yet: Safeguard Actions Under the Agreement on Textiles and Clothing." *The World Economy* 23(1):25-55.

Stout, James V. and Julieta Ugaz-Pereda. (1998). "Western Hemisphere Trading Blocs and Tariff Barriers for U.S. Agricultural Exports." In Mary E. Burfisher and Elizabeth A. Jones (Eds.), *Regional Trade Agreements and U.S. Agriculture.* Economic Research Service, U.S. Department of Agriculture. Agricultural Economic Report No. 771, Washington, DC.

Chapter 16

# Regionalism and Trade Creation: The Case of NAFTA

Dragan Miljkovic
Rodney Paul

## *INTRODUCTION*

Preferential Trade Areas (PTA) were historically limited to arrangements within Western Europe, among developing countries, and trade preferences by developed to developing countries.[1] Because the developing-country arrangements, undertaken principally in Latin America and Africa, were largely ineffective and trade preferences by developed to developing countries were limited, effective PTAs were confined to the two arrangements in Western Europe: the European Community (EC) and the European Free Trade Area (EFTA). The limited role for PTAs meant that the architects of the global trading system, i.e., the General Agreement on Tariffs and Trade (GATT), did not have to fear that regional arrangements might undermine the multilateral process of trade liberalization (Panagariya, 1998).

The United States championed, until recently, a nondiscriminatory global trade regime based on the Most Favored Nation (MFN) clause in Article I of the GATT, which forbids member countries from pursuing discriminatory trade policies against one another. During 1980s the situation changed. The United States felt that the EC was stalling the multilateral trade negotiations process and decided that PTAs were the only means left for keeping the process of trade liberalization afloat. In 1985 the United States went on to conclude a PTA with Israel, and in 1989 a PTA with Canada (CUSTA). Although the Uruguay Round of GATT negotiations was launched in the meantime, the United States went ahead in 1994 with yet another PTA, this time jointly with Canada and Mexico (NAFTA).[2] With the Uruguay Round having been successfully concluded and the multilateral process working well, at least until the WTO conference in Seattle in late 1999, the original rationale for the U.S. pursuit of PTAs has disappeared. But what was originally viewed as a temporary diversion to force the EC to the negotiating table has turned into a race for securing preferentially the neighbors' markets for one's exports.

The effects of regional trade agreements on union members have been studied mostly in the context of Vinerian, i.e., trade creating versus trade diverting, static welfare analysis (e.g., Bhagwati, 1995; Bhagwati, Greenaway, and Panagariya, 1998; Bhagwati and Panagariya, 1996; Jacquemin and Sapir, 1991; Panagariya, 1996; Spilimbergo and Stein, 1996; Summers, 1991; Wonnacott and Lutz, 1989), the implications of differences in transport costs across potential union members (e.g., Krugman, 1991; Frankel, Stein, and Wei, 1995; Wei and Frankel, 1998), the implications of the rules of origin (e.g., Krishna and Krueger, 1995; Krueger, 1993), and nontraditional gains including guaranteed market access, shelter from contingent protection, locking in the reforms, and dispute settlement (e.g., Panagariya, 1996, 1997). While measuring all of the above effects of a PTA on its member countries is impossible simply because some benefits/costs cannot be measured, it is reasonable to assume that creating a PTA is governed by an idea of easing trade barriers and increasing trade flows among the member countries. In other words, a country will enter a regional trade agreement having expectations of increased trade with its new union partners.

Creation of CUSTA and especially NAFTA was subject to bitter discussions and divisions among both politicians and economists in the United States. Both trade agreements were expected to create new trade between the United States and Canada, and between the United States and Canada and Mexico, respectively. The issue arose not about the increased trade among member countries but rather about welfare implications of that increase. Agriculture was one of the sectors in which there was considerable concern about potential effects of these agreements on domestic producers and consumers. No recent paper, however, addressed the question—has new trade in agricultural products been created at all?[3] We define trade creation in agricultural products as a statistically significant positive break in the trend function of the growth in exports and imports between member countries. We determine the time of break, providing one exists, in the post second oil-shock growth trend of real exports and imports between CUSTA and NAFTA member countries from the first quarter of 1980 through the second quarter of 1999, and document the scale of the phenomenon. We further discuss why there is or there is no trade creation in agricultural commodities caused by signing CUSTA or NAFTA in light of trade theory, and policy and political economy arguments.

## DETERMINATION OF THE BREAK PERIOD: METHOD

We utilize recent research on structural change in time-series econometrics that enables us to be explicit about the timing and the significance of the purported breaks. While earlier work imposed restrictive assumptions such as identically and independently distributed, nontrending, or stationary data, these restrictions have been successfully relaxed. The breaks in this study are determined using tests for detecting shifts in the trend function of a dynamic time series developed by Vogelsang (1997). These tests, which allow for serial correlation and have good finite sample power, remain valid whether or not the series is characterized by a unit root.

Therefore it is most useful to know whether such a break period (quarter) *i* even exists, and to the extent that it does, to determine when it occurs. This is done here for the total U.S. agricultural exports to Canada and to Mexico and the total U.S. agricultural imports from Canada and from Mexico. Furthermore, this procedure is repeated for a number of agricultural products based on two-digit Standard Industrial Trade Classification (SITC).[4] The determination of such date then facilitates a more accurate appraisal of the existence (absence) of trade creation caused by signing CUSTA or NAFTA or by some other events preceding or following the signing of these trade agreements.

We begin by examining total U.S. agricultural exports and imports to and from Canada and Mexico (in levels), which we define as the logarithm of real exports and imports. Since exports and imports are clearly trending, structural change involves a break in the linear deterministic trend. The Vogelsang (1997) tests, as we mentioned before, which will be used to determine the existence and timing of the trend breaks, are valid whether or not a unit root is present in a series. The critical values, however, depend on whether the series is stationary or contains a unit root. Therefore, the unit root question must be resolved first, and then the focus can shift to an investigation of trend breaks.

It is by now well known the nonrejection of the unit-root hypothesis can be caused by misspecification of the deterministic trend. Perron (1989) developed tests for unit roots which extend the standard Dickey-Fuller procedure by adding dummy variables for different intercepts and slopes, assuming that the break dates are known a priori. These tests were extended by Banerjee, Lumsdaine, and Stock (1992), Zivot and Andrews (1992), and Andrews (1993) to the case of unknown break dates.

We use a variant of these tests developed in Perron (1997) which allows both a change in the intercept and the slope at time $T_B$. The sequential trend break tests involve regressions of the following form:

$$\Delta X_t = \mu + \theta DU_t + \beta t + \gamma DT_t + \delta D(T_b)_t + \alpha X_{t-1} + \Sigma_{j=1}^{k} c_j \Delta X_{t-j} + \varepsilon_t$$

$$(16.1)$$

where $X$ is the log of real exports and $\Delta X$ is the first difference.[5] The period at which the change in the parameters of the trend function occurs will be referred to as the time of break, or $T_B$. The break dummy variables have the following values: $DU_t = 1$ if $t > T_B$, 0 otherwise; $DT_t = t - T_B$ if $t > T_B$, 0 otherwise; and $D(T_b)_t = 1$ if $t = T_B + 1$, 0 otherwise. Equation (16.1) is estimated sequentially for $T_B = 2, .... T - 1$, where $T$ is the number of observations after adjusting for those "lost" by first-differencing and incorporating the lag length $k$.

The time of break for each series is selected by choosing the value of $T_B$ for which the Dickey-Fuller $t$-statistics (the absolute value of the $t$-statistic for $\alpha$) is maximized. The null hypothesis, that the series $\{X_t\}$ is an integrated process, is tested against the alternative hypothesis that $\{X_t\}$ is trend stationary with a one-time break in the trend function which occurs at an unknown time.

There is considerable evidence suggesting that data-dependent methods for selecting the value of the lag length $k$ are superior to making an a priori choice of a fixed $k$. We follow the procedure suggested by Campbell and Perron (1991) and Ng and Perron (1995) by starting with an upper bound of $k_{max}$ on $k$. If the last lag included in equation (16.1) is significant, then the choice of $k$ is $k_{max}$. If the lag is not significant, then $k$ is reduced by 1. This process continues until the last lag becomes significant and $k$ is determined. If no lags are significant, then $k$ is set to 0. $k_{max}$ is initially set at 16, and the 10 percent value of the asymptotic normal distribution is used to assess the significance of the last lag.[6]

The null hypothesis of a unit root is rejected if the $t$-statistic for $\alpha$ is greater (in absolute value) than the appropriate critical value. Perron (1997, p. 362) provides finite-sample critical values for the lag length selection method described. The unit root null can be rejected for the U.S. total agricultural exports to Canada at 1 percent significance level, while for the other series the null cannot be rejected at any standard significance levels (i.e., 1, 5 or 10 percent). This finding is reported in the first column of Table 16.1.

We now proceed to test for structural change. Among all the tests that Vogelsang (1997) develops in his paper, only the sup Wald (or sup $F_t$) test provides estimates of the break date. Vogelsang (1997) extends the sup Wald test of Andrews (1993) and the mean and exponential Wald tests of Andrews and Ploberger (1994) to permit trending regressors and unit-root errors. The test for trending data consists of estimating the following equation:

$$X_t = \mu + \theta DU_t + \beta t + \gamma DT_t + \Sigma_{j=1}^k c_j X_{t-j} + \varepsilon_t. \tag{16.2}$$

Equation (16.2) is estimated sequentially for each break quarter with 15 percent trimming, i.e., for $0.15T < T_B < 0.85T$, where $T$ is the number of observations.[7] Sup $F_t$ is the maximum, over all possible trend breaks, of two times the standard $F$-statistic for testing $\theta = \gamma = 0$. It is important to understand that the break periods are determined endogenously, with no ex ante preferences given to any particular quarter.[8]

The results of the sup $F_t$ tests are summarized in Table 16.1. As indicated, Vogelsang (1997) tabulates critical values for both stationary and unit-root series. We use the stationary critical values for those countries for which the unit-root null can be rejected at the 10 percent level by the Perron (1997) tests, and the unit-root critical values otherwise. The no-trend-break null hypothesis is rejected in favor of the broken trend alternative for U.S. agricultural exports to Canada, only at 1 percent significance level. The parameter $\theta$ is positive, confirming the evidence of an increase of U.S. agricultural exports to Canada after the break.[9]

Vogelsang (1997) shows that if a series contains a unit root, power can be improved by conducting tests in first differences. We therefore proceed to examine exports (imports) growth (the first difference of the logarithm of real exports and imports) for the three series for which the unit-root null cannot be rejected by the Perron (1997) tests.[10] Since export (import) growth is nontrending, structural change involves a break in the mean of the

TABLE 16.1. Sequential Trend Break Tests (Levels)—Total Agricultural Exports and Imports

| $X_t = \mu + \theta DU_t + \beta t + \gamma DT_t + \Sigma_{j=1}^k c_j X_{t-j} + \varepsilon_t$ | | | |
|---|---|---|---|
| Country | Unit Root | Break Year | Sup$F_T$ |
| U.S. Exports to Canada | $-6.61$^^^ | 1989:3 | 21.40*** |
| U.S. Imports from Canada | $-3.41$ | 1989:1 | 8.66 |
| U.S. Exports to Mexico | $-3.76$ | 1989:1 | 5.38 |
| U.S. Imports from Mexico | $-3.46$ | 1990:3 | 7.76 |

*Note:* ^^^ denotes statistical significance for unit root test (Perron, 1997, p. 362) at 1 percent level (critical values are –6.07, –5.33, and –4.94 at 1, 5, and 10 percent, respectively). ***, **, and * denote statistical significance either using stationary critical values at the 1 percent (17.51), 5 percent (13.29), and 10 percent (11.25) levels or using unit root critical values at the 1 percent (30.36), 5 percent (25.10), and 10 percent (22.29) levels (Vogelsang, 1997, pp. 824-825).

growth rate. This is done by using the sup $F_T$ test for nontrending data, which consists of estimating the following equation:

$$\Delta X_t = \mu + \theta DU_t + \Sigma_{j=1}^{k} c_j X_{t-j} + \varepsilon_t. \qquad (16.3)$$

Equation (16.3) is estimated sequentially for each break period with 15 percent trimming, and sup $F_T$ is the maximum, over all possible trend breaks, of the standard $F$-statistic for testing $\theta = 0$.

The results of the sup $F_T$ tests are summarized in Table 16.2. Assuming that output contains at most one unit root, output exports (imports) growth will not contain a unit root, and stationary critical values can be used. The no-trend-break null hypothesis could not be rejected in any of the three remaining cases at any standard level of significance (1, 5, and 10 percent).

## WHAT HAPPENED TO TRADE CREATION?

Our results clearly indicate that there was no statistically significant break and increase in trade in three out of four cases tested. Then whatever in the world happened with the trade flows between the United States and Canada and Mexico? Viner (1950) noted that since PTAs liberalize trade preferentially, they create new trade between union members. We can see, however, that the U.S. agricultural imports from Canada and both exports and imports to and from Mexico follow the same (positive) trend over the period under consideration and that no new trade was created due to CUSTA or NAFTA.

First we should note that we ran the same model for the total U.S. agricultural exports and imports during the same period and that results indicate that there was no statistically significant break in trend of agricultural ex-

TABLE 16.2. Sequential Trend Break Tests (First Differences)—Total Agricultural Exports and Imports

| $\Delta X_t = \mu + \theta DU_t + \Sigma_{j=1}^{k} c_j X_{t-j} + \varepsilon_t$ | | |
|---|---|---|
| Country | Break Year | Sup$F_T$ |
| U.S. Imports from Canada | 1991:I | 6.73 |
| U.S. Exports to Mexico | 1988:I | 1.52 |
| U.S. Imports from Mexico | 1989:I | 0.18 |

*Note:* Stationary critical values are 13.02, 9.00, and 7.32 at 1, 5, and 10 percent respectively (Vogelsang, 1997, p. 824).

ports or imports. Sup $F_T$ test results for agricultural exports and imports in first differences (both series have a unit root in levels) are 3.007 and 5.357, respectively. Therefore our results, reported in Table 16.1 and Table 16.2, are not affected significantly by some global movement such as GATT or WTO agreements.

Opponents of NAFTA in the United States suggested that Mexico will be the major beneficiary of the agreement for several reasons. First, NAFTA would guarantee to Mexico access to the large U.S. market, i.e., the agreement assures Mexico that if the United States becomes protectionist in the future, its access to the U.S. market will be preserved. It seems, however, that Mexico has the access to the U.S. market guaranteed by WTO agreements anyway. In other words, if U.S. commitment to WTO is credible, this argument is rather weak.

Second, it was believed that Mexico may escape antidumping and safeguards measures by the United States to which other trading partners can be subjected. In practice, however, Mexico experienced administered protection by the United States in a number of cases affecting Mexican exports, actual or potential, to the United States. Examples relevant for agriculture are the special agreements on sugar and orange juice or subsequent restrictions on Mexican tomatoes that allow for the play of administered protection in the event of import surges from Mexico in these sectors. Sup $F_T$ test results for the value of U.S. imports of fruits (1.238) from Mexico indicate that there was no statistically significant break in trend. Sup $F_T$ test result for the value of U.S. imports of sugar and vegetables from Mexico (25.976 and 9.298, respectively) suggest the existence of a break in trend. These breaks in trend occurred, however, during the third quarter of 1989 and 1988, respectively, i.e., long before the NAFTA was signed. Results for fruits, sugar, and vegetables are in first differences due to the existence of a unit root in levels of these series. Also, side agreements on environment and labor standards give the United States new powers to subject Mexico to dispute settlement procedures that can lead to fines of up to 20 million U.S. dollars. Results for all agricultural commodities imported to the United States from Mexico are in accord with the two-digit SITC classification (Table 16.3).

Many believed that NAFTA will lock the reforms in Mexico, making it very difficult for more protection-minded future governments to reverse actions of their predecessors. This means, among other things, that in the case of NAFTA there is a more effective dispute settlement process available to private parties such as business groups and activists or labor unions. The WTO dispute settlement process, by contrast, is available to the governments of member countries only. However, shortly after signing NAFTA it became obvious that it is impossible to lock in all reforms in Mexico. The peso crisis showed that NAFTA does not and cannot guarantee macroeconomic stability. Of course, this crisis became a major concern to U.S. ex-

TABLE 16.3. Sequential Trend Break Tests (Levels and First Differences)—U.S. Agricultural Imports from Mexico Using Two-Digit SITC System

| Product | Unit Root | Break Year | SupF_T | First Diff. SupF_T |
|---|---|---|---|---|
| Animals | $-5.356^{\wedge\wedge}$ | 1988:1 | 5.918 | |
| Grains | $-4.493$ | 1986:3 | 4.001 | 1.230 |
| Fruits | $-3.075$ | 1986:3 | 3.632 | 1.238 |
| Nuts | $-5.281^{\wedge}$ | 1983:3 | 33.451*** | |
| Oilseeds | $-7.905^{\wedge\wedge\wedge}$ | 1988:3 | 3.200 | |
| Vegetables | $-3.291$ | 1988:3 | 2.319 | 9.298** |
| Essential Oils | $-7.815^{\wedge\wedge\wedge}$ | 1988:3 | 1.636 | |
| Seeds | $-6.150^{\wedge\wedge\wedge}$ | 1988:3 | 0.523 | |
| Sugar | $-4.329$ | 1989:3 | 8.664 | 25.976*** |
| Other | $-6.068^{\wedge\wedge}$ | 1988:4 | 33.293*** | |
| Nursery Prod. | $-4.058$ | 1983:1 | 15.348** | 9.356** |
| Beverages | $-3.734$ | 1988:3 | 10.361 | 7.011 |

Note: $^{\wedge\wedge\wedge}$ denotes statistical significance for unit root test (Perron, 1997, p. 362) at 1 percent level (critical values are −6.07, −5.33, and −4.94 at 1, 5, and 10 percent, respectively). ***, **, and * denote statistical significance either using stationary critical values at the 1 percent (17.51), 5 percent (13.29), and 10 percent (11.25) levels or using unit root critical values at the 1 percent (30.36), 5 percent (25.10), and 10 percent (22.29) levels (Vogelsang, 1997, pp. 824-825). ***, **, and * also denote statistical significance using stationary critical values at the 1 percent (13.02), 5 percent (9.00), and 10 percent (7.32) in first differences (Vogelsang, 1997, p. 824).

porters and also brought some smiles to the faces of those who lobbied for introducing the side agreements to protect certain sectors from import surges from Mexico (e.g., sugar or orange juice). The main area where the lock-in argument may apply is trade policy. However, Mexico could have as easily locked in its trade reforms on a multilateral basis by binding tariffs with the WTO at the applied rates. Instead, it chose to bind tariffs at levels much higher than applied rates. Recognizing that it will not be feasible to raise tariffs on the bulk of imports coming from the United States, Mexican authorities may have decided to leave themselves considerable room in the choice of external tariffs in case pressures from domestic industry necessitate a rolling back of trade liberalization. As it turned out, this flexibility was used after the peso crisis with tariffs on 503 items rising from less than 20 percent to 35 percent (Panagariya, 1998). As for the argument of the dispute settlement process being more effective because of its availability to private parties, Levy (1997) shows that the access of private parties can lead governments not to sign agreements that are otherwise beneficial.

Finally, it is known that in unions between a high-tariff country such as Mexico and a low-tariff country such as the United States, losses from the PTA to the former can be considerable. In other words, if a country forms a PTA with another country with substantially lower tariffs than its own, its losses are larger the more it imports from the partner. Panagariya (1997) estimated that the redistributive effect due to NAFTA may be costing Mexico as much as 3.25 billion U.S. dollars per year.

When we summarize, it is clear that there was no trade creation in agricultural products due to NAFTA between the United States and Mexico. It seems that the United States recognized the potential pitfalls of this agreement for its agriculture and protected itself well in advance. While low labor cost, macroeconomic instability, and the peso crisis caused a great deal of nervousness in U.S. agricultural producers, side agreements were designed to protect them from such occurrences. And while Mexico had to rely heavily on the United States as the market for its agricultural products, the United States could maintain its rather marginal trade flows with Mexico or turn to a number of larger and more stable markets.[11] In terms of never-realized opportunity benefits, it seems that Mexico is on the losing side of the deal. Notice, however, that the United States tremendously benefited from NAFTA considering the entire economy: "NAFTA amounted to a 4 percent expansion of the American economy, to include a country that accepted virtually every demand placed upon it in the negotiations and which made virtually all the concessions" (Bergsten, 1997, p. 26). It seems that the deep integration[12] of NAFTA led to a greater payoff to the side with more bargaining power, in this case the United States.

In the case of trade creation between the United States and Canada, the situation is somewhat different. First, there was trade creation in agricultural commodities going one way only, from the United States to Canada. Canada represents the third largest export market for U.S. agricultural products (Japan and the rest of Pacific Rim are the two largest export markets). After the signing of CUSTA,[13] more than half of Canadian agricultural imports were coming from the United States. And although many of the same issues in the Mexican case apply to the Canadian case, major differences lie in the internal stability of Canadian economy and its high national income (actual and potential). Canada has not experienced in its recent history any macroeconomic crisis or breakdown as serious as the ones in Mexico. Its currency has been relatively stable for a long time. Also, national income and market potential in that sense are much higher in Canada than in Mexico. This was a perfect situation for generally risk-averse U.S. agricultural producers and exporters to use the advantages that a PTA brings.

On the other hand, Canadian agricultural exports in the post-CUSTA era continued to follow the same pre-CUSTA trend, i.e., no structural break occurred. Approximately 10 percent of U.S. agricultural imports come from Canada. There might be several possible answers to the question why there was no trade creation from Canada to the United States, and we will offer two of them. One answer might be that some of the major Canadian agricultural exporting commodities such as wheat or barley are produced and exported from the United States as well. The reason for that might be similar climate and quality of soil. These commodities are produced mostly in the provinces of Alberta, Saskatchewan, and Manitoba, which border the states of Montana, North Dakota, and Minnesota, the largest producing states of these commodities. Another reason might be differing food safety standards and the unresolved issue of technical barriers, primarily sanitary and phytosanitary (SPS) measures. This may be most relevant in trade of meats. Notice that this is a very sensitive political issue: "Sovereignty over food safety is a very sensitive political issue. Breakdowns in food safety tend to become extremely emotive consumer issues over which politicians feel particularly vulnerable" (Kerr, 1999, p. 1, Chapter 9).

## REGIONALISM, REGIONALIZATION, AND NAFTA: CONCLUDING REMARKS

The previous discussion illustrates the difference between regionalism and regionalization, which are often believed to represent the same thing. Regionalism is the notion of specific regional policies enacted by governments to promote trade and economic integration. Regionalization is the notion of an increasing share of intraregional international trade arising due to "natural" economic and market forces. NAFTA should be listed under the category of regionalism (Baier and Bergstrand, 1997). The important thing is that regionalism does not imply regionalization. Regionalism is most often politically motivated, disregarding completely or partially economic incentives that agents in an economy may have.

Any concern that agricultural producers might have had in North America due to NAFTA or CUSTA is only partially justified. U.S. agricultural producers are well protected from potential adversity caused by import surges from Mexico via a number of side agreements, while they benefited directly from the expanded exports to Canada. On the other hand, potential benefits to agricultural producers in Mexico and Canada via trade creation were never realized.

The primary difference between NAFTA and other PTAs is that the United States clearly determined outcomes (trading rules) in NAFTA. Other unions, such as the European Union (EU), have a large number of members where no single member dominates the rest of the union. There the rules are set, in most cases, in a way that accommodates more the process of regionalization itself rather than any single country in particular. It will be interesting to see how U.S. engagement and performance in NAFTA will be perceived by the EU countries, Japan, and possibly China, as major players in future multilateral trade negotiations.

## NOTES

1. A PTA is a union between two or more countries in which goods produced within the union are subject to lower trade barriers than the goods produced outside the union. A Free Trade Area (FTA) is a PTA in which member countries do not impose any trade barriers on goods produced within the union but do so on those produced outside the union.

2. NAFTA stands for the North American Free Trade Agreement. However, NAFTA is not a nondiscriminatory free trade area. For instance, in NAFTA, anti-dumping measures can be used by member countries against one another.

3. Papers by Karamera and Koo (1994), and Uhm and Koo (1990) addressed the issue of trade creation and diversion (defined in Vinerian fashion) of agricultural and industrial goods in CUSTA.

4. Quarterly data on U.S. agricultural exports and imports to and from Canada and Mexico for period 1980:I through 1999:II are obtained from the USDA ERS as a courtesy of Carolyn Whitton. Nominal values are deflated by using the US CPI from the International Financial Statistics of the International Monetary Fund.

5. $M$ would denote the log of real imports in an equivalent formulation for the imports.

6. Ng and Perron (1995) use simulations to show that these sequential tests have an advantage over information-based methods since the former produces tests with more robust size properties without much loss of power.

7. Vogelsang (1997) reports critical values for both 1 percent and 15 percent trimming. The 15 percent trimming was used here because it has greater power to detect breaks near the middle of the sample.

8. These tests allow for only one break. While it would be desirable to use the methods developed by Bai and Perron (1998) to investigate multiple structural changes, relatively short time span of data (seventy-eight observations) makes this problematic. In addition, their tests are restricted to stationary and nontrending data.

9. Note that negative value of parameter $\theta$ would indicate a slowdown in U.S. agricultural exports after the break.

10. We do not perform a structural change test on the series for which the unit root null is rejected because, if a series is trend stationary with a break in trend, the tests for structural change have no local asymptotic power.

11. To clarify this point, one should know that about half or more of agricultural imports into Mexico were coming from the United States; that represents between 5

and 8 percent of the total U.S. agricultural exports. By the same token, less than 10 percent of agricultural imports come to the United States from Mexico.

12. Deep integration is a term introduced by Lawrence (1997). In addition to the liberalization of trade among members, it involves coordination, or sometimes complete harmonization, of other policies including competition policies, product standards, regulatory regimes, environmental policies, labor standards, investment codes, and so on.

13. Notice that NAFTA did not change significantly the relationship established between the United States and Canada by CUSTA. That is why we refer to CUSTA primarily when talking about the U.S.-Canada trading relationship.

# REFERENCES

Andrews, D. (1993). "Tests for Parameter Instability and Structural Change with Unknown Change Point." *Econometrica* 61 (July): 821-856.

Andrews, D. and W. Ploberger. (1994). "Optimal Tests When a Nuisance Parameter Is Present Only Under the Alternative." *Econometrica* 62 (November): 1383-1414.

Bai, J. and P. Perron. (1998). "Testing for and Estimation of Multiple Structural Changes." *Econometrica* 66 (July): 817-858.

Baier, S.L. and J.H. Bergstrand. (1997). "International Trade, Regional Free Trade Agreements, and Economic Development." *Review of Development Economics* 1 (June): 153-170.

Banerjee, A., R. Lumsdaine, and J. Stock. (1992). "Recursive and Sequential Tests of the Unit Root and Trend Break Hypothses: Theory and International Evidence." *Journal of Business and Economic Statistics* 10 (July): 271-287.

Bergsten, F. (1997). "American Politics, Global Trade." *Economist* (September 27): 23-26.

Bhagwati, J. (1995). "U.S. Trade Policy: The Infatuation with Free Trade Areas." In *The Dangerous Drift to Preferential Trade Agreements*, J. Bhagwati and A.O. Krueger (Eds.), pp. 1-18, Washington, DC: American Enterprise Institute for Public Policy Research.

Bhagwati, J., D. Greenaway, and A. Panagariya. (1998). "Trading Preferentially: Theory and Policy." *Economic Journal* 108: 1128-1148.

Bhagwati, J. and A. Panagariya. (1996). "The Theory of Preferential Trade Agreements: Historical Evolution and Current Trends." *American Economic Review* 86 (May): 82-87.

Campbell, J.Y. and P. Perron. (1991). "Pitfalls and Opportunities: What Macroeconomists Should Know About Unit Roots." *NBER Macroeconomic Annual* 6: 141-201.

Frankel, J., E. Stein, and S. Wei. (1995). "Trading Blocs and the Americas: The Natural, the Unnatural and the Supernatural." *Journal of Development Economics* 47: 61-96.

Jacquemin, A. and A. Sapir. (1991). "Europe Post-1992: Internal and External Liberalization." *American Economic Review* 81 (May): 166-170.

Karamera, D. and W. Koo. (1994). "Trade Creation and Diversion Effects of U.S.-Canada Free Trade Agreement." *Contemporary Economic Policy* 12(4): 12-45.

Krishna, K. and A.O. Krueger. (1995)."Implementing Free Trade Areas: Rules of Origin and Hidden Protection." In *New Directions in Trade Theory*, A. Deardorff, J. Levinsohn, and R. Stern (Eds.), pp. 149-187, Ann Arbor: University of Michigan Press.

Krueger, A.O. (1993). "Free Trade Agreements as Protectionist Devices: Rules of Origin." *NBER Working Paper No. 4352*, National Bureau of Economic Research, Cambridge, MA.

Krugman, P. (1991). "The Move to Free Trade Zones." *Policy Implications of Trade and Currency Zones*, Symposium Sponsored by the Federal Reserve Bank of Kansas City.

Lawrence, R. (1997). *Regionalism, Multilateralism and Deeper Integration*. Washington, DC: Brookings Institution.

Ng, S. and P. Perron. (1995)."Unit Root Tests in ARMA Models with Data Dependent Methods for the Selection of the Truncation Lag." *Journal of the American Statistical Association* 90 (March): 268-281.

Panagariya, A. (1996)."The Free Trade Area of the Americas: Good for Latin America?" *World Economy* 19 (September): 485-515.

Panagariya, A. (1997)."An Empirical Estimate of Static Welfare Losses to Mexico from NAFTA." Mimeo, Center for International Economics, University of Maryland, College Park, MD.

Panagariya, A. (1998). "The Regionalism Debate: An Overview." Presented at the International Agricultural Trade Consortium Annual Meeting, St. Petersburg, FL, December.

Perron, P. (1989). "The Great Crash, the Oil Price Shock, and the Unit Root Hypothesis." *Econometrica* 57 (November): 1361-1401.

Perron, P. (1997). "Further Evidence on Breaking Trend Functions in Macroeconomic Variables." *Journal of Econometrics* 80 (October): 355-385.

Spilimbergo, A. and E. Stein. (1996). "The Welfare Implications of Trading Blocs Among Countries with Different Endowments." *NBER Working Paper No. 5471*, National Bureau of Economic Research, Cambridge, MA.

Summers, L. (1991). "Regionalism and the World Trading System." *Policy Implications of Trade and Currency Zones*, Symposium Sponsored by the Federal Reserve Bank of Kansas City.

Uhm, I. and W. Koo. (1990). "The Effects of The Canadian-United States Free Trade Agreement on Bilateral Trade Flows of Agricultural and Industrial Products." *Canadian Journal of Agricultural Economics* 38 (December): 991-1004.

Viner, J. (1950). *The Customs Union Issue*. New York: Carnegie Endowment for International Peace.

Vogelsang, T. (1997). "Wald-Type Tests for Detecting Shifts in the Trend Function of a Dynamic Time Series." *Econometric Theory* 13 (December): 818-849.

Wei, Shang-Jin and Jeffrey A. Frankel (1998). "Open Regionalism in a World of Continental Trade Blocs." Staff Papers—International Monetary Fund 45(3): 440-485.

Wonnacott, P. and M. Lutz. (1989). "Is There a Case for Free Trade Areas." In *Free Trade Areas and U.S. Trade Policy,* J. Schott (Ed.), pp. 59-84, Washington, DC: Institute for International Economics.

Zivot, E. and D. Andrews. (1992). "Further Evidence on the Great Crash, the Oil Price Shock, and the Unit Root Hypothesis." *Journal of Business and Economic Statistics* 10 (July): 251-270.

Chapter 17

# Increased Use of Antidumping Weakens Global Trade Liberalization

Anita Regmi

## *INTRODUCTION*

Although countries have continued to move toward further trade liberalization, global trade flow has been increasingly hampered by the use of the WTO antidumping (AD) disciplines as a growing number of countries initiate AD investigations against their trade partners. While antidumping was initially used by only a few industrialized countries, AD use has grown significantly since the 1980s, with many developing countries following the example set by developed countries. Increased use of AD as a restrictive trade practice threatens to roll back many of the gains in trade liberalization made with the implementation of the Uruguay Round Agreement. Furthermore, the principle of broadly including unvalued cost factors in establishing the existence of dumping appears to favor increased use of this practice.

Globally, the number of countries using AD tripled between 1987 and 1997 from seven to twenty-two, with developing countries accounting for about half of the AD cases initiated in 1997, compared with only 16 percent a decade earlier. Data suggests that as economies evolve from a controlled to a more liberal trade regime, particularly those implementing unilateral reforms, the use of antidumping as a border measure rises. This has been illustrated by the increased use of antidumping by many middle-income countries, especially Latin American nations such as Mexico, Argentina, and Brazil.

Antidumping laws provide safety valves that enable governments facing domestic political pressure to maintain barriers to implement trade liberalization. However, many observers suggest that the current WTO AD provisions are too generous. The mechanism by which an antidumping duty is levied under WTO disciplines has been stated as being biased toward finding in favor of dumping and injury claimants. Many of the practices considered unfair in an antidumping investigation would be ruled fair in domestic commerce.

The main problems associated with the current WTO AD provisions arise through the mechanism by which the fair value of a product is determined. WTO disciplines allow the use of a constructed value that incorporates adjustments for transportation, overhead, labor costs, and "normal" profits when actual data are not available. This does not reflect a common business practice of pricing below the average total cost in the short run during periods of market downturn, which is allowed under U.S. antitrust law. Similarly, injury determination in an AD investigation is based only on economic injury to the industry, often determined by whether there is actual or potential decline in output, sales, profits, market share, productivity, return on investments, or utilization capacity. Injury estimation ignores the benefits accruing to consumers and importing companies and therefore discounts the overall national welfare.[1]

In this chapter, WTO AD disciplines, their use by countries, and the implications for global agricultural trade will be discussed. In reviewing the AD rule, the chapter draws upon many of the previous studies conducted in this area. Finally, the implications for global trade will be discussed, taking into consideration the potential for further reforms in trade rules resulting from a new round of WTO negotiations.

## WHAT IS DUMPING?

According to Article VI, GATT 1994 (WTO, 1996), *dumping* is defined as the introduction of a product of one country into the commerce of another country at less than its *normal value*. A product is considered to be introduced at less than its normal value if the price of the product is less than the comparable price for the product, in the ordinary course of trade, when destined for domestic consumption in the exporting country. When such a price is not available, normal value may be computed using a comparable price for the product exported to any third country, or by considering the cost of production for the product taking into account additional selling expenses and profits. *AD duties can be levied* if it is determined that dumping causes or threatens to cause material injury to an established industry in the country or retards the establishment of a domestic industry. To offset the dumping, the importing country can levy AD duties not greater than the *margin of dumping* (difference between the export price and the normal value).

The WTO AD law requires that AD investigations be terminated if dumping margins are found to be *de minimis,* which is defined as less than 2 percent of the export price. Similarly, AD investigations are to be terminated if the volume of the dumped product from a country is less than 3 percent of the total imports for the product. However, the AD law allows the practice of cumulating imports from various countries when the particular group of countries is simultaneously subject to an AD investigation. In this case, the de minimis volume for the group collectively is 7 percent of the total imports of the like product.[2] Finally, the WTO AD law establishes five years as the duration of antidumping duties, unless a prior review determines that the expiry of the AD duties will lead to continuation or recurrence of dumping and injury.

U.S. AD law was changed under the Uruguay Round Agreements Acts (effective January 1, 1995) to make it consistent with WTO AD disciplines. According to the U.S. law, an AD duty shall be imposed, in addition to any other duty, if two conditions are met. First, the Department of Commerce (DOC) must determine that "a class or kind of foreign merchandise is being, or is likely to be, sold in the United States at less than its fair value" (USDOC, 2000). Second, the U.S. International Trade Commission (ITC) must determine that "an industry in the United States is materially injured, or is threatened with material injury, or the establishment of an industry in the United States is materially retarded, by reason of imports of that merchandise" (USDOC, 2000). If the two conditions are met, an AD order is issued imposing antidumping duties equal to the amount by which normal value exceeds the export price for the merchandise (the dumping margin). In the United States, AD investigation may be self-initiated by the DOC or may be initiated as a result of a petition filed by an interested party. Consistent with WTO disciplines, U.S. AD law requires DOC and ITC to conduct "sunset reviews" no later than five years after issuance of an AD order to determine whether revoking the order would likely lead to continuation or recurrence of dumping and material injury.

## USE OF ANTIDUMPING BY COUNTRIES

Although antidumping was initially used by only a few industrialized countries, AD use has grown significantly since the 1980s, with many developing countries following the example set by developed countries. The number of countries using AD tripled between 1987 and 1997 from seven to twenty-two with AD use by nontraditional users, such as Mexico, Argentina, Brazil, and Korea, increasing considerably between 1987-1997 (Table 17.1). Although the United States, Australia, and the European Union together account for over half of all AD cases (2,196) investigated during 1987 and 1997, AD use by traditional users has slowed down in recent years compared with the early 1990s. The decline in AD use by industrialized countries has been attributed to general economic prosperity and a growing realization that AD use had not served their national interest (Finger, 1998). Moreover, the United States and the European Union have faced increasing

TABLE 17.1. Antidumping Investigations by Reporting Country

| Traditional Users | 1987 | 1988 | 1989 | 1990 | 1991 | 1992 | 1993 | 1994 | 1995 | 1996 | 1997 | Total |
|---|---|---|---|---|---|---|---|---|---|---|---|---|
| Australia | 22 | 16 | 21 | 47 | 68 | 71 | 59 | 15 | 5 | 17 | 42 | 383 |
| Canada | 31 | 15 | 13 | 15 | 11 | 46 | 25 | 2 | 11 | 5 | 14 | 188 |
| EC | 28 | 27 | 18 | 48 | 29 | 42 | 21 | 43 | 33 | 25 | 41 | 355 |
| New Zealand | 0 | 9 | 1 | 1 | 9 | 14 | 0 | 6 | 10 | 4 | 5 | 59 |
| United States | 15 | 40 | 24 | 34 | 63 | 83 | 32 | 48 | 14 | 22 | 16 | 391 |
| Total | 96 | 107 | 77 | 145 | 180 | 256 | 137 | 114 | 73 | 73 | 118 | 1,376 |
| New Established Users | | | | | | | | | | | | |
| Argentina | 0 | 0 | 0 | 0 | 1 | 14 | 27 | 17 | 27 | 22 | 15 | 123 |
| Brazil | 0 | 1 | 1 | 2 | 7 | 9 | 34 | 9 | 5 | 18 | 11 | 97 |
| Colombia | 0 | 0 | 0 | 0 | 2 | 3 | 6 | 3 | 4 | 1 | 1 | 20 |
| Korea | 1 | 0 | 1 | 5 | 0 | 5 | 5 | 4 | 4 | 13 | 15 | 53 |
| Mexico | 18 | 11 | 7 | 11 | 9 | 26 | 70 | 22 | 4 | 4 | 6 | 188 |
| Poland | 0 | 0 | 0 | 0 | 24 | 0 | 0 | 0 | 0 | 0 | 1 | 25 |
| Total | 19 | 12 | 9 | 18 | 43 | 57 | 142 | 55 | 44 | 58 | 49 | 506 |
| Infrequent Developed Country Users | | | | | | | | | | | | |
| Austria | 0 | 0 | 0 | 0 | 0 | 5 | 4 | 0 | 0 | 0 | 0 | 9 |
| Finland | 5 | 5 | 2 | 0 | 1 | 0 | 0 | 0 | 0 | 0 | 0 | 13 |
| Japan | 0 | 0 | 0 | 0 | 3 | 0 | 0 | 1 | 0 | 0 | 0 | 4 |
| Singapore | 0 | 0 | 0 | 0 | 0 | 0 | 0 | 2 | 0 | 0 | 0 | 2 |

| | | | | | | | | | | | | |
|---|---|---|---|---|---|---|---|---|---|---|---|---|
| Sweden | 0 | 0 | 8 | 2 | 1 | 0 | 0 | 0 | 0 | 0 | 0 | 11 |
| Total | 5 | 5 | 10 | 2 | 5 | 5 | 4 | 3 | 0 | 0 | 0 | 39 |
| Emerging Developing Country Users | | | | | | | | | | | | |
| Chile | 0 | 0 | 0 | 0 | 0 | 0 | 1 | 1 | 4 | 3 | 0 | 9 |
| Costa Rica | 0 | 0 | 0 | 0 | 0 | 0 | 0 | 0 | 0 | 4 | 1 | 5 |
| Guatemala | 0 | 0 | 0 | 0 | 0 | 0 | 0 | 0 | 0 | 1 | 0 | 1 |
| India | 0 | 0 | 0 | 0 | 0 | 8 | 0 | 7 | 6 | 21 | 13 | 55 |
| Indonesia | 0 | 0 | 0 | 0 | 0 | 0 | 0 | 0 | 0 | 9 | 4 | 13 |
| Israel | 0 | 0 | 0 | 0 | 0 | 0 | 1 | 1 | 5 | 6 | 3 | 16 |
| Malaysia | 0 | 0 | 0 | 0 | 0 | 0 | 0 | 0 | 3 | 2 | 8 | 13 |
| Philippines | 0 | 0 | 0 | 0 | 0 | 0 | 1 | 7 | 0 | 1 | 2 | 11 |
| Peru | 0 | 0 | 0 | 0 | 0 | 0 | 0 | 3 | 2 | 7 | 2 | 14 |
| South Africa | 0 | 0 | 0 | 0 | 0 | 0 | 0 | 16 | 16 | 33 | 23 | 88 |
| Thailand | 0 | 0 | 0 | 0 | 0 | 0 | 3 | 0 | 0 | 1 | 2 | 6 |
| Turkey | 0 | 0 | 0 | 0 | 0 | 0 | 7 | 21 | 0 | 0 | 4 | 32 |
| Venezuela | 0 | 0 | 0 | 0 | 0 | 0 | 3 | 0 | 3 | 2 | 4 | 12 |
| Total | 0 | 0 | 0 | 0 | 0 | 8 | 16 | 56 | 39 | 90 | 66 | 275 |
| Total | 120 | 124 | 96 | 165 | 228 | 326 | 299 | 228 | 156 | 221 | 233 | 2,196 |

*Source:* Miranda, Torres, and Ruiz (1998).

pressure from domestic importing industries that pay the price of AD protection. The use by developing countries, on the other hand, is on the rise (Figure 17.1). Unless AD laws are amended in the next round of WTO negotiations, there appears to be a "race to the bottom"[3] as developing countries try to match the restrictions in place in developed countries.

Data suggests that as economies evolve from a controlled to a more liberal trade regime, particularly those implementing unilateral reforms, the use of antidumping as a border measure rises. This may be due to the fact that antidumping laws provide safety valves that enable governments to implement trade liberalization with less domestic political opposition. The knowledge that trade concessions resulting from trade agreements are in fact "escapable" may also facilitate trade negotiations (Skyes, 1991). As illustrated in the figure, developing country use of AD measure dramatically increased beginning in 1990. This coincides with major trade policy changes undertaken in developing countries, including those undertaken in Latin America to replace the import substitution policies that led to debt crises in the 1980s (Miranda, Torres, and Ruiz, 1998). In late 1980s and early 1990s, Mexico, Argentina, and Brazil undertook massive unilateral trade liberalization, significantly exceeding their GATT obligations. For example, in 1987, the year Mexico activated its AD system, maximum tariff levels in Mexico were reduced to 20 percent, well below the 50 percent bound rate under GATT. Initiation of a similar import liberalization policy in 1984 (Pyo, Kim, and Cheong, 1996) precedes Korea's aggressive AD use. Developing country use of AD measures received further boosts in the 1990s with India's movement toward freer trade in 1991 and with South Africa's

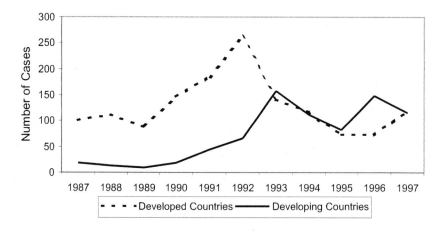

FIGURE 17.1. Antidumping investigations by countries. (*Source:* Miranda, Torres, and Ruiz, 1998.)

participation in the global economy since the dismantling of its apartheid regime. Although South Africa had previously never used AD measures, the number of AD investigations initiated between 1994 and 1997 by South Africa totaled eighty-etight. This accounts for almost 11 percent of the total number of investigations initiated during the same period. Despite the reforms implemented since 1991, India continues to restrict imports of a vast number of consumer goods, including many agricultural and food products. Given the recent WTO ruling against India, on a case brought up by the United States, the government of India is expected to accelerate the phaseout of all import restrictions.[4] Considering the pattern exhibited by developing countries, the resulting liberal import regime in India will likely lead to further expansion in its use of AD.

Comparison of the use of AD provisions on the basis of absolute number of investigations does not take into consideration the relative share of global trade of individual countries (Miranda, Torres, and Ruiz, 1998). Countries with a large share of global exports have a greater probability of having more AD investigations initiated against them. Therefore, Miranda, Torres, and Ruiz (1998) developed a trade-weighted index, which is a ratio of a country's share of total number of investigations to its share of global trade. Using this index, countries that use AD are compared based on the ratio of their share of the total number of AD investigations to their share of global imports (Table 17.2). Although the United States has been the most frequent user of AD in absolute numbers, based on the trade-weighted criterion it

TABLE 17.2. Top Ten Users of Antidumping

| Country | Number of Investigations | Trade-Weighted Ratio[1] | Trade-Weighted Ranking |
|---|---|---|---|
| United States | 391 | 0.7 | Australia |
| Australia | 383 | 8.3 | Argentina |
| European Community | 355 | 0.7 | New Zealand |
| Canada | 188 | 1.5 | South Africa |
| Mexico | 188 | 3.0 | Mexico |
| Argentina | 123 | 7.4 | Brazil |
| Brazil | 97 | 2.4 | Peru |
| South Africa | 88 | 4.2 | India |
| New Zealand | 59 | 5.7 | Costa Rica |
| India | 55 | 2.1 | Colombia |

*Source:* Miranda, Torres, and Ruiz, 1998.
[1] Trade-weighted ratio = share of total investigation/share of global import value

ranks nineteenth in its intensity of AD use. Using the trade-weighted index, Australia ranks as the most intensive user of AD, with its share of investigations eight times larger than its share of global imports.

Similarly, countries affected by AD investigations are compared based on the ratio of their share of total number of investigations and global exports. Although China has had the greatest number of AD investigations launched against it, when relative share of global exports are taken into consideration, China is the sixteenth most intensely investigated country. Likewise, the United States ranks as the second most frequently investigated country in absolute terms. However, when relative import share is taken into consideration, the United States ranks eighty-fifth among ninety-nine countries that were subject to AD investigations between 1987 and 1997. Using the trade-weighted criterion, Miranda, Torres, and Ruiz (1998) have demonstrated that the transition economies, followed by developing countries, are most intensively affected by AD investigations (Table 17.3). The ratio of the share of investigations to the share of exports was greater than one for forty-six out of ninety-nine countries. These were almost all transition and developing countries, while developed and oil-exporting countries generally had a ratio of less than one. This finding highlights the fact that developing countries are most intensely affected by AD investigations, which explains their desire to change the current provision to make it more difficult to impose AD duties.

TABLE 17.3. Top Ten Countries Most Affected by Antidumping

| Country | Number of Investigations | Trade-Weighted Ratio[1] | Trade-Weighted Ranking |
|---|---|---|---|
| China | 247 | 2.9 | Turkmenistan |
| United States | 188 | 0.5 | Mozambique |
| Korea | 139 | 1.9 | Kazakstan |
| Japan | 133 | 0.6 | Ukraine |
| Brazil | 105 | 4.0 | Belarus |
| Taiwan | 100 | 1.6 | Romania |
| Germany | 92 | 0.7 | Lithuania |
| Thailand | 62 | 2.0 | Latvia |
| India | 57 | 3.1 | Estonia |
| United Kingdom | 54 | 0.9 | Egypt |

*Source:* Miranda, Torres, and Ruiz, 1998.
[1]Trade-weighted ratio = share of total investigations/share of global export value

## EXAMINATION OF THE CURRENT AD LAW

Remedial trade measures such as AD are said to have been created to protect domestic industries from predatory pricing by foreign firms. However, various studies have indicated that the imposition of AD duties results in net welfare loss in the importing country (Gallaway, Blonigen, and Flynn 1999; Canning and Vroomen, 1996; Devault, 1996; USITC, 1995; Webb, 1992). In fact, WTO criteria for antidumping has often been criticized as economically baseless (Finger, 1998; Lindsey, 1999), and a 1996 OECD review indicated that about 90 percent of the instances of import sales found to be unfair under antidumping investigations would have been considered fair if practiced in domestic commerce. Using criteria developed in antitrust literature, Shin (1994) has illustrated that less than 10 percent of 282 U.S. AD cases examined indicated the possibilities of behavior consistent with dumping motivated by predatory pricing. Economists have also questioned whether the economic realities of a firm can allow dumping to be undertaken. Some have debated that the occurrence of long-term dumping is economically inconceivable since no firm can sustain selling below the cost of production for a long period, and that short-run intermittent dumping may not necessarily cause injury (Krishna, 1997). Finally, Prusa (1996) demonstrates that AD duties may not be the best tool to shield domestic producers from foreign imports. Due to trade diversion from third countries (whose products are not subject to AD duties) total imports may not decline after the imposition of AD duties. For example, in a U.S. AD case (USITC, 1995), although lamb meat prices increased 10 percent after the imposition of AD duties on lamb meat from New Zealand, its overall supply did not decrease with imports from Australia largely replacing those from New Zealand.

A major shortcoming of current WTO AD disciplines identified by previous studies is the mechanism by which the normal value of a product is determined. The rules of the current AD disciplines do not detail the circumstances under which comparison between exported and home market prices may take place, often resulting in comparison between apples and oranges. Products sold in home and foreign markets may differ on the basis of packaging, different commodity and currency market situations under which they are sold, and quality differences (Rowat, 1990; Lindsey, 1999). When no home market or third country prices are available, a constructed value can be used to determine the normal value of a product. However, a constructed value may also be used if these prices are deemed as being unrepresentative, allowing significant bias to occur in the investigative process. When calculating "constructed value," AD disciplines require that adjustments be made for transportation, overhead, labor costs, and normal profits. This may not reflect the reality that firms often base their price decisions on short-run variable costs. This is especially true for agriculture where the value of family land and labor may be difficult to incorporate when estimating the cost of production. In fact, pricing below the average cost is considered economically rational behavior in the short run during periods of market downturns. This pricing behavior is allowed under the U.S. domestic antitrust law, except when proven that low prices had a predatory purpose (Rowat, 1990).

In order for a country to implement AD duties, material injury to a domestic industry from dumping must be established in addition to the existence of dumping. Just like establishing the existence of dumping, WTO disciplines used in establishing the existence of injury are not definite regarding the criteria to be used. Injury to the domestic industry can be estimated based on whether there is actual or potential decline in output, sales, profits, market share, productivity, return on investments, or utilization capacity (Krishna, 1997). The definition of injury does not take into consideration the overall national welfare. Cheap imports capable of injuring specific import competing firms benefit the companies that use them and the consumers that buy and use products from these companies. These benefits accruing to importing companies and consumers are excluded from AD injury estimation. A study conducted by the USITC in 1991 stated that the economy-wide effects of a simultaneous removal of outstanding AD/CVD orders (AD comprised of 163 of 239 orders) would result in a welfare gain to the U.S. economy of $1.59 billion. Since this study did not incorporate the impacts of revoked, suspended, or terminated cases, the actual welfare gain from the removal of AD/CVD orders could be potentially higher than $1.59 billion. Individual case studies conducted by USITC demonstrate that imports can decline substantially following AD/CVD petition filing, and may also continue to decline after revoking of an AD/CVD order.

## AGRICULTURE AND ANTIDUMPING

AD investigations on agricultural products have a greater probability of dumping and injury finding due to the prevailing wider and more frequent price variations, especially among perishable products (University of Michigan, 1980). Agricultural prices vary not only annually and seasonally, but also weekly, daily, and even hourly. Since most agricultural products are harvested seasonally, producers experience seasonal high and low prices. Additionally, weather and other agronomic factors can affect regional yields, which in turn affect commodity prices. Given the length of time required to produce any agricultural products, most producers are unable to adjust their outputs to price changes. At least a year is required for producers to change hog production, three years to change beef supply, and five to ten years for growers to change production plans for tree crops (Tomek and

Robinson, 1981). Agricultural supply curve in the short run is therefore inelastic. Producers of storable commodities such as grains may be able to adjust supply to a certain extent depending on the prevailing storage cost, while suppliers of perishable products have little alternative other than selling at prevailing market price, which at times may be below cost.

Assuming perfectly competitive market conditions and profit-maximizing firm behavior, the supply curve of an agricultural commodity can be given by its marginal cost curve, $S$ in Figure 17.2. Under the assumption that the producer is a price taker, $Q$ level of product will be sold at market price $P_1$. However, if the market price falls to $P_2$, $Q_2$ level of product will be sold when the producer is able to adjust the level of firm output. This is unlikely in agriculture where short-run supply curves are inelastic and may be depicted by $QQ'$.[5] For a harvest-ready crop, a farmer will choose to sell $Q$ level of product if $P_2$ is greater than the variable cost of harvesting and marketing. This action will allow the farmer to recover at least some of the sunken expenses. Selling below cost of production, therefore, is often the economically optimal action for many farmers.

Thus, economic necessity and not predatory pricing behavior often lead to selling agricultural products below their cost of production. Although predatory pricing behavior cannot be ruled out, it is highly unlikely among

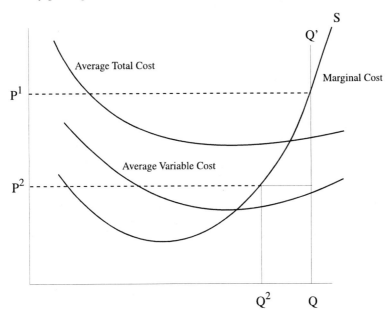

FIGURE 17.2. Agricultural commodity supply curve.

agricultural commodities. For successful predatory pricing, producers must be able to discriminate between market sectors, charging high prices in inelastic markets and lower prices in elastic markets. This may not be possible in agriculture, which is generally characterized by numerous producers with little or no control over product prices. In addition, to successfully discriminate between domestic and foreign markets, the seller must operate in a domestic market where foreign competition is restricted.

Specific problems arising due to the implementation of AD rules on agricultural products have been described by Palmeter (1989). Historically, AD cases on agricultural products have been initiated primarily against perishable fruit and vegetable products rather than on bulk commodities. Palmeter illustrates the problems encountered in estimating the fair value of products that are produced by many different producers, are highly perishable, and whose prices vary both seasonally and also within the course of a day depending on freshness. Using U.S. investigations, Palmeter highlights how the procedures used in establishing dumping and material injury have differed from one case to the other, giving credence to claims of being biased toward findings favorable to dumping and injury claimants. In a case involving vegetables from Mexico, home market price was estimated based on a sample of thirty-four growers representing approximately 15 percent of imports, and for a case involving potatoes from Canada, the sample included only ten growers representing an unknown share of total imports. In addition, DOC has used different pricing mechanisms in establishing dumping in cases involving perishable products. Typically DOC determined the fair value in the exporting country by calculating the weighted average home market price over a period of at least six months. Individual export prices were then compared to this weighted average price to determine dumping.[6] However, given the difficulty in estimating prices for agricultural products, DOC has used innovative methods on an ad hoc basis. These have included using the average export price over variable time periods (rather than individual price) for comparison with the fair value. Furthermore, price comparison has ranged over a wide period, including comparing the daily export price to the daily third-country price used to represent fair value (instead of weighted six-month average third country price [Palmeter, 1989]).

Antidumping cases on agriculture have also raised questions regarding the extent to which an industry is entitled to complain about imports that are further processed products of a domestically produced good. For example, the case brought up by U.S. citrus growers (Florida Citrus Mutual) against Brazilian frozen concentrated orange juice was opposed by several U.S. companies (including the National Juice Products Association, Procter & Gamble, Tropicana, and Coca-Cola) on the basis that Florida Citrus Mutual did not produce a like product and that a majority of the industry did not support the investigation (Braga and Silber, 1993). However, after six frozen concentrated orange juice processors were added to the petition, DOC found sufficient grounds to continue the investigation, stating that majority support in the domestic industry is not necessary for AD investigation.[7] This argument was later substantiated, with the court stating that the DOC was not "required to dismiss petitions that are not proven to be affirmatively supported by the domestic industry." This ruling raises questions about the economic justifications of imposing AD duties on a value-added product to protect a small segment of domestic raw material producers. Additionally, to what extent is a segment of agricultural industry such as grape producers more entitled to complain about imports of processed products such as raisins, jelly, or wine, than a nonagricultural industry such as producers of wire is to complain about imports of coat hangers (Palmeter, 1989).

## POTENTIAL IMPACTS ON FUTURE
## AGRICULTURAL TRADE

With increased use by developing countries, AD investigations on agricultural products are in danger of experiencing escalations. This is partly due to the fact that developing economies rely to a greater extent on agriculture than industrial economies do. Many of these countries, such as India and China, currently continue to restrict food and agricultural imports through very high tariffs, quantitative restrictions, licensing requirements, and parastatal import controls. As these countries further liberalize agricultural trade to meet their WTO obligations, they will likely increasingly resort to using AD measure to protect their domestic industries. This is already evident by the fact that while agriculture accounted for about 6 percent of the total number of AD cases launched between 1987 and 1997, it accounted for 10 percent of the total cases among the newly established AD users, including 96 percent of all cases for Poland (Table 17.4). Similarly, the share of agriculture in total AD cases investigated is high for some developed countries, such as New Zealand, Australia, and Canada, that are major agricultural exporters.[8]

Agriculture also remains especially vulnerable to AD investigations given the current rule by which the normal value of a product is calculated based on estimates of total production costs adjusted for marketing, handling, and an appropriate share of profits. This can prove to be particularly harmful to U.S. agriculture, which accounts for about 13 percent of global agricultural exports, and repeatedly suffers from very low market prices. Since many of the developing countries, especially Latin American, directly compete with the United States for agricultural markets, the United States

TABLE 17.4. Antidumping Investigations by Sector (1987-1997)

| | Agriculture Products | | Forest Products | | Textile & Footwear | | Base Metals | | Other Products | | Total | |
|---|---|---|---|---|---|---|---|---|---|---|---|---|
| | # | % | # | % | # | % | # | % | # | % | # | % |
| **Traditional Users** | | | | | | | | | | | | |
| Australia | 34 | 9 | 32 | 8 | 18 | 5 | 16 | 4 | 281 | 74 | 381 | 17 |
| Canada | 17 | 9 | 20 | 11 | 17 | 9 | 87 | 47 | 44 | 24 | 185 | 8 |
| EC | 5 | 1 | 14 | 4 | 63 | 18 | 85 | 24 | 188 | 53 | 355 | 16 |
| New Zealand | 13 | 22 | 0 | 0 | 10 | 17 | 4 | 7 | 32 | 54 | 59 | 3 |
| United States | 14 | 4 | 9 | 2 | 10 | 3 | 178 | 46 | 176 | 45 | 289 | 18 |
| Total | 83 | 6 | 75 | 5 | 118 | 9 | 370 | 27 | 723 | 53 | 1,369 | 100 |
| **New Established Users** | | | | | | | | | | | | |
| Argentina | 2 | 2 | 8 | 7 | 3 | 2 | 33 | 27 | 77 | 63 | 123 | 6 |
| Brazil | 13 | 13 | 0 | 0 | 7 | 7 | 24 | 25 | 53 | 55 | 97 | 4 |
| Colombia | 2 | 10 | 0 | 0 | 0 | 0 | 5 | 25 | 13 | 65 | 20 | 1 |
| Korea | 0 | 0 | 4 | 8 | 0 | 0 | 5 | 9 | 44 | 83 | 53 | 2 |
| Mexico | 8 | 4 | 6 | 3 | 23 | 14 | 64 | 34 | 84 | 45 | 188 | 9 |
| Poland | 24 | 96 | 0 | 0 | 0 | 0 | 0 | 0 | 1 | 4 | 25 | 1 |
| Total | 49 | 10 | 18 | 4 | 36 | 7 | 131 | 26 | 272 | 54 | 506 | 100 |
| **Infrequent Developed Country Users** | | | | | | | | | | | | |
| Austria | 0 | 0 | 0 | 0 | 0 | 0 | 0 | 0 | 9 | 100 | 9 | 0 |
| Finland | 0 | 0 | 2 | 18 | 0 | 0 | 1 | 9 | 8 | 73 | 11 | 1 |
| Japan | 0 | 0 | 0 | 0 | 1 | 25 | 3 | 75 | 0 | 0 | 4 | 0 |
| Singapore | 0 | 0 | 0 | 0 | 0 | 0 | 2 | 100 | 0 | 0 | 2 | 0 |

| | | | | | | | | | | | | |
|---|---|---|---|---|---|---|---|---|---|---|---|---|
| Sweden | 0 | 0 | 4 | 36 | 0 | 0 | 0 | 0 | 7 | 64 | 11 | 1 |
| Total | 0 | 0 | 6 | 16 | 1 | 3 | 6 | 16 | 24 | 65 | 37 | 100 |
| **Emerging Developing Country Users** | | | | | | | | | | | | |
| Chile | 2 | 22 | 0 | 0 | 1 | 11 | 2 | 22 | 4 | 44 | 9 | 0 |
| Costa Rica | 3 | 60 | 0 | 0 | 0 | 0 | 0 | 0 | 2 | 40 | 5 | 0 |
| Guatemala | 0 | 0 | 0 | 0 | 0 | 0 | 0 | 0 | 1 | 100 | 1 | 0 |
| India | 0 | 0 | 3 | 6 | 3 | 6 | 6 | 12 | 38 | 76 | 50 | 2 |
| Indonesia | 0 | 0 | 3 | 23 | 2 | 15 | 6 | 46 | 2 | 15 | 13 | 1 |
| Israel | 1 | 7 | 6 | 40 | 0 | 0 | 2 | 13 | 6 | 40 | 15 | 1 |
| Malaysia | 0 | 0 | 10 | 77 | 0 | 0 | 0 | 0 | 3 | 23 | 13 | 1 |
| Peru | 3 | 21 | 0 | 0 | 5 | 36 | 0 | 0 | 6 | 43 | 14 | 1 |
| Philippines | 0 | 0 | 0 | 0 | 2 | 18 | 3 | 27 | 6 | 55 | 11 | 1 |
| South Africa | 0 | 0 | 10 | 11 | 3 | 3 | 15 | 17 | 60 | 68 | 88 | 4 |
| Thailand | 0 | 0 | 0 | 0 | 0 | 0 | 2 | 33 | 4 | 67 | 6 | 0 |
| Turkey | 0 | 0 | 4 | 13 | 10 | 31 | 3 | 9 | 15 | 47 | 32 | 1 |
| Venezuela | 0 | 0 | 1 | 8 | 2 | 17 | 6 | 50 | 3 | 25 | 12 | 1 |
| Total | 9 | 3 | 37 | 14 | 28 | 10 | 45 | 17 | 150 | 56 | 269 | 100 |
| Total | 141 | 6 | 136 | 6 | 183 | 8 | 552 | 25 | 1,169 | 54 | 2,181 | 100 |
| Traditional % | 83 | 59 | 75 | 1 | 118 | 64 | 370 | 67 | 723 | 62 | 1,369 | 63 |
| New Established % | 49 | 35 | 18 | 13 | 36 | 20 | 131 | 24 | 272 | 23 | 506 | 23 |
| Infrequent % | 0 | 0 | 6 | 4 | 1 | 1 | 6 | 1 | 24 | 2 | 37 | 2 |
| Emerging % | 9 | 6 | 37 | 27 | 28 | 15 | 45 | 8 | 150 | 13 | 269 | 12 |

*Source:* Miranda, Torres, and Ruiz (1998).

can be expected to be the target of increased AD investigations on agriculture. This again has been proven by the fact that, based on U.S. Department of Commerce data, about a quarter of all cases launched against the United States by two of the United States' major competitors, Canada and Mexico have been on agriculture (USDOC, 2000). This is disproportionately high considering the fact that agricultural products account for about only 5 percent of total U.S. exports to Canada and about 8 percent of total U.S. exports to Mexico (USITC, 1999).

Developing countries, which have been subject to many AD investigations by the United States and the European Union, repeatedly complained that the expectations of increased market access due to the implementation of the Uruguay Round were not realized because of frequent use of AD measures. These countries proposed that the WTO AD disciplines be revised during the new round of WTO negotiations. With strong support from Japan, South Korea, and many other countries, it was agreed at the November 2001 Doha Ministerial that the new round of WTO negotiations would aim at clarifying and improving disciplines under the Agreements of the Implementations of Article VI of GATT 1994; that is the antidumping measure. As discussed earlier, the provision of AD measures in multilateral trade rules encourages greater trade liberalization by providing a safety valve for governments implementing trade policy reforms. Nevertheless, as previously described in this chapter, the current provisions are considered by many observers to be too generous and, unless tightened, can lead to rampant abuse by an increasing number of countries. In addition to the economic justifications for tightening the rules, an agreement on tightening the WTO AD rule may make the climate more amenable for developing countries to accept difficult reforms in global trading rules in other sectors during future round of WTO negotiations.

With AD on the agenda in future trade negotiations, reform efforts should be directed toward establishing clearer guidelines for estimating fair values, including taking into consideration the special circumstances encountered in investigations involving agricultural products. Additionally, injury determination should incorporate economy-wide potential benefits and costs resulting from the imposition of AD duties. Past research on AD has generally focused on nonagricultural products, and the few works on agriculture have primarily discussed the legal aspects of AD disciplines. With growing use of AD on agricultural products and the potential for reform in current WTO AD disciplines, more research is necessary to examine the impacts of AD on agricultural products and also to detail the reform areas to make WTO AD disciplines better suited for agricultural products.

## NOTES

1. It is important to note that reforms aimed at clarifying and improving disciplines under the Agreements on Implementation of Article VI of the Uruguay Round agreement were agreed upon at Doha.

2. This practice has been often criticized as being unfair to small exporters whose exports would not have caused injury if they had not been cumulated with exports of other countries.

3. As termed by Rowat, 1990.

4. In April 1999, a WTO Dispute Settlement Panel ruled that India's continued use of quantitative restrictions on imports of a wide range of consumer goods under the Balance of Payments provisions of GATT Article XVIII:B was inconsistent with the GATT guidelines and called for their removal.

5. Note that as agents will accumulate stocks of storable commodities at low prices, the supply curve for such commodities will not plunge all the way to the x-axis.

6. Under the modified U.S. AD laws (effective January, 1995), price comparisons in AD investigations have to be either weighted average with weighted average, or individual with individual. However, exceptions are allowed when these comparisons do not account for a pattern of prices that differ significantly among purchasers, regions, or time periods (Wilson, 1995).

7. Under the Uruguay Round Agreements Act (effective January 1, 1995), U.S. law states that if less than 25 percent of the domestic industry support a petition, no AD investigation will be initiated (Wilson, 1995).

8. Based on FAO data, agriculture's share of total exports are the following: 8 percent for the world, Australia 26 percent, Canada 7 percent, New Zealand 47 percent, and the United States 8 percent.

## REFERENCES

Braga, C. A. P. and S. D. Silber. (1993). "Brazilian Frozen Concentrated Orange Juice: The Folly of Unfair Trade Cases." In *Antidumping: How It Works and Who Gets Hurt*, J. M. Finger (Ed.), (pp. 83-102). Ann Arbor, MI: The University of Michigan Press.

Canning, P. N. and H. Vroomen. (1996) "Measuring Welfare Impacts from Equilibrium Supply and Demand Curves: An Antidumping Case Study." *American Journal of Agricultural Economics* 78:1026-1033.

Devault, J. M. (1996). "The Welfare Effects of U.S. Antidumping Duties." *Open Economies Review* 7:19-33.

Finger, M. J. (1998). "GATT Experience with Safeguards: Making Economic and Political Sense of the Possibilities That the GATT Allows to Restrict Imports." Working Paper #201700. Washington, DC: The World Bank.

Gallaway, M. P., B. A. Blonigen, and J. E. Flynn. (1999). "Welfare Costs of the U.S. Antidumping and Countervailing Duty Laws." *Journal of International Economics* 49:211-244.

Krishna, R. (1997). "Antidumping in Law and Practice." Policy Research Working Paper # 1823, The World Bank.

Lindsey, B. (1999). "The U.S. Antidumping Law: Rhetoric versus Reality." Center for Trade Policy Studies, CATO Institute, Washington, DC.

Miranda, J., R. A. Torres, and M. Ruiz. (1998). "The International Use of Antidumping: 1987-1997." *Journal of World Trade* 32(5):5-71.

OECD. (1996). "Trade and Competition, Frictions After the Uruguay Round: Note by the Secretariat." Economics Department, Paris.

Palmeter, D. N. (1989). "Agriculture and Trade Regulation: Issues in the Application of U.S. Antidumping and Countervailing Duty Laws." *Journal of World Trade Law* 23:47-68.

Prusa, T. J. (1996). "The Trade Effects of U.S. Antidumping Actions," Working Paper 5440, National Bureau of Economic Research, Cambridge, MA.

Pyo, H. K., K. Kim, and I. Cheong. (1996). "Foreign Import Restrictions, WTO Commitments, and Welfare Effects: The Case of Republic of Korea." *Asian Development Review* 14:21-43.

Rowat, M. D. (1990). "Protectionist Tilts in Antidumping Legislation of Developed Countries and the LDC Response: Is the "Race to the Bottom" Inevitable?" *Journal of World Trade* 24:5-29.

Shin, J. H. (1994). "Antidumping Law and Foreign Firm Behavior: An Empirical Analysis," Unpublished Dissertation, Yale University.

Skyes, A. O. (1991). "Protectionism as a 'Safeguard': A Positive Analysis of the GATT 'Escape Clause' with Normative Speculations." *The University of Chicago Law Review* 58:255-305.

Tomek, W. G. and K. L. Robinson. (1981). *Agricultural Product Prices.* Ithaca, New York: Cornell University Press.

University of Michigan. (1980). "Applying Antidumping Law to Perishable Agricultural Goods." *Michigan Law Review* 80:524-561.

USDOC. (2000). "Foreign Cases Against U.S. Firms." International Trade Administration, Import Administration. <http://ia.ita.doc.gov/>.

USITC. (1995). "The Economic Effects of Antidumping and Countervailing Duty Orders and Suspension Agreements." United States International Trade Commission, Publication 2900, Washington, DC.

USITC. (1999). "U.S. Export Data." <http://dataweb.usitc.gov/scripts/>.

Webb, M. (1992). "The Ambiguous Consequences of Antidumping Laws." *Economic Inquiry* 30:437-448.

Wilson, A. (1995). "Antidumping and the Uruguay Round: An Overview." CRS Report for the Congress, 95-192 E, Congressional Research Service, The Library of Congress, Washington, DC.

World Trade Organization (WTO). (1996). *The Results of the Uruguay Round of Multilateral Trade Negotiations: The Legal Texts.* Geneva, Switzerland.

# Chapter 18

# Implementation of the SPS Agreement

Suzanne D. Thornsbury
Tara M. Minton

Prior to 1995, technical restrictions on agricultural trade were governed by the original GATT Articles (primarily Article XX) and by the 1979 Tokyo Round Agreement on Technical Barriers to Trade (TBT Agreement). Although these codes required that measures not be applied in a discriminatory manner, create unnecessary obstacles to trade, or act as disguised trade restrictions, it was generally agreed that they were not effective in preventing the misuse of technical barriers (Roberts, 1998). In response to these concerns as well as others, the Agreement on the Application of Sanitary and Phytosanitary Measures (SPS Agreement) was negotiated during the 1986-1994 Uruguay Round. The SPS Agreement governs regulations designed to mitigate potential negative externalities associated with the movement of products across national borders that might adversely impact the life or health of humans, animals, or plants.

When legitimate externalities or other market failures are addressed, such technical barriers have the potential to increase national welfare, even without consideration of terms-of-trade effects. Governments may also impose technical barriers to isolate domestic producers from international competition. In these cases under small-country assumptions, technical barriers are welfare-decreasing policies. The goal of the SPS Agreement is to protect the legitimate rights of importing countries with respect to national health and safety without providing a loophole for countries to avoid other trade-liberalizing disciplines of the Uruguay Round Agreements.

In contrast to the Agreement on Agriculture, which sets numerical limits and commitments in the areas of market access, domestic support, and export subsidies, provisions of the SPS Agreement provide guidelines for government behavior in implementing technical measures. Guidelines are designed to help identify when such barriers act as disguised protection, but

We would like to thank the University of Florida's Center for International Business and Economic Research for support of this research and Sandra Vokaty of IICA Caribbean Regional Center for additional comments.

the legitimacy of specific SPS measures is still evaluated on a case-by-case basis.

After approximately five years, it is possible to comment on the implementation of the SPS Agreement. Although not inclusive of all provisions, this chapter focuses on institutional and regulatory standard-setting changes that were implemented at the end of the Uruguay Round. The dispute settlement process, the role of the SPS committee, and the notification process are three important components within the institutional framework. The standard-setting guidelines include encouragement of harmonization, scientific risk assessment, and regionalization. Emerging issues for the next round of negotiations can then be evaluated within this context.

## INSTITUTIONAL CHANGES

In any policy arena, the institutional structure has a decisive impact on outcomes. The Uruguay Round Agreements implemented some significant changes in institutional structure that impact the multilateral governance of SPS regulations. The concept of the GATT as a single undertaking is new with the Uruguay Round Agreements. Prior to 1995, members were able to sign the basic GATT Agreement without signing numerous side agreements (such as the 1979 TBT Agreement) that were also negotiated among subsets of members. Since 1995, all members are bound by the full set of GATT/ WTO obligations, thus helping to clarify the legal relationship between the side agreements and obligations of the GATT. Within this single framework, additional changes were implemented including revised dispute settlement procedures, creation of an SPS committee, and increased transparency through equivalence and notification.

### Dispute Settlement Process

At the end of the Uruguay Round negotiations, creation of the WTO and revised dispute settlement procedures were seen as critical components of institutional strengthening for all GATT disciplines. The dispute settlement procedures are likely to be of particular relevance for technical barriers since the SPS Agreement provides guidelines for government behavior, as opposed to numerical limits on policy adoption. The establishment of the WTO creates a formal institution to govern and monitor the administration of these multilateral trade agreements as was proposed, but not implemented, at the 1947 Geneva Conference.

One of the goals of the Uruguay Round negotiations was to establish a more efficient and credible dispute settlement mechanism based on an agreed-upon set of procedures. As an important step toward this goal, the previous consensus-based dispute settlement system was converted to a quasi-judicial system governed by the WTO. Under the new system, members have the right to initiate a panel hearing without full consensus; strict time limits are set for each stage in the dispute process; panel reports can only be rejected by unanimous vote; and panel rulings are to be binding on all members. The changes in dispute resolution have moved GATT away from a system of "soft law" where legal norms do not effectively compel compliance toward one of "hard law" where there is a relatively high expectation of compliance with legal norms (Abbott, 1997). A priori, supporters viewed conversion as a tool that should increase the likelihood of compliance and the credibility of the enforcement mechanisms.

Critics argued, however, that the movement of GATT dispute resolution toward a system of hard law might not enhance the long-run prospects for freer trade. The prospect of an increase in the number of disputes could cause governments to resist making new commitments or lead to a breakdown in negotiations. The costs to pursue formal dispute settlement under the WTO could prove prohibitive for many small and developing countries, thus making access to informal dispute mechanisms critical. Further, the new hard laws may not be designed to decrease trade barriers; for example, the SPS Agreement specifically authorizes members to set their own acceptable level of risk, which could potentially legitimize reduced trade.

Since implementation, the WTO has witnessed an increase in the number of complaints relating to the SPS Agreement. Among all countries, formal WTO complaints related to nine distinct SPS issues have been filed under the dispute settlement procedures since 1995, whereas during the prior forty-seven years of GATT there were virtually no trade disputes over SPS measures that reached formal dispute settlement. Three of the disputes have advanced to WTO panel and Appellate Body rulings: the EU-U.S./Canada Hormones case, the Australia-Canada Salmon case, and the Japan-U.S. Varietal Testing case (Roberts, 1998). While the institutional framework has provided positive opportunities for discussion and formal resolution, outcomes of the three cases have been somewhat disappointing in terms of increasing agricultural trade.

In the Hormones dispute, the complainants were authorized to introduce retaliation in the form of increased tariffs on EU exports after the EU failed to either modify its measure to comply with the final Appellate Body ruling or to offer acceptable compensating trade concessions. As of October 2000, after one and a half years of retaliation, negotiations were still underway for the EU to comply with the WTO ruling. The United States would then reduce its $116.8 million per year sanctioned duties levied against food items from the EU (WTO, 1999c). In the Salmon case, the revised measures notified by Australia have been challenged by Canada in regard to their compliance. As of February 2000, Australia had exceeded the reasonable period of time for implementation, causing panel recommendations for the Dispute Settlement Body to request that Australia bring its measures into conformity. Of the three cases, Varietal Testing has the most potential for a positive movement toward freer trade. In October 1998, Japan, due to its varietal testing procedures, was found to be in violation of WTO obligations. Prior to the ruling, Japan imposed stringent phytosanitary requirements on different varieties of several produce items, for example apples, cherries, and peaches, requiring separate testing be completed for each variety. The United States, a major apple producer, felt these variety tests severely restricted its apple trade and violated provisions of the SPS Agreement (Calvin and Krissoff, 1998). The Dispute Settlement Body decision was upheld by the Appellate Body in February 1999. Japan notified the WTO of its intentions to comply with the Appellate Body ruling in January 2000; however, by fall 2000, the ruling had yet to be implemented (WTO, 2000a).

### Role of the SPS Committee

Article 12 of the SPS Agreement established a committee, composed of representatives from all members, to administer a number of tasks related to the Agreement. The SPS committee provides a forum for discussion, information exchange, and informal resolution regarding sanitary and phytosanitary issues. The committee reviews the implementation of the SPS Agreement, monitors the process of harmonization among member nations, and strives for consistency in the applications of various concepts brought forth in the SPS Agreement. Acting as a liaison, the committee provides coordination among other international organizations in the field of sanitary and phytosanitary regulations. Last, the committee may make recommendations regarding the future of the SPS Agreement.

An important contribution of the SPS committee is its role in clarification of the Agreement through rulings and interpretations. Extensive discussion on particular issues has helped to draw attention to specific trade concerns and to avoid potential conflicts (WTO, 1999a). For example, the ability of members to set a prohibitively low level of acceptable risk will be limited by the nondiscrimination policy of GATT, which requires that domestic and imported products be subject to the same requirements. A consensus that emerged in the SPS committee is that the objective is to avoid arbitrary or unjustified distinctions between the levels used across regulatory measures (Greifer, 1998). In conducting its own internal review, the SPS committee noted that a substantial part of its time was devoted to the discussion of specific implementation problems. A number of SPS-related trade issues were resolved through formal discussion of the committee. For example, U.S. concerns over South Korean procedures for establishing shelf-life restrictions on numerous agricultural products were debated in the SPS committee and a compromise was reached.

### Notification Process

In addition to the forum provided by the SPS committee, movement toward a more transparent process was implemented through recognition of equivalence, establishment of enquiry points, and notification guidelines. Transparency that permits greater public scrutiny of regulations could raise the political costs of using technical regulations as disguised protection for domestic agriculture (Sykes, 1995). Equivalency obliges members to accept the equivalence of other members' SPS measures, if they achieve the same level of risk protection. In order to determine if the level of protection is the same, members are encouraged to provide access for testing, inspection of facilities, or other relevant procedures (Art. 4, Para. 1).

Each member is required to establish an enquiry point that can provide answers to questions from potential trading partners regarding SPS regulations, testing, inspection, and/or other procedures, risk assessment, and participation in other agreements. More recently members have been encouraged, but not required, to publish such information on the Web to allow broader and more timely access.

Notification requirements provide an opportunity for comment before regulatory changes are enacted and have proven to be a catalyst for effective implementation of the SPS Agreement as countries publicize measures under consideration. Prompt publication is required if the proposed change puts in place restrictions that are not substantially the same as an international standard and if the measures may have a significant impact on trade (Art. 7). Hence, the ex ante opportunity for members to comment on proposed regulatory changes has been enhanced by the SPS Agreement. The transparency requirements are obligatory for all members, including least-developed countries, except when urgent problems arise. Over 1,100 SPS notifications were filed during the first five years of Agreement implementation. For example, notification by the EU of a future decrease in the acceptable maximum level of aflatoxin residues resulted in a large number of comments from both developed and developing countries that may have had difficulty learning about proposed changes prior to implementation of the SPS Agreement (Roberts, Orden, and Josling, 1999).

An increasing number of WTO members have begun to notify their SPS measures. By March 1999, fifty-nine members had submitted notifications, ninety-one members had established National Notification Authorities, and 100 members had established National Enquiry Points (WTO, 1999a). For calendar year 1999, the SPS committee reported the circulation of ninety-five SPS notifications including thirty-two from high income, fifty-nine from middle-income, and four from low-income countries (Table 18.1). Notifications from middle- and lower-income countries have steadily been increasing over the implementation period.[1]

Of the total number of SPS notifications prior to 1999, eighty-three were challenged in the informal forum provided by the WTO's SPS committee. Additional disputes were resolved through bilateral negotiations. For example, between October 1996 and September 1998, the U.S. Animal and Plant Health Inspection Service (APHIS) reported that 112 restrictions related to plant and animal health were resolved through bilateral technical exchange. Allowances are made for ex poste emergency notifications (Annex B) and provisional measures in cases where relevant scientific evidence is deemed insufficient (Art. 5.7), often referred to as the "precautionary principle." Concerns over interpretation of the precautionary principle were raised in both the Hormones and Varietal Testing cases and continue to be debated. The EU considers the precautionary principle a central part of their decision-making paradigm and has argued to extend its scope to cover insufficient, inconclusive, or uncertain evidence and those cases where there are reasonable grounds for concern (WTO, 2000c).

### STANDARD-SETTING PROCEDURES

Along with the changes to institutional structure, the Uruguay Round Agreements also contain a number of provisions governing the setting of regulatory standards. Provisions include encouragement of harmonization based on international standards, the requirement for scientific risk assessment for standards, and the recognition of regional areas not necessarily constrained by national boundaries.

TABLE 18.1. SPS Notifications Circulated, 1999

| WTO Members | Measures Notified (no.) | Members Notifying (%) |
| --- | --- | --- |
| High Income | 32 | 42 |
| Upper-Middle Income | 25 | 54 |
| Lower-Middle Income | 34 | 38 |
| Low Income | 4 | 10 |
| Total | 95 | |

*Source:* World Trade Organization (2000d) and authors' calculations

### International Standards

Uniform harmonization with international standards is not an explicit requirement of the SPS Agreement, yet it is obviously important to implementation success (Josling, 1998). Members are encouraged to base their SPS regulations on guidelines established by recognized international organizations in order "to harmonize sanitary and phytosanitary measures on as wide a basis as possible" (Art. 3, Para. 1).[2] Recognizing that the international guidelines may not reflect the preferences and/or needs for externality mitigation within every nation, the SPS Agreement also allows members to set their own (presumably higher) level of protection if it is based upon a risk assessment that incorporates available scientific evidence.

Movement toward harmonization could reduce the risk members face when trying to access a new market. It could also potentially lower the costs associated with applying required treatments or processes; for example, if total volume treated increases. In addition to these anticipated benefits, there were some a priori concerns over the ability to set acceptable regulatory standards in an international forum. Critics feared that, in order to reach a negotiated consensus, internationally adopted standards would reflect the least common denominator, or lowest level of protection. The emphasis on reaching consensus might also act to ensure delay or avoidance of discussion on controversial matters. Although raised initially, such concerns were addressed, at least in part, by allowing members to set their own higher levels of protection, and have not been substantiated in the implementation phase.

There are no readily available quantitative measures indicating the number of members who have revised national regulations to comply with international standards or the resulting level of market access achieved. The relevant standard-setting bodies have been increasingly active in cooperative roles with WTO. The SPS committee specifically recognized the (1) intensification and simplification of the Codex standard-setting procedures; (2) increased activity in the Office International des Epizooties; and (3) revision of the International Plant Protection Convention (WTO, 1999a).

### Scientific Standards

The use of science as criteria for policy evaluation is unique to the SPS Agreement. It arguably holds governments to a higher standard of accountability when compared to criteria used as the basis for implementing other, non-SPS, technical measures such as those covered by the TBT Agreement (Roberts, 1998). In addition, scientific criteria can provide for greater stability of expectations among trading partners concerning trade and in designing domestic regulations (Bhagwati, 1996). While its current specifications are more stringent than criteria negotiated in previous GATT Rounds, there remains substantial room for interpretation by members. Other than requiring a risk assessment, the SPS Agreement does not provide guidelines for identifying acceptable scientific procedures. Concurrence may be difficult to achieve since science is, in practice, usually the findings and explanations proposed by various experts in the field of concern and, over time, criteria in use will change as scientific and risk assessment methodology evolves.

As technology for detection and measurement of physical occurrences becomes more sophisticated, the potential for disagreement between opposing scientific views can increase. For example, the methodology for assessment of biological hazards, such as plant pests and microbial pathogens, is relatively new and much more complex than that for toxicological and chemical risks. There are costs associated not only with the development of detection methodology, but also with implementation. Developing countries often do not have the human and technical resources necessary to detect and measures the increasing number of hazards that have been identified internationally.

The results of a risk assessment depend critically on the probabilities, both objective and subjective, associated with specific events and outcomes. Decision makers must choose among policy options when they are uncertain about the probabilities to associate with specific outcomes. They are missing information about the consequences of their actions that could be known if more precise scientific knowledge was available. In such cases, decision makers often exhibit a cautious approach and can be shown to prefer situations where they have more knowledge. In particular, probabilities for low-probability, high-consequence events are often overestimated. These events are relevant for SPS issues as, for example, the probability of introducing a nonindigenous pest as a result of relaxing a technical measure may be quite low, but the economic consequences of such an event could be significant for domestic crop or livestock producers.[3]

Once a low-probability, high-consequence event has occurred, scientific experts may agree that reoccurrence is unlikely, yet public concerns about the probability of reoccurrence are likely to be biased upward, and the demand for stricter regulations may increase. Differences between scientific assessment and public perception may prove difficult to resolve as WTO members debate the role of consumer sovereignty. In addition to positive risk assessment, there are two normative issues that must be addressed in setting an appropriate level of risk protection: whose interests should be protected and what level of risk is acceptable. Article 5.5 requires members to set a consistent level of appropriate risk among policies. Although not defined in the Agreement, the SPS committee has provided a forum for debate on this issue and has developed a draft interpretation of consistency.

The legitimacy of specific technical barriers remains subject to challenge on scientific or other procedural grounds (for example, not-least-trade-restrictive). Empirical evidence concerning the extent of questionable technical measures in international agricultural trade is very difficult to assess, as no current comprehensive data source is available. A 1996 USDA survey of technical barriers to U.S. agricultural exports identified 302 questionable barriers imposed by sixty-one countries and two regional trading blocks that had an estimated trade impact of $4.9 billion.[4] Since the cross-sectional survey was not repeated, there is no direct evidence concerning either escalation or reductions in these numbers.

### Regionalization

Article 6 of the Agreement obliges members to adapt their regulations to conditions in regional areas not necessarily constrained by geographical areas defined by national borders. Members are obliged to recognize the existence of pest- or disease-free areas within an exporting nation. In addition, pest-free zones can be recognized which include relevant geographic areas of multiple countries, for example, the Caribbean. This regionalization criterion has opened up some previously restricted areas to agricultural trade. In review of the operation and implementation of the SPS Agreement, the SPS committee noted the critical importance for international trade in agricultural products of recognizing pest- or disease-free areas or areas of low pest or disease prevalence.

For example, prior to the Uruguay Round Agreements, cattle, swine, sheep, and some meat from countries where foot-and-mouth disease (FMD) was present were banned from entry into the United States. Revised, post–Uruguay Round U.S. regulations allow imports from low-risk regions within a country. By 1997, beef from Argentina was initially allowed into the United States under a series of procedural and inspection protocols established for low-risk regions. In 2000, FMD was again detected in parts of Argentina and a ban on exports to the United States was reinstated. Under the SPS Agreement, such an emergency measure is considered temporary until the potential infestation can be resolved.

## ISSUES FOR THE NEXT ROUND

Article 20 of the Agreement on Agriculture committed members to initiating continued negotiations one year before the end of the implementation period. The mandate for negotiations includes taking into account the following:

1. The experience to that date from implementing the reduction commitments
2. The effects of the reduction commitments on world trade in agriculture
3. Nontrade concerns such as special and differential treatment to developing-country members
4. The objective to establish a fair and market-oriented agricultural trading system
5. Further commitments necessary to achieve the long-term objectives listed in the preamble to the Agreement

In preparation for the next round of negotiations, a series of ministerial conferences were held. Although the November 30-December 3, 1999, WTO Ministerial meetings in Seattle ended without agreement on an agenda for the next round, a preliminary agenda for agriculture was drafted. In this early draft there was no consensus to open or renegotiate the SPS Agreement, and the preliminary positions of specific countries did not indicate a strong movement to do so. However, as in the initial Uruguay Round negotiations, discussion with regard to other agricultural trade and institutional issues may have significant impacts on interpretation and further implementation of the SPS Agreement. The EU position on agriculture specifically mentions the SPS Agreement, particularly as it relates to food safety issues and asks for confirmation that nondiscriminatory science-based measures designed to achieve the determined level of safety are in conformity with the SPS Agreement (WTO, 1999d). A timetable was set for continued negotiations on agriculture with proposals on member objectives due by the end of 2000, with some flexibility for revisions before March 2001 as input into the current round.

The use of biotechnology would likely, at least in part, be governed by the disciplines of the SPS Agreement. Prior to the Ministerial meeting in Seattle, the Canadian government proposed establishment of a working group on biotechnology. Such a group would be a cross-committee and thus could incorporate issues governed by multiple agreements. Although the working group was not launched, biotechnology is likely to be a contentious topic of debate in any agenda on agriculture and may have significant impacts on interpretation and enforcement of the SPS Agreement.

A number of issues related to biotechnology are likely to arise as genetically modified organisms (GMOs) continue to gain awareness in the public arena. Although scientists from some countries have found GMOs to be safe, consumer and environmental groups have voiced opposition. The EU and Japan have both raised concerns over regulatory procedures and proposed labeling of GMOs. Currently, labeling of GMOs is not required in the United States for products that are found to be fundamentally the same as the non-GMO variety (Kelch, Simone, and Madell, 1998). Continued debate over GMOs will force a reexamination of some fundamental principles of the SPS Agreement including the role of the precautionary principle and the definition of science.

Equal access and full participation by developing and least-developed countries are issues of ongoing concern. In review of implementation issues, the SPS committee recognized a continuing need for enhanced technical assistance through the international organizations as well as continued bilateral exchange. An increasing concern of least-developed country members is their declining share in world trade, investment, and output (WTO, 2000b). The ability to implement and/or document compliance with SPS regulatory restrictions has been suggested as a contributing factor to this marginalization. Inconsistent standards for domestic products could provide a window for low-quality imports, which cannot be discriminated against under provisions of the SPS Agreement. Conversely, Roberts, Orden, and Josling (1999) caution against extensive use of claims for special and differential treatment by developing countries. An overuse of such claims could reinforce perceptions that the products are unable to meet standards and may actually hamper market access. The need for domestic institutional building is clear; technical assistance and access to funding are likely to be important components of the next round.

An additional concern remains about the ability of developing and least-developed countries to participate fully in the benefits of the SPS Agreement. Real opportunities to participate in international standard-setting bodies, the WTO, and numerous committees, including the SPS committee, may be limited by resource constraints. Membership lists for WTO, Codex, OIE, and IPPC indicate that both the number and the percentage of members in each of the organizations were higher for low- and middle-income countries in almost every case (Table 18.2). With the exception of OIE, the greatest number of members was in the low-income category. OIE is an in-

TABLE 18.2. Membership in WTO and International Standard-Setting Bodies, 1999 (Percentage of Total in Parentheses)

| Members | Organization | | | | |
|---|---|---|---|---|---|
| | WTO | Codex | OIE | IPPC | Total |
| High Income | 36 (72) | 33 (66) | 26 (52) | 33 (66) | 50 (100) |
| Upper-Middle | 24 (67) | 25 (69) | 23 (64) | 31 (86) | 36 (100) |
| Lower-Middle | 32 (56) | 40 (70) | 32 (56) | 46 (81) | 57 (100) |
| Low Income | 42 (67) | 54 (86) | 26 (41) | 53 (84) | 63 (100) |

*Source:* WTO (1999b) and authors' calculations

dependent organization that charges annual membership fees, whereas no fees are charged by the IPPC and Codex that are associated with parent international organizations.[5] However, membership in an organization is not always indicative of active participation as evidenced by a recent survey of Caribbean countries undertaken by the Inter-American Institute for Cooperation on Agriculture (IICA). Survey results indicate that there is a much higher percentage of membership in Codex and IPPC than active participation. Decisions undertaken by these bodies also have trade impacts on countries that do not participate. Impacts may be trade-restricting when standards are set at levels that resource-constrained countries cannot meet technologically. There is also potential for positive impacts as promulgated standards may serve to legitimize current or slightly modified practices as acceptable to potential importers.

## CONCLUSION

The Uruguay Round negotiations ended with substantial institutional reforms, both for GATT in general and with regard to agricultural trade in particular. Creation of the WTO provided a formal organization to oversee the GATT agreements and strengthened the compliance and enforcement mechanisms. The Agreement on Agriculture brought agricultural trade discipline more in concordance with the underlying principles of the GATT by requiring specific numerical commitments on market access, domestic support, and export subsidies. Technical barriers to trade were explicitly addressed in the SPS and TBT Agreements, designed to recognize the rights of governments to use technical barriers as a tool to correct negative externalities, while seeking to limit the ability to intentionally misuse such measures. The SPS Agreement was seen by most as a positive step toward freer trade in agricultural products.

As WTO members drafted their positions for the Ministerial meetings in Seattle, there was not a strong agenda to reopen the SPS Agreement for further negotiations, albeit with some areas of concern as well as recognition of the need to continue with progress that had begun. In an internal review, the SPS committee concluded that the Agreement had contributed to improving international trading relationships. Increasing use of the transparency requirements, adaptation of regulations to regional conditions, progress in the recognized standard-setting bodies, and continued conflict resolution through informal negotiations and discussions are areas of subtle and steady progress toward freer trade. Areas of concern include enhanced participation by developing countries, the practical application of the equivalence principal, and enforcement of formal Dispute Settlement Body rulings. A consensus seems to be emerging that the SPS Agreement has been arguably successful and that future progress can be best achieved through continued refinement of the existing structure, the work of the SPS committee, and the WTO dispute settlement procedures.

## NOTES

1. Roberts, Orden, and Josling (1999) reported notifications from 81, 54, 44, and 6 percent of high, upper-middle, lower-middle, and low-income countries from 1995 through June 1999, respectively.

2. Recognized international organizations are the Codex Alimentarius Commission (Codex) for food additives, veterinary drug and pesticide residues, contaminants, methods of analysis and sampling, and codes and guidelines of hygienic practices; the International Office of Epizootics (OIE) for animal health; and the International Plant Protection Convention (IPPC) for plant health.

3. Importation of a given product may result in a pest infestation that increases the costs to domestic producers but lowers the market equilibrium price faced by domestic consumers. Orden and Romano (1996) illustrate this dichotomy among outcomes from a SPS policy decision for a long-standing dispute over the import of HASS avocados into the United States from Mexico and conclude, in part, that even when an SPS trade barrier is based on sound scientific analysis, it may not be the preferred economic policy in terms of maximizing national welfare.

4. Barriers identified in the USDA survey include an expert consensus-based view of 1996 measures that decreased, or potentially decreased, U.S. agricultural exports to specified export markets and appeared to violate at least one of the disciplines of the WTO Agreements. Of the 302 measures identified, over 86 percent were considered questionable because they potentially violate a principle of the SPS Agreement, 9 percent were potentially disciplined by the TBT Agreement, and 5 percent served more than one purpose and so might be covered by multiple agreements, or appeared to violate another provision of GATT 1994 (Thornsbury et al., 1999).

5. The Codex Alimentarius Commission is a subsidiary body of two United Nations organizations, the World Health Organization and the Food and Agriculture Organization. The International Plant Protection Convention is a subsidiary body of the Food and Agriculture Organization.

## REFERENCES

Abbott, Frederick L. (1997). "The Intersection of Law and Trade in the WTO System: Economics and the Transition to a Hard Law System." In *Understanding Technical Barriers to Agricultural Trade,* David Orden and Donna Roberts (Eds.). St. Paul, Minnesota: University of Minnesota, Department of Applied Economics, International Agricultural Trade Research Consortium, pp. 33-48.

Bhagwati, Jagdish (1996). "Trade and the Environment: Exploring the Critical Linkages." In *Agriculture, Trade and the Environment,* Maury Bredahl, Nicole Ballenger, John Dunmore, and Terry Roe (Eds.). Boulder: Westview Press, pp. 13-22.

Calvin, Linda and Barry Krissoff (1998). "Technical Barriers to Trade: A Case Study of Phytosanitary Barriers and U.S.-Japanese Apple Trade." *Journal of Agricultural and Resource Economics.* 23(2):351-366.

Greifer, John (1998). "Setting the Appropriate Level of Protection in Trade." *USDA ORACBA News.* Washington, DC, January-February, pp. 1-4.

Josling, Timothy (1998). *Agricultural Trade Policy: Completing the Reform.* Washington, DC: Institute for International Economics.

Kelch, David, Mark Simone, and Mary Lisa Madell (1998). "Biotechnology in Agriculture Confronts Agreements in the WTO." *Agriculture in the WTO.* Economic Research Service/USDA WRS-98-4, December, Washington, DC, pp. 34-35.

Orden, David and Eduardo Romano (1996). "The Avocado Dispute and Other Technical Barriers to Agricultural Trade Under NAFTA." Invited paper presented at the conference NAFTA and Agriculture: Is the Experiment Working, San Antonio, TX, November.

Roberts, Donna (1998). "Preliminary Assessment of the Effects of the WTO Agreement on Sanitary and Phytosanitary Trade Regulations." *Journal of International Economic Law.* 2:377-405.

Roberts, Donna, David Orden, and Tim Josling (1999). "WTO Disciplines on Sanitary and Phytosanitary Barriers to Agricultural Trade: Progress, Prospects, and Implications for Developing Countries." Invited paper presented at the 1999 World Bank Global Conference on Agriculture and the New Trade Agenda from a Development Perspective: Interests and Options in the WTO Negotiations, Geneva, October 1-2.

Sykes, Alan O. (1995). *Product Standards for Internationally Integrated Goods Markets.* Washington, DC: The Brookings Institute.

Thornsbury, Suzanne, Donna Roberts, Kate DeRemer, and David Orden (1999). "A First Step in Understanding Technical Barriers to Agricultural Trade." In *Food Security, Diversification, and Resource Management: Refocusing the Role of Agriculture?* G.H. Peters and J. Von Braun (Eds.). Brookfield, Vermont: Ashgate, pp. 453-463.

World Trade Organization, Committee on Sanitary and Phytosanitary Measures (WTO) (1999a). "Review of the Operation and Implementation of the Agreement on the Application of Sanitary and Phytosanitary Measures." G/SPS/12, Geneva, March 11.

World Trade Organization, Committee on Sanitary and Phytosanitary Measures (WTO) (1999b). "Membership in WTO and International Standard-Setting Bodies." G/SPS/GEN/49/Rev.1, Geneva, May 7.

World Trade Organization (WTO) (1999c). "European Communities—Measures Concerning Meat and Meat Products (Hormones)." WT/DS26/21, Geneva, July 15.

World Trade Organization, General Council (WTO) (1999d). "EC Approach on Agriculture." WT/GC/W/273, Geneva, July 27.

World Trade Organization (WTO) (2000a). "Japan-Measures Affecting Agricultural Products." Status Report by Japan. WT/DS76/11, January 17.

World Trade Organization (WTO) (2000b). "Declaration of the Ministers of Trade of the Least-Developed Countries, Seattle, November 20, 1999." WT/L/343, Geneva, February 9.

World Trade Organization, Committee on Sanitary and Phytosanitary Measures (WTO) (2000c). "Communication from the Commission on the Precautionary Principle." G/SPS/GEN/168, Geneva, March 14.

World Trade Organization, Committee on Sanitary and Phytosanitary Measures (WTO) (2000d). "Documents and Notifications Circulated During 1999." G/SPS/GEN/171, Geneva, March 15.

Chapter 19

# Modeling Impacts of the Macroeconomic and Political Environment on Long-Term Prospects for Agricultural World Markets

Martin von Lampe

## *INTRODUCTION*

The developments and prospects of regional and global agricultural markets have been of particular interest for policymakers and economists for a long time. High prices on international markets both in the 1970s and the 1990s gave cause for concern about possible global food shortages and options to improve the world food situation, resulting in the World Food Summit held in November 1996 in Rome, Italy (FAO, 1996). A number of scientists feared that production could no longer keep pace with the rapidly growing demand for food due to limited resources. On the other hand, high levels of support for agricultural producers in many industrialized countries, and large surpluses that could be exported only with the use of subsidies, resulted in increasingly distorted markets and high budgetary costs. International efforts aiming to liberalize markets resulted in the Uruguay Round Agreement on Agriculture (URAA), where the international community agreed to put bounds on import barriers, domestic support, and export subsidies for the first time (Henrichsmeyer and Witzke, 1994, pp. 592; WTO, undated), and to do further steps toward more liberalized markets in later negotiations that have now started. In line with the international agreements, individual countries revised their agricultural policies.

In this context, projections of agricultural markets are important for several reasons: Political decision makers are interested in what they should be prepared for, and they also need to know about the impact of possible policy changes on markets and trade. In addition, they need information on the sensitivity of market developments with respect to the main driving factors on supply and demand.

A number of models are employed for outlook purposes, and some of them are also used for sensitivity analyses. These models, however, mainly focus on the medium-term perspective, and a number of them mainly concentrate on price policies. Others, designed for longer projection horizons, take macroeconomic conditions into consideration, but neglect the impact of agricultural policies to a certain degree.

Following an overview on these models, this chapter discusses the World Agricultural Trade Simulation (WATSIM) model. Developed at the University of Bonn, it incorporates both macroeconomic, sectoral and policy variables in a differentiated way and allows for detailed long-term projections and simulations of agricultural markets.

## BRIEF OVERVIEW OF OTHER AGRICULTURAL MARKET MODELS

Over the last decades several multimarket multiregion partial equilibrium models have been developed and applied to project and simulate agricultural markets.[1] The Organization for Economic Cooperation and Development's (OECD) AGLINK model, the USDA's Country-Link System (CLS), and the Food and Agricultural Policy Research Institute's (FAPRI) Global Modeling System (GMS) are used for medium-term outlooks and focus on agricultural policy issues. The main aim of these models is to draw up the annual baselines published by the respective institutions.

The AGLINK model (OECD, 1998) distinguishes between thirteen commodity markets in twelve regions, mainly OECD member countries. Used to combine consistently the projections of the OECD member countries' national authorities, the main focus is to put on markets of particular interest for the foreign trade of these regions, i.e., grains, oilseeds, meat, and milk products. Dynamic formulations allow the model to calculate annual time series. An asymmetric structure takes into account specific properties of the different regions and commodity markets.

The CLS (Landes, 1998) distinguishes between twenty-four agricultural commodities and forty-six regions plus an exogenous Rest of World aggregate. It consists of a set of stand-alone country models with individual depths of differentiation, but uniformly defined interfaces with respect to the world market. A linking procedure allows to solve the models simultaneously. The CLS considers a broad set of policy measures including tariffs, quotas, etc., and differentiates between gross imports and exports using an Armington-approach.

Similar to the CLS, the GMS (Chaudhary and Premakumar, 1996; Westhoff and Young, 2000) is a set of different models designed over specific sets of commodities and differentiated at the regional level. In contrast to the CLS, however, the GMS submodels are not formally linked with each other, but jointly used within the modeling groups. Apart from this difference, a number of similarities with the CLS are observed, including the broad set of both commodities and regions and the representation of various policy measures.

Given the fact that both the CLS and the GMS are actually sets of models, their usability for policy simulations is restricted for practical reasons. The individual models of both systems, however, are also used for policy analyses. As with the AGLINK model, simulations of specific shift factors are not the aim of CLS nor of GMS.

Two other models are used for long-term projections, namely the Food and Agricultural Organization of the United Nations' (FAO) World Food Model (WFM) and the International Food Policy Research Institute's (IFPRI) International Model for Policy Analysis of Agricultural Commodities and Trade (IMPACT). The WFM (FAO, 1998) covers the markets for thirteen food commodities, but distinguishes between 146 countries and country groups. It is mainly used in the context of expert-based long-term projections of world food markets and food security and is supplemented by separate projections for a number of other markets to give a largely complete picture of the nutritional prospects. The dynamic structure of the WFM allows for the calculation of time series projections and adjustment paths following certain shocks.

IMPACT (Rosegrant, Agcaoili-Sombilla, and Perez, 1995) distinguishes between seventeen agricultural commodities and thirty-six regions, many of which are individual countries. Projections of supply, demand, and net trade on agricultural markets are based on various macroeconomic variables. Given the focus on food security issues, a malnutrition index is calculated from the energy consumption modeled endogenously and additional exogenous data on a number of socioeconomic variables.

The representation of policy variables is limited to some price wedges in both the WFM and the IMPACT models making them incapable to incorporate other forms of market intervention. This excludes detailed policy simulations, and may also lead to problems in market projections.

For the analysis of long-term developments of agricultural world markets and trade, the WATSIM model should therefore meet a number of requirements: Apart from the capability to describe the basic trends on both the supply and the demand side of agricultural markets, processes of economic adjustment of supply and demand should be realistically portrayed. The interactions between different markets closely related to each other are particularly important, and should be included. The principal policy variables, such as tariffs, floor prices, quotas, set-aside regimes, and export restrictions should be part of the formal model, since these variables are known to have a major influence on the development of regional and world markets. And finally, the main factors influencing long-term developments of supply and demand, as well as the quantitative relationship between supply and demand and these variables, should be explicitly formulated in the model. This would allow for the determination of the specific impact of these variables.

## OUTLINE OF THE WATSIM MODELING SYSTEM

In general, the WATSIM model represents a standard partial equilibrium approach.[2] Based on a comprehensive and consistent database, which contains detailed data for some 100 agricultural product markets at single-country level and is fed from a broad set of data sources (including FAOSTAT, PS&D, World Development Indicators, PSE/CSE Estimations, and others), the aggregation level of the simulation model can be determined in a flexible way according to the purpose of the model. For this study, the model was specified for twenty-nine interdependent commodity markets in fifteen regions. Each regional market is represented by several iso-elastic supply and demand functions, describing the impact of price changes on market quantities. Regional markets are cleared via net trade positions meeting on the world markets. To bring global markets to a new equilibrium, world market prices, which are linked to the regional market and incentive prices via price transmission functions, are adjusted. Apart from political border measures, these functions also include a transport cost term depending on the regional net trade. Figure 19.1 shows the principal structure of the partial equilibrium model.

In order to make the model capable of market adjustments and substitutions across commodities both on the supply and the demand side, the respective functions include a full set of calibrated cross-price elasticities. In this context, the proper representation of the livestock-feed relationships is of particular importance. The appropriate calibration of available parameters ensures that the regional feed markets are balanced in terms of total feed energy use.

### Policy Representation

A subset of policy measures reflected in the model enter the price transmission functions. Given the aim of the model to represent policies in a way that is as close to real world as possible, not only ad valorem and specific tariffs are considered. Direct product-related payments and other subsidies also alter incentive prices, and administrated minimum market prices partly prevent changes of world market prices from being transmitted to domestic markets. Assuming that average yields are independent from prices,[3] premium levels paid per hectare of harvested area or per livestock unit are transformed to price equivalents and included in the price transmission functions as well.

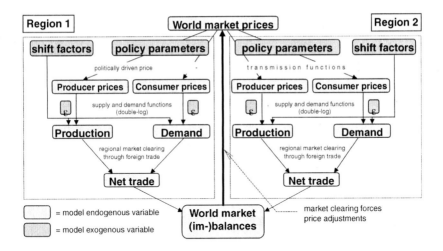

FIGURE 19.1. Schematic structure of the partial equilibrium model WATSIM. (*Source:* Modified from von Lampe, 1999, p. 17.)

Other measures directly influence supply and trade quantities. Production quotas set upper bounds to production quantities and are modeled by price-independent fixation of supply variables. Compulsory set-aside of land reduces harvested areas and is taken into account by adjusting the production areas for each affected commodity. Export restrictions limit the exports of certain commodities whenever subsidies are necessary, i.e., as long as foreign trade prices plus some price volatility adjustments[4] are lower than domestic prices, exports are bounded by the limits set by the URAA. If domestic market prices are fixed (e.g., by an intervention system), administrated stocks are modeled to buffer the excess supply; otherwise, market prices are allowed to decline until the domestic markets are cleared.

Most of the data on policy measures are taken from the OECD, supplemented by various other sources. For most developing countries, however, no policy information are included due to data limitations, implying the assumptions of full price transmission and unchanged policies in both the baseline projections and the simulations.

### *Representation of Shift Factors*

Apart from policy measures, a number of socioeconomic and sectoral variables influence agricultural markets that become more important the longer the time horizon of model simulations is. In order to make the WATSIM model able to represent and simulate these developments, a set of exogenous shift factors are explicitly included. These key variables are considered to be the main quantifiable driving forces for supply and demand on agricultural markets.

### *Supply Side*

Land availability for crop production is strongly limited and partly even decreasing in most regions of the industrialized world, but also in several developing countries in Asia, under the pressure of land needs for other purposes (e.g., urbanization, industrialization, and infrastructure). For most regions, however, an increase in the cropping index is expected, a development much more important with respect to the increase of agricultural production than the cultivation of new land. Projections on both total land availability and cropping indices are based on time series analysis, explicitly taking into account the impact of urbanization.

The share of irrigated land is explicitly considered to formulate crop production developments as an important factor with respect to both cropping indices and crop yields. Projections on irrigation are based on time series analysis and checked against other studies. Linear relationships are assumed between the irrigation share and both cropping indices and average crop yields [eqs. (19.1) and (19.2)]. Corresponding parameters are derived from various studies.

$$HRVI_{r,IRSH_r}^{trend,t} = HRVI_{r,IRSH_r^{bas}}^{trend,t} * \frac{100 + IRSH_r^t * \frac{HADI_r^t}{100}}{100 + IRSH_r^{bas} * \frac{HADI_r^t}{100}} \qquad (19.1)$$

$$YIEL_{i,r,IRSH_r^t}^{trend,t} = YIEL_{i,r,IRSH_r^{bas}}^{trend,t} * \frac{100 + IRSH_r^t * \frac{YADI_{i,r}}{100}}{100 + IRSH_r^{bas} * \frac{YADI_{i,r}}{100}} \qquad (19.2)$$

Where:

| | | | |
|------|-----------------------------------------------------|-------|---------------------------------------|
| *HRVI* | harvesting index (total harvested area relative to total land used for crops) | *trend* | value of variable affected by other factors |
| *IRSH* | irrigation share | *t* | time index |
| *HADI* | advantage coefficient of the harvesting index due to irrigation | *i* | product index |
| *YIEL* | average crop yield | *r* | regional index |
| *YADI* | advantage coefficient of average crop yield due to irrigation | *bas* | base year |

The changes both in feed regimes and feed efficiency is crucial to determine the future needs of the livestock sectors. Feed requirement parameters estimated as marketable feed energy intake per kg of animal product were projected via time series analysis and expert judgment. Their use in the feed energy balances is presented by eq. (19.3):

$$\sum_{f} FEED_{f,r}^{t} * ENER_{f} = \sum_{l} PROD_{l,r}^{t} * NEED_{l,r}^{t} \qquad (19.3)$$

Where:

| | | | |
|---|---|---|---|
| *FEED* | feed use quantity | *t* | time index |
| *ENER* | energy content of feed stuff | *f* | index for feed products |
| *PROD* | production quantity | *l* | index for livestock products |
| *NEED* | energy requirement per unit of livestock product | *r* | regional index |

Technical progress, particularly due to modern breeding and biotechnology, holds a large potential impact on productivity. This aspect is difficult to quantify and is captured mainly by trend estimation and supplemented by expert judgment on yield developments.

### Demand Side

By modeling human consumption as per capita demand, the consideration of population growth, projected by the UN, becomes straightforward. Further growth in income, approximated as real gross domestic product per capita, is one of the most important driving forces for the development of consumption growth on agricultural markets, particularly in the developing countries. Like in many other models that are synthetic, one of the characteristics of the WATSIM model is that the steering parameters, and therefore income elasticities as well, are mainly taken from external sources, such as from other models or from literature. They are not estimated within or explicitly for the model. While many models work with constant elasticity functions, others employ linear functions to determine consumption quantities.

Two major problems are encountered with such a procedure. First, the assumption of constant income elasticities becomes implausible particularly in the long run. Not only does the share of food expenditures with respect to total expenditures decrease with rising income (Engel's law, see, e.g., Henrichsmeyer, Gans, and Evers, 1986, p. 577), a fact that implies income elasticities for food to be smaller than unity, but income elasticities themselves also decrease with a rising income (Henrichsmeyer and Witzke, 1991, p. 298). Since it is believed that considerable gains in per capita income can be expected within the next twenty or thirty years in many developing countries, particularly in Asia, the use of constant income elasticities may lead to an overestimation of the demand for agricultural products.

Second, the validity of the income elasticities found in literature is questionable even for the base year used for projections. The parameters published, e.g., by FAO (1989) or in the documentation of the SWOPSIM database of 1992 by the USDA (Sullivan et al., 1992; see also Gardiner et al., 1989), were mainly taken from other studies and models. Hence, the time span relevant for the estimation of those parameters often further dates back several years. Between 1980 and 1994, however, the gross domestic product per capita in China,[5] for example, increased by more than 180 percent in real terms (The World Bank, 1998).

Hence, the use of income elasticities published in older studies, and not adapted as suggested above, may lead to a biased specification of the impact of income on demand particularly in the long-term simulation model. In medium-term calculations this problem is limited if the income situations in the different regions differ only slightly from the situation at present, and if the parameters are adjusted for income changes since the estimation period.

Therefore, the WATSIM model does not use constant elasticities of income. Instead, cross sample estimations over the set of regional aggregates in WATSIM were done to derive a functional relationship between the income elasticities and per capita income. As in the model itself, per capita income is approximated by the gross domestic product divided by total population. In the estimation model, the per capita income in 1980 is used as the explanatory variable for the different regional income elasticities. For the sake of simplicity a semi-logarithmic function is assumed [eq. (19.4)], i.e., an increase of per capita income by 10 percent leads to the same absolute change in the income elasticity for all income levels and regions.

$$\eta_{i,r} = a_{0,i} + b_i * \ln PCI_r + \varepsilon_{i,r} \qquad (19.4)$$

Where:

| | | | |
|---|---|---|---|
| $\eta$ | income elasticity of demand | $i$ | product index |
| $a_0,b$ | parameters to be estimated | $r$ | regional index |
| $PCI$ | per capita income, 1980 [GDP per capita, US$ (1987)] | | |
| $\varepsilon$ | error term | | |

Although the coefficients of determination vary widely across commodities, the estimations generally show better results for commodities with a high share in direct food consumption than for products where direct consumption only plays a minor role. Estimation results for cereals and meat are shown in Table 19.1. For specific products and regions, however, estimation errors can be considerable. Therefore, the estimation errors are interpreted as systematic indicators for regional specificities and added to the constant term, yielding a regional constant:

$$\eta_{i,r} = a_{i,r} + b_i * \ln PCI_r$$

$$\text{with } a_{i,r} = a_{0,i} + \varepsilon_{i,r} = \eta_{i,r}^{1980} - b_i * \ln PCI_r^{1980} \qquad (19.5)$$

Where:

$a_{i,r}$     regional constant parameter
Other variables and indices already declared

This procedure should be interpreted as follows: While the regional comparison results into product specific parameters describing the effect of income changes on the income elasticities, the functional relationship between per capita income and the income elasticities fully makes use of the information in the parameters actually found in literature, which are assumed to be valid for 1980. Hence, income elasticities are made dependent on growing income in a plausible manner, using the information available from literature.

The linear relationship between income elasticities and the logarithm of per capita income given by equation (19.5) implies that the logarithm of per

TABLE 19.1. Results of OLS Estimation of Income Elasticities Depending on Per Capita Income for Cereals and Meat

| Product | $a_0$ | b | $R^2$ |
|---|---|---|---|
| Wheat | 1.4100 *(9.61)* | −0.1598 *(−8.71)* | 85.37% |
| Barley | 0.1811 *(2.03)* | −0.0253 *(−2.27)* | 28.35% |
| Maize | 0.4081 *(3.22)* | −0.0573 *(−3.61)* | 50.07% |
| Other Cereals | 0.1828 *(2.03)* | −0.0255 *(−2.26)* | 28.27% |
| Rice | 0.9167 *(4.93)* | −0.0806 *(−3.47)* | 48.07% |
| Beef | 1.2478 *(11.79)* | −0.1114 *(−8.42)* | 84.51% |
| Pig Meat | 1.1541 *(8.79)* | −0.0967 *(−5.89)* | 72.77% |
| Other Meat | 1.3606 *(9.19)* | −0.1214 *(−6.56)* | 76.81% |
| Poultry | 1.7985 *(11.40)* | −0.1631 *(−8.27)* | 84.03% |

*Source:* von Lampe, 1999, appendix 3. Reprinted by permission.

capita consumption is a function quadratic in the logarithm of per capita income, and is as follows:

$$\ln HCPC^t_{i,r} = c^t_{i,r} + a_{i,r} * \ln PCI^t_r + 0.5 * b_i * \left(\ln PCI^t_r\right)^2 \quad (19.6)$$

Where:

| | |
|---|---|
| $HCPC$ | per capita human consumption |
| $c^t_{i,r}$ | parameter describing the change of per capita demand due to other factors |

Other variables and indices already declared

To give an example, Figure 19.2 shows the actual development as well as the income-corrected development of per capita pig meat consumption in China from 1961 to 1995. Given the assumptions on economic growth in China (an average geometric growth rate for the per capita income of 6.2 percent p.a. between 1994 and 2005, and 4.3 percent p.a. between 2005 and 2020), and holding all other factors like prices fixed, the average consumer would be expected to consume some 46.7 kg of pig meat in the projection year 2020. In contrast, assuming constant income elasticities taken from literature, i.e., 0.68 for poultry, would result in a per capita consumption of 58 kg/cap. in 2020. The difference of 24 percent makes it obvious that the way of income representation is crucial for the projection results.

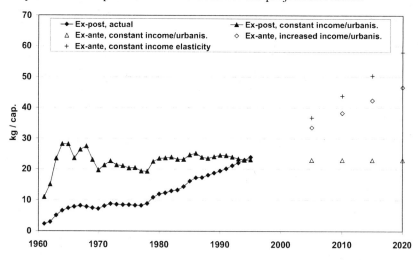

FIGURE 19.2. Actual and income-projected development of per capita pig meat consumption in China, 1961-2020. (*Source:* WATSIM database, own estimations)

Independent of income growth, urbanization has proved to have significant impacts on consumption patterns: Wheat, sugar, and particularly animal products are consumed more in urban areas than in rural areas even with per capita income held fixed, while rice, coarse grains, starchy products, and pulses are consumed less. Again, a linear relationship is assumed between the share of urban population and consumption quantities at a given level of per capita income:

$$HCPC_{i,r}^{t,urb(t)} = HCPC_{i,r}^{t,urb(bas)} * \left(1 + URBS_{i,r} * \left(URB_r^t - URB_r^{bas}\right)\right) \quad (19.7)$$

Where:

| | | | |
|---|---|---|---|
| HCPC | per capita human consumption | $i$ | product index |
| URBS | urbanization shift parameter | $r$ | regional index |
| URB | urban share of total population | $t$ | time index |
| | | $bas$ | base year value |
| | | $urb(t)$ | data associated with actual share of urbanization |
| | | $urb(bas)$ | data associated with base year share of urbanization |

Urbanization shift parameters are derived from surveys on Asian consumption patterns. Like the projections on total population, urbanization figures were taken from the World Population Assessment.

## MODEL RESULTS: PROSPECTS AND SENSITIVITY OF AGRICULTURAL WORLD MARKETS

### Assumptions

Baseline projections of the agricultural markets were done with respect to a number of assumptions both on macroeconomic and sectoral variables and on agricultural policy measures. The former are mainly drawn from other studies. In particular, the world population prospects on urban and total population published by the United Nations (1996), the income prospects (real GDP) estimated by the International Monetary Fund (1998, 1999), and inflation rate estimates by the World Bank (1996) were taken into account. Assumptions on the longer-term development of real GDP are based on FAPRI (1998). Trend-based projections include those for land availability and irrigation of agricultural land and were cross-checked with available literature such as Alexandratos (1995). Impacts of changes in the development of irrigation are accounted for in the first impact analysis.

With respect to agricultural policies, the baseline represents a "status-quo" projection, i.e., existing and agreed policies are assumed to remain unchanged for the projection period. In particular, this relates to the results of the 1993 Uruguay Round Agreement on Agriculture (URAA), the 1992 reform of the Common Agricultural Policy (CAP) of the European Union, and the 1996 Federal Agriculture Improvement and Reform (FAIR) Act of the United States. The EU Agenda 2000 reform is accounted for in the second impact analysis of the paper.

### Main Results of the Baseline Projection

Driven by rising incomes and growing population numbers, meat demand in developing regions will increase significantly within the next decades. High consumption growth rates are projected especially for poultry and pig meat. Meat demand in industrialized countries is projected to grow following a more moderate pace due to the high levels of per capita consumption already observed and the low population growth in these regions.

Part of the additional developing countries' meat demand will have to be imported, mainly from the industrialized countries. Net imports to South and South East Asia are projected to grow from 286,000 t in 1994 to 4.9 million (Mio.) t by 2020. Similarly, net imports to Africa and West Asia are projected to increase from 1.1 Mio. t to 6.1 Mio. t within the same period. The United States is projected to become the largest meat exporter, with an excess supply of almost 7 Mio. t by 2020, mainly in poultry. The EU, restricted in the medium term due to the GATT commitments on subsidized exports, could increase net meat exports by 2020 to more than 4.6 Mio. t, a large part of which again would be poultry. Australia and New Zealand should increase their meat exports as well, although at more moderate rates. The projected growth of meat trade would imply an adequate expansion of transport capacities both between regions and within the developing countries, as well as more or less liberal import policies.

Given the rapidly growing livestock production, feed demand for cereals is projected to increase in most regions, and food consumption will add to the grain demand growth particularly in developing regions. Similar to meat, a growing share of the demand in the developing regions will be provided by foreign imports according to the model results. The Mediterranean region of North Africa and West Asia will remain the largest grain importer, with the deficit growing from roughly 40 Mio. t in the mid-1990s to some 70 Mio. t by 2020. Sub-Saharan Africa, China, the Rest of Asia, and Latin America are also projected to increase net cereal imports. The United States, on the other hand, will remain the world's largest grain exporter with an excess supply doubling to reach 140 Mio. t by 2020. The EU could also provide large net exports to the world grain market by 2020. In the medium term, in contrast, exports are limited due to the GATT restrictions, since unsubsidized exports are projected to be possible only to some extent on the wheat market under pre-Agenda 2000 conditions. Other growing cereal exporters include Canada, Australia, and New Zealand.

Figure 19.3 shows the projected world market price changes between 1994 and 2020. In real terms, world prices in the year 2020 are projected to be similar to 1994 prices for beef, pig meat, and poultry, while prices for sheep and goat meat are estimated to slightly increase. World grain prices, in contrast, are projected to further decline in real terms, with average growth rates ranging between –1.05 percent and –2.10 percent p.a. for the period 1994-2005, and between –0.84 percent and –1.78 percent p.a. for the period 2005-2020. Similar to the meat markets, prices for most cereals develop more favorably in the long term than in the medium term. The only exception is wheat, mainly because the EU export restriction is expected to become unbinding in the long run, and growth in Chinese food demand is projected to cease by the end of the projection period. Although wheat exports are projected to be possible without subsidies to some extent already by 2005, intervention purchases of wheat and coarse grains are necessary on a large scale. Given a 10 percent set-aside rate and the unchanged Common Agricultural Policy (CAP), intervention stocks of wheat and coarse grains would increase by some 7 Mio. t each in 2005, and in 2020 purchases of coarse grains would still be necessary with 2.3 Mio. t, both making evident the need for adjustments in the EU policy.

These results indicate that global agricultural production appears to be able to keep pace with global demand even with decreasing real prices. Compared to the developments in the past decades, however, the projections are significantly more favorable for exporters. Between 1950 and the mid-1990s average world market prices for wheat, maize, and rice declined with average rates of –2.20 percent, –2.82 percent, and –2.15 percent p.a., respectively, rates that were between 0.7 and 1.8 percentage points lower than the projections for the 1994-2020 period. Similarly, real meat prices are projected to not decline anymore. Hence, the baseline results also indicate less favorable food prices for the poorest parts of the global society.

### Results of the Impact Analyses

The baseline projections discussed above are subject to significant uncertainties in three main areas: First, the assumptions on the macroeconomic and sectoral environment can change significantly whenever new data become available. On the demand side, income projections have proved to be very difficult and uncertain (compare for example the pros-

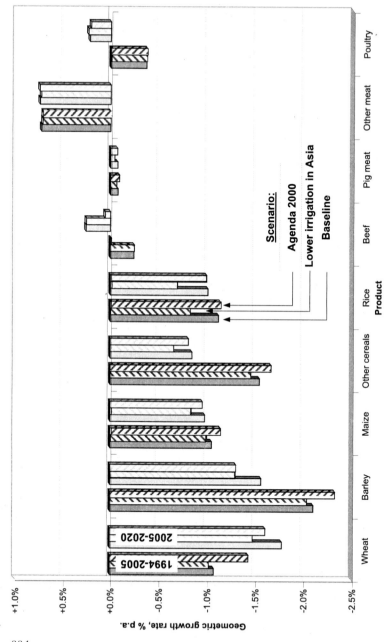

FIGURE 19.3. Projected development of real world market prices for cereals and meat, 1994-2020. (*Source:* WATSIM simulation results)

pects for Asia published in IMF, 1999, with those in IMF, 1998). On the supply side, productivity growth depends on a number of variables including irrigation and modern breeding technologies (e.g., genetically manipulated
organisms), the projections of which appear to be extremely difficult. See
Evenson, Rosegrant, 1995 for a discussion of this issue).

Second, agricultural policies are already changing compared to the assumptions in the baseline. The European Union has decided on a new reform of the CAP, and further negotiations on the liberalization of regional
and international commodity markets are on their way.

Third, the model results neglect the impact of a number of factors that
can be interpreted as "random" variables. In particular, deviations from average weather conditions are not considered at all.

Although impact analyses were done to show how market prospects depend on sectoral and policy variables, weather and other random variables
remain untouched in this discussion. They should, however, be kept in mind
by interpreting the model outcomes as means of probability distributions
rather than exact forecasts.

### Macroeconomic and Sectoral Shifts:
### The Impact of Reduced Irrigation Growth in Asia

As an important production factor, irrigation has been expanded considerably in many Asian countries, and thus has contributed to production
growth to a large extent, both in terms of multicropping capacities and average crop yields. Though more moderately than in the past, further increases
in the irrigation shares are assumed in the baseline for the Asian regions. To
analyze the impact of this development, a scenario with lower growth in irrigation shares is carried out with the WATSIM model. More specifically,
growth rates for the 1994 to 2020 period were cut by half in China, India, the
ASEAN, and the Rest of Asia region (which excludes Japan).

While the impact of reduced irrigation on harvested areas is limited and
partly compensated by higher crop prices, lower average yields would result
in about 2 percent reduction in Chinese wheat and rice production, respectively, by the end of the projection period. For India, wheat and rice production would be lower by more than 4 percent. Similar impacts are calculated
for the smaller ASEAN and Rest of Asia regions. Due to higher prices, large
parts of the reduction in rice production would be offset by lower consumption in these regions, given that the bulk of world's rice consumption takes
place in these four regions. In contrast, net wheat imports to China, India,
and the Rest of Asia would increase by some 7.5 Mio. t by 2020, compared
to the baseline.

By 2020, international rice prices would be more than 8 percent above
baseline level, given the lower irrigation in Asia (see Figure 19.3). Simi-

larly, wheat and barley prices would be 5 percent higher. With the exception of rice (+3 percent), the impact on grain prices would be much smaller in the medium term, i.e., by 2005. The improved market situation would allow the EU to cut intervention purchases of wheat by half, thereby to increase net wheat exports by another 3 Mio. t. Therefore, the impact on international market prices is largely buffered. Intervention purchases of coarse grains would be more or less unaffected in the medium term, but cut by 40 percent in 2020. Both regional and global meat markets would be largely unaffected by changes in the irrigation regime, with slightly higher world market prices for pig meat and poultry compared to the baseline.

### The Political Environment: The EU Agenda 2000 Reform

With the Berlin Agreement of March 1999, the EU has done another step to prepare for future challenges. The reform of the Common Agricultural Policy is only one part of the Agenda 2000 package. It defines, however, significant changes in the political environment for the agricultural sector. Without going into details, the main points of this reform can be characterized as first, the reduction of administrated price levels for cereals, beef, and milk products, second, the increase of area and headage related compensation payments for cereals and cattle, third, the decrease of direct subsidies for oilseed area to the level provided for cereals, and fourth, the increase of the total milk quota (for a more comprehensive discussion of the Agenda 2000 reform see Knaster, 1999).

The simulation results indicate that the impact of Agenda 2000 on international markets are much less significant in the long run than in the medium term. Particularly the meat markets are affected only marginally, with beef prices being slightly increased because of reduced EU exports in the medium term due to lower domestic prices (see Figure 19.3). In the long run, the EU can be expected to further increase meat exports, particularly pig meat and poultry, due to reduced feed prices. The impact on international prices is only very small, however.

On cereal markets, the impact is more significant. This is particularly due to the fact that the price reduction mainly affects coarse grain markets, while the reduction of domestic wheat prices is largely dampened by export opportunities. The need for large intervention purchases will strongly drop, and the amount of unsubsidized wheat exports is modeled to increase significantly. Production of wheat can be expected to be stimulated in favor of coarse grains and particularly of oilseeds. Within the coarse grains group, corn production will be favored somewhat in the long run compared to barley and other cereals.

International cereal prices show a much stronger reaction in the medium term compared to the long run since the increase in EU wheat exports in the medium term is supported by both a larger surplus and the drop in intervention purchases, while the latter is no longer relevant in the long run.

## FURTHER DEVELOPMENTS

The calculations done with the WATSIM model still suffer from some problems related to the model structure. Given that the changes in the agricultural policies are not only related to the level of policy measures, but also to the shift toward more specific measures, the structure of the WATSIM model is currently being refined in two aspects: First, the net-trade approach is being dropped. An Armington approach is used to differentiate between gross imports and gross exports of each region. This will improve not only the representation of restrictions on subsidized exports, but also the consideration of import tariffs. Second, given the market access commitments agreed upon in the Uruguay Round Agreement on Agriculture, tariff rate quotas become more important in regional policy regimes. Their explicit representation in the model is therefore considered an important task, particularly in view of the upcoming new round of multilateral trade negotiations.

## SUMMARY AND CONCLUSION

Given the current problems on global agricultural markets—persistent malnutrition in numerous countries on the one hand, and large imbalances due to protectionist and supportive policy measures causing both trade distortions and high budgetary costs in a number of industrialized countries—sound analyses of market prospects are important. To enhance transparency of projections, and to assess the various influencing factors, the detailed consideration of both macroeconomic and sectoral data and policy measures in the models used for this task is necessary.

The World Agricultural Trade Simulation modeling system is designed to specify baseline projections and simulations of agricultural commodity markets depending on both the macroeconomic and the political environment. On the macroeconomic side, apart from population developments, changes in urbanization and per capita income are considered to be important driving forces for food demand. It is argued, however, that the use of income elasticities from literature is problematic unless they are adjusted to changes in the economic situation. Similarly, feed intensities and feed efficiency, as well as irrigation capacities, are considered in the projection and simulation of production developments.

The results show that market prospects on both the regional and the global level strongly depend on a number of "exogenous" factors. In general, gradual changes in the macroeconomic and sectoral development will have an increasing impact on markets over time, resulting in larger differences in the market outcomes in the long run than in the medium term. This is true, for instance, in the case of irrigation capacities, which may significantly influence production levels and hence net trade positions, e.g., in Asia, where irrigation plays a major role in agriculture. While a reduced growth in Asian irrigation capacities would affect international rice markets mainly via higher prices, the region's net imports of wheat would be increased significantly.

The direct impact of policy measures, in contrast, is basically a medium-term one, particularly if their changes are considered to be completed at some point of time. This, as well as the fact that indirect consequences from policy changes such as changes in research and investment behavior with additional long-term impacts on markets are not taken into account, are simplifications of the model analyses that have to be kept in mind.

## NOTES

1. Another comprehensive review on applied models of agricultural trade with a somewhat different focus can be found in van Tongeren and van Meijl, 1999.

2. For a more comprehensive discussion of the methodological concept of WATSIM see von Lampe, 1999.

3. This assumption is based on the expectation that, while rising prices will increase yields at the field level, this impact is more or less offset on the sectoral and regional level by the expansion to marginal areas.

4. Given the comparative-static characteristics of the model, price volatility cannot be reflected explicitly. The impact of fluctuating prices within and across years on the opportunity for unsubsidized exports is taken into account, however, by the assumption that these are possible even if average prices on world markets are below domestic prices by a certain percentage which is derived from sensitivity analyses.

5. Including Taiwan and Hong Kong.

## REFERENCES

Alexandratos, N. (ed.) (1995): *World Agriculture: Towards 2010—An FAO Study.* Chichester: Food and Agriculture Organization of the United Nations (FAO) and John Wiley and Sons.

Chaudhary S. and Premakumar, V. (1996): *FAPRI Agricultural Policy Modeling: An Integrated Systems Approach—An Introduction to FAPRI.* Ames, IA: Food and Agricultural Policy Research Institute, Center for Agricultural and Rural Development, Iowa State University.

Evenson, R.E. and Rosegrant, M.W. (1995): *Developing Productivity (Non-Price Yield and Area) Projections for Commodity Market Modeling.* Washington, DC: International Food Policy Research Institute. Mimeo.

FAO (1989): *Income Elasticities of Demand for Agricultural Products, Estimated from Household Consumption and Budget Surveys.* Rome: Food and Agriculture Organization of the United Nations.

FAO (1996): "FAO Regional Conference for Europe Will Discuss Food Security, World Food Summit and Environmental Problems." Press Release 96/10, April 25, 1996, Rome.

FAO (1998): World Food Model—Technical Documentation. Paper prepared for the World Food Outlook Conference, Rome, May 28-29, 1998. Rome: Food and Agriculture Organization of the United Nations, Commodities and Trade Division.

FAPRI (1998): *World Agricultural Outlook 1998.* FAPRI Staff Report #2-1998. Available at <http://www.fapri.org/outlook98/outlook98_pdf.htm>.

Gardiner, Walter H., Roningen, V. O., and Liu, K. (1989): "Elasticities in the Trade Liberalization Database." Staff Report No. AGEC 89-20, Agriculture and Trade Analysis Division, Economic Research Service, United States Department of Agriculture, May 1989.

Henrichsmeyer, W., Gans, O., and Evers, I. (1986): *Einführung in die Volkswirtschaftslehre,* Seventh edition. Stuttgart: Ulmer.

Henrichsmeyer, W. and Witzke, H.P. (1991): *Agrarpolitik Band 1—Agrarökonomische Grundlagen.* Stuttgart: Ulmer.

Henrichsmeyer, W. and Witzke, H.P. (1994): *Agrarpolitik Band 2—Bewertung und Willensbildung.* Stuttgart: Ulmer.

IMF (1998): *World Economic Outlook May 1998.* On the Internet under <http://www.imf.org/external/pubs/ft/weo/weo0598/index.htm>.

IMF (1999): *World Economic Outlook May 1999.* On the Internet under <http://www.imf.org/external/pubs/ft/weo/1999/01/index.htm>.

Knaster, B. (1999): *The Final Agenda 2000 Agreement on Agriculture: An Assessment.* Agricultural and Resource Economics, Discussion Paper 99-01, Institute for Agricultural Policy, Bonn University. On the Internet under <http://www. agp.uni-bonn.de/agpo/publ/dispap/dispap_e.htm>.

Landes, R. (1998): *The Country-Link System.* Unpublished slide presentation of the Markets and Trade Economics Division, Economic Research Service, US Department of Agriculture.

OECD (1998): *AGLINK General Characteristics.* Information file for national agencies using the model. Paris: Organization for Economic Cooperation and Development.

Rosegrant, M.W., Agcaoili-Sombilla, M., and Perez, N.D. (1995): *Global Food Projections to 2020: Implications for Investment.* Food, Agriculture, and the Environment Discussion Paper 5. Washington, DC: International Food Policy Research Institute.

Sullivan, J., Roningen, V., Leetmaa, S., and Gray, D. (1992). *A 1989 Global Database for the Static World Policy Simulation (SWOPSIM) Modeling Framework.* ERS Staff Report No. AGES 9215. Washington, DC: Agriculture and Trade Analysis Division, Economic Research Service, U.S. Department of Agriculture.

UN (1996): *World Population Prospects: The 1996 Revision.* New York: United Nations, Population Division of the Department for Economic and Social Information and Policy Analysis.

van Tongeren, F. and van Meijl, H. (1999): Review of Applied Models of International Trade in Agriculture and Related Resource and Environmental Modelling. FAIR6 CT 98-4148 Interim Report No. 1. The Hague: Agricultural Economics Research Institute (LEI).

von Lampe, M. (1999): *A Modelling Concept for the Long-Term Projection and Simulation of Agricultural World Market Developments—World Agricultural Trade Simulation Model WATSIM.* Dissertation. Bonn, Aachen: Shaker Verlag.

Westhoff, P. and Young, R. (2000): *The Status of FAPRI's EU modeling effort.* Paper presented at the 65th Seminar of the European Association of Agricultural Economists EAAE, Bonn, March.

World Bank, The (1996): *Commodity Markets and the Developing Countries—A World Bank Quarterly.* November.

World Bank, The (1998): *World Development Indicators 1998 on CD-ROM.* Washington, DC: The World Bank.

World Trade Organization (WTO) (undated): *A Summary of the Final Act of the Uruguay Round.* On the Internet under <www.wto.org/english/docs_e/legal_e/ursum_e.htm>.

Chapter 20

# Price Volatility: A Bitter Pill of Trade Liberalization in Agriculture?

Robert D. Weaver
William C. Natcher

## *INTRODUCTION*

Public sector involvement in the agricultural markets has historically included import quotas, export subsidies, price supports, and supply-oriented measures such as input restrictions or inventory management. From a sector basis, these interventions have most often attempted to enhance prices or to control (symmetrically or asymmetrically) the range of their variation. Since the institution of sector-targeted policies during the Great Depression, agricultural and food markets have evolved substantially, as has the distribution of market power in those markets. Over the past few decades, the wisdom of these roles of government in agricultural markets has been brought to question by their high financial burdens and less than expected efficacy in achieving goals. Further, the high costs of the associated distortions of agricultural trade among nations has been recognized. Together this evolution of thinking concerning the role of government in markets has led to a slow, though persistent progression of changes in policies resulting in an increasingly private, decentralized market orientation at both national and international levels.

At the national level, the United States implemented the Federal Agricultural Improvement Act (FAIR) in 1996 that was heralded as a significant shift in U.S. agricultural policy establishing U.S. commitment to decouple agricultural policy. Translated, this implied moving to policies that do not distort economic decisions. FAIR followed on the heels of the Food Security Act of 1985 and the Food, Agriculture, Conservation, and Trade Act (FACTA) of 1990 that had also attempted to reduce government involvement in the agricultural sector and move prices to competitive market determination. Each of these policies targeted U.S. farm programs that had evolved over several decades to achieve various goals for enhancing or stabilizing farm level prices through a combination of instruments and actions.

These policies were exemplary of similar types of policies implemented in other nations that impacted economic decisions including spatial arbitrage and trade. The basket of policy goals included setting of limit prices, price management through public storage and trade transactions, subsidizing private on-farm storage, supply control, subsidization of desirable environmental outputs, and protection from import competition.

Combined, this set of distortions became the target of trade reforms and international trade liberalization. The General Agreement on Trade and Tariffs (GATT), Uruguay Round, the North American Free Trade Agreement (NAFTA), and the formation of the European Union took further steps to open domestic markets to regional and global competition.

Through the course of this series of policy changes, concerns have been raised that increased price volatility will result or has resulted. For example, extensive farm press and extension coverage (e.g., *Agri Finance,* 1997; Yonkers and Dunn, 1996) suggested that price volatility would increase or has dramatically increased as a result of FAIR. Similar accusations have been made concerning the implications of taking down government storage programs, and opening of trade.

Within this context, it is of interest to assess the implications of deregulation of agricultural markets through national and trade reforms on price volatility. Domestic policy reforms have significantly reduced the role of government through decoupling of sector policies and dramatic reduction in government storage. International trade policy changes have exposed some agricultural markets to external sources of volatility. The net effect on price volatility of the convolution of these changes is unclear from a theoretical perspective and deserves empirical examination.

In this chapter, we examine the track record over the past several decades of price volatility. Livestock, grain, and dairy market data are considered across the United States. In particular, we consider first, the record of price volatility based on univariate autoregressive conditional heteroskedasticity (GARCH) estimates (Engle, 1982; Bollerslev, 1986). Next, we examine both parametric and nonparametric evidence of shifts and changes in the levels of price volatility.

## APPROACH

The implications of trade liberalization for price volatility have received very little attention in the literature. Crain and Lee (1996) considered the implications of changes in U.S. domestic agricultural policy across three eras of post-World War II farm programs: quota dominated, mandatory programs (January 1950-April 1964); acreage control, voluntary programs

(April 1964-December 1985); and increasingly market oriented programs (December 1985-1997). Based on natural volatility estimates for spot and futures wheat prices across these program eras, Crain and Lee (1996) found evidence that volatility had changed significantly across program regimes. They concluded that wheat price volatility was higher in the 1964-1985 policy regime than the 1985-1993 regime while mandatory and long-term land diversion programs were associated with reduced volatility in prices. They also reported evidence that suggested that low loan rate levels were associated with high levels of volatility. These results provide support for press and extension observations concerning changes in volatility that would be associated with FAIR. The authors are unaware of other published considerations of the impacts of policy changes on price volatility.

While Crain and Lee's (1996) study is suggestive of a role of government programs in altering the volatility of prices, the market setting for agricultural commodities is complicated simultaneously by farm, tax, macro, and trade policies of both the United States and its trading partners. Further, changes in government programs analyzed by Crain and Lee involved numerous changes impacting incentives and constraints affecting private sector production, storage, trade, and utilization decisions. The confluence of these policies and their differential implementations suggest that a less structured approach to assessing changes in price volatility is of interest.

To proceed, we evaluate how price volatility has changed over time for commodities exposed to different levels of policy intervention and trade. We consider price data over the past several decades for commodities with three distinct histories of experience in national and trade policy. Empirical evidence of changes in price volatility is based on weekly U.S. prices drawn for grains (#2 yellow corn and #1 yellow soybeans) and beef (feeder and live cattle), and monthly prices for dairy products (40 lb blocks and 500 lb barrels of cheddar cheese, and the base price of raw milk). While both of these grains are heavily traded, corn was affected by direct U.S. government supply control policies. Alternatively, the beef markets have not been distorted by domestic policies, however, numerous non-tariff barriers and tariffs have affected the trade of beef. The experience for dairy products and milk has been of a high degree of domestic regulation of markets and direct intervention through government stock acquisition to stabilize prices. Details on data are summarized in Table 20.1 and presented graphically in Figure 20.1.

### Measuring Price Volatility

To analyze price volatility, a variety of approaches are available including natural volatility based on a moving average of past variance estimates, conditional volatility based on parameterized time series models, and im-

TABLE 20.1. Data Description

| Commodity | Description | Units |
|-----------|-------------|-------|
| Corn | #2 Yellow, Chicago | Cents/bu |
| Soybeans | #1 Yellow, Central Illinois | Cents/bu |
| Feeder Cattle | Oklahoma City Cash Price | Cents/lb |
| Live Cattle | Texas/Oklahoma Cash Price | Cents/lb |
| BFP | Basic Formula Price (AMS) | $/cwt |
| Cheddar Cheese | Cheddar #40 Blocks Midwest | Cents/lb |

plied volatility based on futures and options market information. In this chapter, we consider conditional volatility estimates.

To begin, first define a price series as $\{P_t\}$. By connotation, price volatility suggests uncertainty and, in fact, it is the unanticipated change in price that economists argue results in real effects. For this reason, a direct measure of price volatility can be based on a measure of the deviation of price from a stationary prediction. For $I(1)$ price series, such a measure is provided by the change in price over time. More generally, the deviation from a transformation to achieve stationarity would provide a measure of such an unanticipated change in price. A parametric approach to estimate price volatility is to derive an estimate of conditional variance from a more general model of the data generation process. An important alternative that has been shown to be capable of representing a wide variety of discrete and continuous dynamic processes is the GARCH model. For example, a GARCH(1,1) model for a nonstationary series $I(1)$ can be written:

$$\Delta P_t = b_0 + \beta_1 \Delta P_{t-1} + \mu_t \sim N(0, h_t)$$
$$h_t = \alpha_0 + \alpha_1 h_{t-1} + \phi_1 \varepsilon_{t-1}^2 \qquad (20.1)$$

where $\Delta P \equiv P_t - P_{t-1}$, $\mu_t$ is a stochastic term distributed normally with a mean zero and a conditional variance $h_t$. The estimate of $h_t$ drawn from (20.1) provides a conditional second moment that varies over each observation without concern for specification of trading intervals.

While (20.1) provides one example of a GARCH process, in general, we would expect a more general form of GARCH to be optimal. First, we could expect that the conditional mean may be determined by a longer AR structure. In (20.1), only an AR(1) is assumed. Second, typically the order of the GARCH process is unknown. Thus, in this chapter we present condition volatility estimates based on a more general AR*(r)*-GARCH*(p,q)* form.

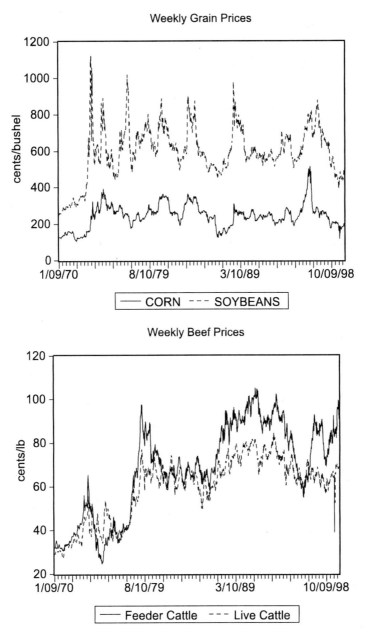

FIGURE 20.1. Price series

Monthly Basic Formula Price

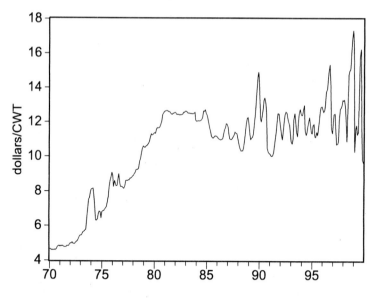

Monthly Cheddar Cheese Price (Blocks)

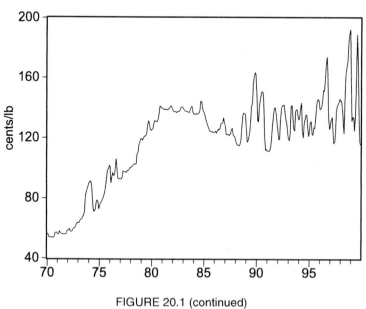

FIGURE 20.1 (continued)

This specification allows us to simultaneously explore optimal models of the conditional mean and conditional variance. Weaver and Natcher (2000) labeled this a "two-moment" approach to estimation of GARCH models. That is, the approach explicitly models both the first and second moments of price. To proceed, the AR($r$)-GARCH($p,q$) model is defined as:

$$R_{it} = \beta(X) + \sum_{j=1}^{r} \delta_j R_{it-j} + \mu_{it}$$

$$\mu_{it} \sim N(0, h_{it}) \qquad\qquad (20.2)$$

$$h_{it} = \alpha_i(X) + + \sum_{j=1}^{p} \gamma_{ij} h_{it-j} + \sum_{j=1}^{q} \phi_{ij} \mu_{it-j}^2$$

where $R_{it} \equiv P_{it}$ for a stationary process and $R_{it} \equiv \Delta P_{it}$ for a nonstationary process. Within this notation, the estimation problem is one of simultaneously choosing $(\beta, \delta, r, \alpha, \gamma, \phi, p, q)$. An important caveat follows from the ARMA form of the GARCH element of the model, namely only a best fitting, parsimonious pair $(p, q)$ can be found, not a unique pair. We implement an approach that chooses these parameters to maximize their likelihood.

### *Examining Evidence of Shifts in Price Volatility or Changes in Regimes*

To consider shifts in the evolution of price volatility over time, both nonparametric and parametric approaches are available. Nonparametric approaches are free of specification decisions and allow the identification of change points in a sequence of random variables. These approaches rely on a search for point of change by examining stability of sample estimators as the sample is changed within a particular realization. Under a maintained hypothesis that the realization being examined was generated by an independently identically distributed process, these sample estimators should not change as the sample is changed (e.g., Inclan and Tiao, 1994). Here, to examine evidence of change in price volatility, we use the iterative cumulative sum of squares (ICSS) test proposed by Inclan and Tiao. This approach is based on the sample estimator of cumulative sum of squares

$$C_k = \sum_{t=1}^{k} \tilde{z}_t^2$$

for a subsample of $k$ observations. The underlying process $\{\tilde{z}_t\}$ is maintained as a series of independent random variables with

$$E[\tilde{z}_t] = 0 \text{ and } E[\tilde{z}_t^2] = \sigma_t^2.$$

A centered cumulative sum of squares (CSS) is defined as:

$$D_k = \frac{C_k}{C_T} - \frac{k}{T}, \; k = 1, ..., T. \tag{20.3}$$

Inclan and Tiao (1994) note that $D_k$ will fluctuate around 0 for a constant variance series but in the event of a structural change, $D_k$ will extend beyond some predetermined confidence interval. The critical values for this confidence interval are provided by Inclan and Tiao.

While the CSS over the entire sample is able to identify the existence of a single change point, the process must be repeated over subsamples to identify multiple change points. For example, if a point change is observed at $\tau T$ where $\tau \in (0,1)$, then this point is used to partition the sample into two subsamples: $t_0 - \tau T$ and $(\tau T + 1) - T$. The CSS is then estimated over both subsamples to identify additional point changes. The process is repeated until no new change points are identified.

Along with the ICSS nonparametric approach to test for changes in the conditional second moment, we also employed the parametric approach of incorporating dummy variables into an optimal AR$(r)$- GARCH$(p,q)$ model as arguments in the vector $X$. The dummy variables coincide with specific events such as the agricultural acts of 1990 and 1996. This type of Chow test (Chow, 1960) relies on a priori identification of specific points in time when change occurs. The power of this type of test is conditional on the validity of the specification of the underlying parametric model.

## EMPIRICAL RESULTS:
## THE HISTORY OF PRICE VOLATILITY
## AND EVIDENCE OF CHANGE

### Characteristics of the Price Series

Figure 20.1 presents the price series examined for the period 1970 to 2000. These graphics are noteworthy because they highlight the temporal variation experienced by each of the products across a spectrum of differing domestic or trade regimes. Comparing corn and soybean prices, it is of interest to note the absence of any visually apparent shift in temporal variation. While a spike occurred in the FAIR implementation period in 1996, the prices reverted to the trading range experienced over nearly a decade. Looking back further, the period before 1989 appears slightly more temporally variable, despite substantial government corn stocks held out of the market for price stabilization purposes although soybean prices appear more temporally variable than corn prices that were directly impacted by domestic policies. Both beef price series exhibited similar patterns of temporal variation, though feeder cattle price variation was greater. Again, a shift in volatility around 1988 appears. The dairy price series illustrate a different story. In both series, temporal variation before the late 1980s was dampened by government stock management. The increased temporal variation after government stocks were nearly eliminated was substantially greater.

### Nonparametric Evidence of Shifts in Price Volatility

The results for the ICSS test reported in Tables 20.2a and 20.2b indicate numerous points in the three-decade period when a shift in price volatility occurred. Recall for this approach we use squared change in price over time as a measure of price volatility. Noteworthy is the fact that multiple change points were found. However, all change points fall after 1988, a point in time after which many market reforms began to be implemented.[1]

The results from the ICCS test indicate that FACTA had minimal impact on the volatility for the commodities analyzed. Although all of the commodities experienced a change in volatility around the FACTA event period only live cattle encountered a change approximately after the passing of the act (7/91). For the other commodities, the changes occurred prior to FACTA or a few years following suggesting the change point was the result of other economic factors.

The results of the ICSS during the FAIR period suggest that FAIR had more of an impact on price volatility compared to FACTA. Tables 20.2a and 20.2b illustrate that corn, cheddar cheese, and the BFP all experienced a change in volatility in October 1996. This date closely corresponds to the passing of FAIR. Also note that both dairy products appear closely related in terms of dates of volatility changes. This result is sensible considering that the BFP is a function of cheddar cheese prices. Finally, Figure 20.2 provides a graph of the CSS and the centered CSS for the BFP. This highlights the change in volatility occurring around October of 1996.

FAIR does not appear to have strongly impacted the volatility of soybeans, live cattle, or feeder cattle. Soybeans and feeder cattle were found to have a change in price volatility during the summer of 1997, a change that might be attributable to adjustments rather than direct effects of FAIR.

These results are consistent with the level of public involvement in these markets. The dairy and corn markets have been directly impacted by public

TABLE 20.2a. Iterative Cumulative Sum of Square Results for Agricultural Products

| Commodity | Change Points |
| --- | --- |
| Soybeans | 8/88, 10/89, 6/93, 7/97 |
| Corn | 8/88, 8/89, 3/96, 10/96, 6/99 |
| Live Cattle | 10/88, 7/91, 8/99 |
| Feeder Cattle | 5/97 |

TABLE 20.2b. Iterative Cumulative Sum of Square Results for Dairy Products

| Commodity | Change Points |
| --- | --- |
| BFP | 6/89, 10/96, 1/99 |
| Cheddar Cheese Blocks | 7/88, 10/96, 12/98 |

involvement while soybeans and the beef markets have only been indirectly impacted. However, it is curious that dairy products show a change associated with FAIR, although they were not directly targeted by changed implemented by FAIR.

Finally, an interesting relationship appears to exist between corn and soybeans. As previously noted, corn has been directly impacted by public intervention while soybeans have been essentially regulation free. While these markets were characterized by dramatically differing levels of government intervention, they were both found to have similar timing of changes in price volatility during the late 1980s. This confirms the suggestion by economists that the effects of government intervention are not restricted to targeted markets when many markets are highly interrelated through a web of arbitrage.

## GARCH Results

Based on the ICSS results, dummy variables were incorporated into the conditional variances of corn, cheddar cheese, and the BFP. These commodities were chosen from the set of commodities given the nonparametric results. A dummy variable representing October 1996 was included where the value was equal to one from 10/96 onward and zero elsewhere. The results from the dummy variable analysis are presented in Table 20.3. The table indicates that the results are consistent with the ICSS results for cheddar cheese and the BFP. That is, the dummy variable is significant at the 5 percent level. Alternatively, inconsistency is found for corn since the dummy

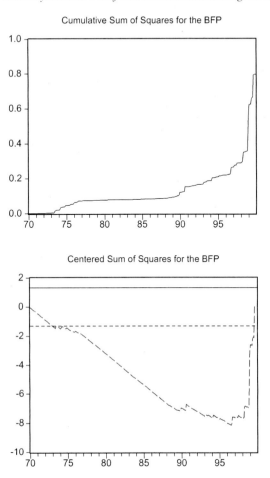

FIGURE 20.2. Iterative Cumulative Sum of Squares Based on the BFP Price Series

variable is found insignificant. Figure 20.3 presents estimated price volatility for the periods analyzed. As is clear from these figures, no apparent shift in the level of price volatility is visible for corn or soybeans. However, temporary increases in volatility are apparent for corn in late 1988 and the fall of 1996. For soybeans these spikes are much less pronounced. For the beef markets, a steady variation in volatility over time is apparent, though for feeder cattle the trend of increased volatility suggested by the estimated GARCH parameters is illustrated. Consistent with parameter estimates for dairy products, an increase after the late 1980s is apparent.

TABLE 20.3. GARCH ICSS Regime Results: Corn, BFP, and Cheddar Cheese

| Optimal Model | Corn AR(1) GARCH(1,1) | BFP AR(1) GARCH(2,1) | Cheddar Cheese AR(1) GARCH(1,2) |
|---|---|---|---|
| AR(1) | .2162 | .5478 | .4982 |
|  | 8.7129 | 11.2515 | 8.7243 |
| ARCH(0) | 1.6977 | .0015 | .0552 |
|  | 7.0829 | 4.6973 | 2.6748 |
| ARCH(1) | .3645 | .1970 | .6392 |
|  | 15.8269 | 4.0553 | 6.4225 |
| ARCH(2) |  |  | −.5482 |
|  |  |  | −5.7219 |
| GARCH(1) | .6597 | .3184 | .9279 |
|  | 37.9455 | 1.0793 | 59.9839 |
| GARCH(2) |  | .4813 |  |
|  |  | 1.8336 |  |
| Dummy | 1.4886 | .3258 | 9.2161 |
| 10/96 | 1.4617 | 3.6605 | 3.2088 |

## *CONCLUSION*

The results presented here consider whether changes in price volatility have been associated with the series of domestic and trade policy reforms that have affected agricultural markets over the past three decades. Based on analysis of prices of corn, soybeans, beef, and dairy for the past three decades we find a number of noteworthy conclusions. First, price levels for each commodity appear to be $I(1)$ while the volatility series are $I(0)$. This suggests that price changes have moments that are time independent. Evidence based on both parametric and nonparametric methods supports the conclusion that price volatility has varied over time, and that depending on the commodity shifts in the level of price volatility has occurred. However, with few exceptions no evidence was found to link changes in price volatility to changes in domestic or trade policy regimes. This may be due to the convolution of domestic and trade policy changes during these periods as well as the fact that their implementation was not instantaneous. It is particularly interesting that the most dramatic changes in price volatility appear for dairy products where the government role in stock management was withdrawn during the late 1980s and early 1990s, though trade liberalization in these products has not as yet appeared as a market force.

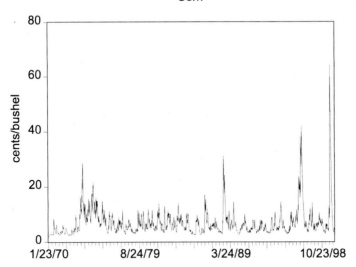

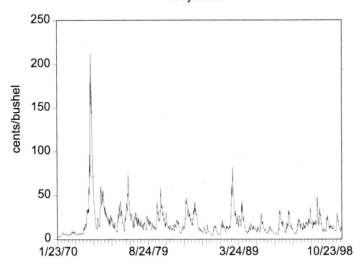

FIGURE 20.3. Conditional Volatility Estimates

### Weekly Conditional Volatility Estimates
### Live Cattle

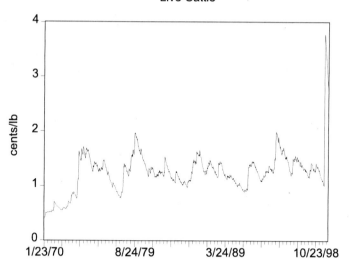

### Weekly Conditional Volatility Estimates
### Feeder Cattle

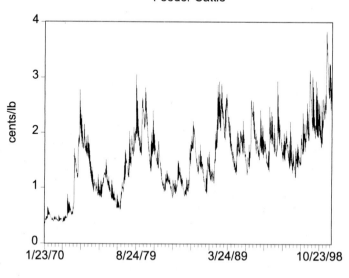

FIGURE 20.3 *(continued)*

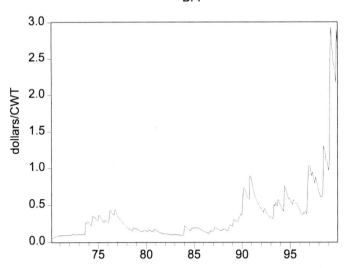

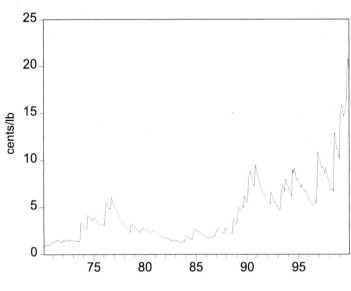

FIGURE 20.3 *(continued)*

## NOTE

1. The results in Table 20.2a and 20.2b were generated based on a structured search. First, following the logic of Crain and Lee (1996), the dates of the Food, Agriculture, Conservation, and Trade Act (FACTA), and the Federal Agriculture Improvement and Reform Act (FAIR) were used as possible break points (identified as 1990 and 1996, respectively). Since the interest is in determining whether a change in volatility occurred as a result of these policies, an event window was formed which encompasses both acts. The window consists of the period 1988-1999. Recall that FACTA was passed in 1990 so two additional years of data prior to 1990 were included to initiate the ICSS algorithm. Given this event window, the ICSS was estimated for the period using the three-step iterative approach outlined by Inclan and Tiao (1994).

## REFERENCES

*Agri Finance* (1997). "1996 Farm Bill Increases Risk." p. 24

Bollerslev, T. (1986). "Generalized Autoregressive Conditional Heteroscedasticity." *Journal of Econometrics,* 31: 307-327

Chow, G.C. (1960). "Tests of Equality Between Sets of Coefficients in Two Linear Regressions." *Econometrica,* 28: 591-605

Crain, S.J. and J.H. Lee. (1996). "Volatility in Wheat Spot and Futures Markets, 1950-1993: Government Farm Programs, Seasonality, and Causality." *The Journal of Finance,* 51: 325-343

Engle, R.F. (1982). "Autoregressive Conditional Heteroscedasticity with Estimates of the Variance of United Kingdom Inflation." *Econometrica,* 50: 987-1007

Inclan, C. and G.C. Tiao. (1994). "Use of Cumulative Sums of Squares for Retrospective Detection of Changes of Variance." *Journal of the American Statistical Association,* 89: 913-923.

Yonkers, R.D. and J.W. Dunn. (1996). "Managing Price Risk in the Absence of Government Support Programs." *Farm Economics.* Pennsylvania State University, November.

Chapter 21

# Challenges and Prospects for Agricultural Trade Policy in the New Millennium

P. Lynn Kennedy
Won W. Koo

## *CHALLENGES FOR AGRICULTURAL TRADE POLICY*

The new millennium holds significant challenges for agricultural trade policy. Perhaps no event highlights this point better than the success of special-interest groups to successfully influence the trade talks in Seattle. Yet before too much emphasis is placed on Seattle, there are many other roadblocks to future agricultural trade negotiations.

The overall objective of the next round of agricultural talks will be to continue the progress made in the Uruguay Round. This implies negotiations on improved market access, additional constraints on export subsidies, and revision of the rules for domestic support. However, diverse country perspectives exist in addition to a number of additional issues that are directly related to the Uruguay Round Agreement on Agriculture core agenda. These issues include such topics as administration of TRQs, state trading, and export restrictions. Additional topics include health and food safety, and environmental issues relating to agriculture and biotechnology, regional trade agreements, and preferential trade arrangements.

### *European Union*

From the perspective of the European Union, there are two main impediments to progress in agricultural trade negotiations. First, reaching agreement among members regarding modifications of the Common Agricultural Policy (CAP) to aid in the addition of new member countries without further budget pressure will be a difficult task. Previous changes to the CAP through the Agenda 2000 were designed to prepare the Union for further expansion of its membership. However, Agenda 2000 provides little in the way of flexibility for further liberalization of EU agricultural trade in the WTO negotiations.

The second obstacle involves the EU's stance on multifunctionality. The view of agriculture as a multifunctional activity, yielding various public goods in addition to food and fiber, with further trade liberalization is difficult to reconcile with the stance of other WTO countries. Achieving some efficient means to ensure the desired supply of public goods will be a major issue.

There are other European concerns associated with agriculture, such as food safety and animal welfare. It is likely that these will complicate the European Union negotiating position. Because these issues are of major public concern to the European citizenry, reaching an agreement with other trading partners on these issues may not be politically feasible.

### United States

There are several domestic policies that could limit the willingness of the United States to push for the reduction of agricultural trade barriers. Trade barriers that protect the U.S. sugar or peanut sectors are examples of policies that are inconsistent with open borders. The current U.S. position is not clear as to the level of trade liberalization favored by the United States.

A second set of policies that may impede the willingness of the United States to strongly support trade liberalization involves indirect export subsidies. While the United States supports the elimination of export subsidies as one of its objectives for the next round, it has maintained or expanded its use of government subsidies for foreign market promotion, foreign market credit guarantees, and food aid. Some argue that each of these measures has some export subsidy component. The willingness of the United States to support the elimination of export subsidies may not be quite as strong if its own policies are challenged as contributing to the problem.

In a similar manner, the U.S. position on domestic support has raised controversy. While calling for reductions in trade-linked farm subsidies, the United States experienced a dramatic increase in its expenditures for farm subsidies. Supporters of trade reform have identified this as inconsistent with the United States subsidy programs. Finally, the United States has identified state trading enterprises (STEs) as an important concern. While the United States lists its Commodity Credit Corporation (CCC) as an STE, it is not clear to what degree the United States would be willing to accept external disciplines on the CCC.

### Developing Countries

The ability of developing counties to identify and defend their interests through multilateral trade negotiations has been inhibited due to resource constraints and small market size. Because of this there was a bias to follow practices established in industrial countries. The results are not suited to the needs of developing countries. Approaches must be developed to guarantee that rules account for the limitations and needs of all members. In addition, assistance promised to these countries through the agreement must be kept.

A key question for developing countries is whether to support initiation of a new round of negotiations. Further, if they are to support a new round, the scope and focus of the round must be determined. From a developing country perspective, the next round must focus on areas that are strongly trade related and where there is strong likelihood of reaching agreement.

### Environmental Issues

Given that unresolved environmental issues will likely slow the pace of agricultural trade liberalization, there is a clear need for international cooperation on a wide range of environmental issues. As international environmental agreements increases, an efficient infrastructure to handle interactions between these agreements and trade rules must be developed. The outcome should yield a framework where the use of policy instruments targeted directly toward the externalities at the heart of environmental problems are encouraged, as opposed to trade measures that typically offer indirect solutions.

## PROSPECTS FOR AGRICULTURAL TRADE POLICY

The current proposals of the WTO members are quite diverse, yet it appears that most are aimed at achieving the long-term objective of establishing a fairer, more market-oriented agricultural trading system and procedures. The identification and reduction of trade-distorting market access, export competition, and domestic support established in the Uruguay Round Agreement on Agriculture established the infrastructure necessary for long-term reform. The challenge participants face is to build on these pillars through the further elimination of inefficient trade-distorting policies.

While most countries are committed to working through the WTO to eliminate trade-distorting policies, they are also committed to addressing non-trade concerns such as food security, rural development, environmental protection, and conservation. However, while it may not be possible to meet these objectives through non-trade-distorting means, they should be achieved through non-price means. Programs should be targeted to the specific intended objective without creating additional distortions. A key criteria is that the cost of achieving these objectives should not be passed on to other countries or result in unfair competition.

Achieving an agreement of this nature will increase market-orientation and furnish producers with greater opportunities to compete in the world market. Domestic policies structured according to these principles will not only remove a source of trade distortion, they will eliminate government policies that result in production inefficiencies. Producers will benefit from expanded economic opportunities. Consumers will gain access to a wider variety of products at more competitive prices. Adoption of this type of policies will diminish food security concerns through increased access to food and greater purchasing power.

While most members desire a fair, market-oriented agricultural trading system, they are also committed to addressing various non-trade concerns. While these objectives are not mutually exclusive, the latter is easiest to achieve by foregoing the former. Restrictions imposed by the Uruguay Round Agreement on Agriculture have brought the world agricultural community closer to achieving these joint objectives. However, the commitment of all members is necessary for this vision to be fully realized.

Policy analysts can provide leadership in this area through the identification of policy alternatives that address the various non-trade concerns in a manner consistent with a fair, market-oriented agricultural trading system. Identification of politically acceptable policy alternatives of this nature could provide members with the necessary incentive for the further elimination of inefficient trade-distorting policies. As a result, members would have the freedom to address various concerns, such as food security, rural development, or environmental protection and conservation, as a sovereign nation, while allowing the agricultural trading system to function in an efficient manner.

# Index

Page numbers followed by the letter "i" indicate illustrations; those followed by the letter "t" indicate tables.

African, Caribbean and Pacific
    Countries (ACP)
  rice trade, 148
  sugar industry, 156
  TRQs, 41
Agenda 2000
  CAP, 19, 60-61, 67
  coarse grain trade, 204
  future trade negotiations, 357
  WATSIM projection, 332, 336-337
Aggregate measurement of support (AMS)
  coarse grain trade, 205-206, 207
  EU rice trade, 148
  Korean rice trade, 147
  U.S. rice trade, 147-148
  wheat trade, 185
  WTO measurement, 15-16, 45, 47
AGLINK, OECD model, 322, 323
Agreement on Agriculture. *See*
    Uruguay Round Agreement
    on Agriculture (URAA)
Agreement on Government
    Procurement (AGP), NGOs
    demand, 122
Agreement on Technical Barriers to
    Trade (Tokyo, 1979), 47-48
Agreement on Textiles and Clothing,
    263
Agricultural and Fishery Marketing
    Cooperation (AFMC), Korean
    rice trade, 147
Agricultural market
  AGLINK model, 322, 323
  CGE model, 107, 245-250, 261
  CLS model, 322-323
  factors, 23-24, 24i, 25i
  global wheat trade model, 188-194,
    191t, 192t, 194t

Agricultural market *(continued)*
  GMS model, 322-323
  IMPACT model, 323
  market conditions, 75
  public sector role, 341, 342
  WATSIM model, 322
  WATSIM projections, 332-333,
    334t, 335-337
  World Food Model (WFM), 323
Agriculture
  antidumping (AD) practices, 295,
    296-299, 297i, 302
  and trade liberalization, 118-119,
    342
Agriculture and Agri-Food Canada, 16
Algeria, wheat trade, 180t, 181, 184
Allocation, TRQs, 40-41
Amber box policy
  coarse grain trade, 205
  Japanese policy, 128
  U.S. rice trade, 147-148
American Lands Alliance,
    environmental demands,
    119-124
Andean Pact
  as RTA, 260
  RTAs/FTAA analysis, 262, 262t,
    264t, 266, 266t, 267t, 268
Animal welfare
  EU policy, 64, 65-66, 67, 358
  Fourth Ministerial Conference
    (Qatar, 2001), 49, 302
Antidumping (AD) laws, 289,
    295-296
Antidumping (AD) practices. *See also*
    Dumping
  policy development, 292, 302
  URAA, 287

APEC food system (AFS), CGE
    simulation, 250, 254-256, 255i
Archer Daniels Midland (ADM), coarse
    grains trade, 210
Argentina
    antidumping (AD) practices, 289,
        290t, 292, 293t, 300t
    coarse grains trade, 198
    meat exports, 81
    wheat trade, 177, 178, 179, 180t,
        181-182, 193
    wheat trade policy, 184, 186
Article 20, URAA, 19-20, 47, 313-314
Asia Pacific Economic Cooperation
    (APEC) Region
    AFS, 250, 254-255
    CGE simulation, 254-256, 255i
    GMO, 49
    as RTA, 2
    service sector restrictions, 97-98
Australia. *See also* Oceania
    antidumping (AD) practices, 289,
        290t, 293t, 294, 299, 300t
    biotech labels, 210
    Cairns Group member, 129
    CGE projection, 95, 96i
    coarse grains trade, 198, 208
    cotton trade, 217
    Pacific-5 member, 250, 254
    rice trade, 143t, 146, 150
    salmon case, 307-308
    sugar industry, 156, 157t, 166t, 167,
        171
    wheat trade, 177, 178, 179, 180t,
        181-182, 193
Australian Productivity Commission,
    restrictiveness index, 97
Australian Wheat Board (AWB)
    barley trade, 208
    export credits, 184
    LTAs, 185
Austria
    antidumping (AD) practices, 290t, 300t
    rice trade, 148


Back-office services, developing
    countries, 98
Barley, world market, 197, 198, 204,
    206, 208, 210-211

Barley malt, 197, 202, 204, 206, 210
Barshefsky, Charlene
    environment trade policy, 124
    WTO negotiations, 1, 77, 78-79, 85
Basel Convention, hazardous waste,
    120
Beef
    dumping complaints, 14-15
    food safety, 30, 48
    hormone treatment controversy, 48,
        65, 307
    multinational corporations, 26
    price supports, 9
    price volatility, 343, 344t, 345i,
        349-350, 350t, 352, 354i
    supply management, 27
    United States policy, 81
Beet sugar, U.S. production, 159, 160i,
    161
Bhagwati, Jagdish, 113, 115
Biotechnology
    EU policy, 127
    future trade issues, 2, 3, 357
    Japanese policy, 128
    labeling requirements, 121, 131
    Ministerial conferences, 30
    rice trade negotiations, 151-152
    SPS Agreement, 314-315
    United States policy, 77, 121, 125,
        133, 173
    wheat industry, 186
Blair House agreement
    blue box policy, 46
    U.S.-EU, 39, 46, 47, 57
Blandford, David, 5
Blue box polices
    Blair House Agreement, 46
    EU policy, 58
    Fourth Ministerial Conference
        (Qatar, 2001), 45
    and Japanese policy, 128
    WTO, 15
Bound rate tariffs
    coarse grains trade, 197, 201-202
    developing countries, 92-93
    EU, 58
    Fourth Ministerial Conference
        (Qatar, 2001), 39, 41
    rice trade negotiations, 142t-143t,
        149

Bound rate tariffs *(continued)*
    United States, 77
    URAA, 36-37, 39, 245
Bovine Spongiform Encephalopathy
    (BSE), EU policy, 66. *See*
    *also* "Mad cow" disease
Brazil
    antidumping (AD) practices, 289,
        290t, 292, 293t, 294t, 300t
    Cairns Group, 129
    coarse grains trade, 198, 202
    domestic supports, 132
    rice trade, 142t, 144
    sugar industry, 156, 157t, 166t, 167,
        171, 173
    wheat trade, 179, 180t, 181, 184
Break period analysis
    description of, 275-278, 277t, 278t
    PTAs, 278-283
Bush, George W., trade policy, 72, 74
Buy and sell (SBS) auction, Japanese
    rice trade, 145


Cairns Group of Agricultural Fair
    Traders
    agricultural export policy, 3, 4
    environmental trade policy, 129,
        132
    Fourth Ministerial Conference
        (Qatar, 2001), 38, 40, 42, 43,
        45, 48, 49
    members of, 129
    Ministerial conferences, 29
    rice trade negotiations, 151
    vision statement, 21
Canada
    agricultural policy, 9-11, 13
    agricultural trade policy, 76, 79
    antidumping (AD) practices, 290t,
        293t, 299, 300t, 302
    beef-hormone dispute, 48, 307
    coarse grains trade, 198, 199, 208
    dumping complaints, 14-15
    Fourth Ministerial Conference
        (Qatar, 2001), 35, 40, 42,
        43-44, 45, 49
    NISA program, 16
    PTA break period analysis, 275,
        277t, 278, 278t, 281-282

Canada *(continued)*
    as PTAs, 273
    RTAs/FTAA analysis, 262, 262t,
        267t, 268, 269
    salmon case, 307
    sugar industry, 156, 157t
    supply management, 27-28
    upward harmonization, 115
    wheat trade, 10i, 24, 177, 178, 179,
        180t, 181-182, 183-184, 186,
        193
Canada Transportation Act (CTA),
    Canadian grain industry, 184
Canada-U.S. Trade Agreement
    (CUSTA)
    break period analysis, 275, 278,
        281-282
    creation of, 185, 273, 274
    policy harmonization, 9-11, 28
Canadian Wheat Board (CWB)
    barley trade, 208
    export credits, 184
    supply management, 43
    and WTO, 13-14
Cane sugar, U.S. production, 159, 160i,
    161
Cargill Incorporated, international
    trade, 26, 44
Caribbean Common Market
    (CARICOM), RTA, 260
Cartagena Protocol on Biosafety, 120,
    121, 125, 128
Casava, coarse grains trade, 198
Center for International Environmental
    Law, environmental demands,
    119-124
Central America, rice trade, 149, 150
Central American Common Market
    (CACM), RTA, 260
Centrally Planned Economies (CPE),
    wheat sales, 24
Chile
    antidumping (AD) practices, 291t,
        301t
    FTAs, 260
    Pacific-5 member, 250, 254
    RTAs/FTAA analysis, 262, 262t,
        264t, 265, 267t, 268

China
  agricultural export policy, 3
  antidumping (AD) practices, 294,
    294t, 299
  CGE projection, 95, 96i, 246, 250,
    252-254, 253i
  coarse grains trade, 198, 208, 209
  cotton trade, 215, 217
  FDI restrictions, 97-98
  Fourth Ministerial Conference
    (Qatar, 2001), 36
  global wheat trade model, 188-194,
    191t, 192t, 194t
  rice trade, 141, 142t, 146, 149, 150,
    151
  STE policy, 30
  sugar industry, 156, 157t, 166t, 167,
    171
  wheat trade, 24, 25i, 177, 178, 179,
    180t, 181, 182, 186, 187,
    187t, 191-194
  WTO accession, 187-194, 209,
    252-254, 253i
"Civil society" coalition, on trade
    liberalization, 113, 119-124
Clinton, Bill, environmental trade
    policy, 125, 130
Clothing
  GATT, 89
  tariffs on, 94
Coarse grains. *See also* Grain
  commodity policy, 5
  export subsidies, 228
  world market, 141, 197, 198-199
  WTO negotiations, 197-198
Colombia
  antidumping (AD) practices, 290t,
    300t
  coarse grains trade, 201
  export value commitments, 229t
  export volume commitments, 230t
Commodities
  break period analysis, 280t
  GTAP-4, 247-248, 248t
  RTAs/FTAA analysis, 262, 262t
  SPS Agreement, 307-308
  WATSIM, 324, 329t
Commodity Credit Corporation (CCC)
  coarse grains program, 207
  STE role, 13, 83-84, 358

Commodity Credit Corporation (CCC)
    *(continued)*
  U.S. sugar programs, 160
  U.S. wheat trade, 160
Common Agricultural Policy (CAP)
  coarse grains trade, 199
  early history, 55-57
  EU policy, 4, 16, 46
  future trade negotiations, 357
  URAA, 19-21
  WATSIM assumptions, 332
Community Nutrition Institute,
    environmental demands,
    119-124
Composition effect, environmental
    impact, 115, 116, 117
Computable General Equilibrium
    (CGE)
  industrial tariffs, 95-96, 96i
  modeling technique, 107, 245-250,
    261
  RTAs/FTAA analysis, 259-260,
    261-269, 262t
  U.S. negotiating strategy, 250-256,
    251i, 253i, 255i
Concessional programs, United States,
    82t, 83
Consumer's Choice Council,
    environmental demands,
    119-124
Convention on International Trade in
    Endangered Species, 120
Corn
  GMO, 198
  price volatility, 343, 344t, 345i,
    349-351, 350t, 352t, 353i
  world market, 197, 198, 199,
    202-203, 204, 206, 210
Cotton
  commodity policy, 5
  supply management, 28
  Turkey, 26-27
  world market, 216-221, 216i, 219i,
    220i
Council on Environmental Quality,
    environmental assessment,
    122
Countervailing duties
  green box policies, 15
  international trade, 14, 26
  United States policy, 81-82, 83-84

Country-Link System (CLS), USDA
    model, 322-323
Credit programs, United States, 82t, 83
Crow Rate transportation subsidy,
    Canadian agricultural policy,
    9-10
Cuba, sugar industry, 156, 157t


Dairy products
    Agenda 2000, 60, 61
    export subsidies, 228
    price volatility, 342, 343, 346i,
        349-351, 350t, 351i, 352, 355i
    supply management, 28
de Gorter, Harry, 5
*De minimis,* GATT definition, 288
Decoupling
    agricultural policy, 17-18, 22, 23i
    environmental policy, 132
Defenders of Wildlife, environmental
    demands, 119-124
Developing countries. *See also* Less
    developed countries (LDC)
    and agricultural trade liberalization,
        4, 90-93, 91t, 92i
    antidumping (AD) practices, 289,
        291t, 292, 292i, 299, 301t,
        302
    CGE projection, 95, 96i
    cotton trade, 215
    development policies, 100-101
    environmental trade policy, 129-130
    Fourth Ministerial Conference
        (Qatar, 2001), 36, 38-39
    industrial production, 93-97, 94i,
        95t, 96i
    IPRs, 103-104
    negotiating asymmetries, 102
    rice trade, 148, 149
    service sector, 97-100
    SPS Agreement, 315-316
    sugar policies, 166t, 167
    and United States, 77-78
    WTO participation, 106-107, 200,
        358-359
"Dirty tariffication," 58, 92, 201, 245
Disaster relief payments, WTO policy,
    14

Dispute settlements
    SPS Agreement, 307, 308-310
    URAA mechanism, 306-307, 316
Dohlman, Erik, 5
Domestic services markets, GATS,
    99-100
Domestic supports
    agricultural prices, 23, 24, 25i
    CGE projections, 246, 250, 251i,
        251-252
    coarse grain sector, 197, 205-207
    developing countries, 91-92
    EU, 59, 132
    Fourth Ministerial Conference
        (Qatar, 2001), 37, 45-47
    Japanese policy, 128, 132
    Ministerial conferences, 30
    Norwegian policy, 132
    rice industry, 141, 142t-143t, 149,
        151
    sugar policies, 166, 166t, 171
    United States, 77, 125, 358
    URAA, 245
    wheat industry, 24, 186
Dominican Republic, sugar trade, 26
Dumping. *See also* Antidumping (AD)
    practices
    GATT definition, 288
    international trade, 14-15, 26
    United States policy, 81, 83-84


Earthjustice Legal Defense Fund,
    environmental demands,
    119-124
Ecolabels
    EU policy, 127
    Japanese policy, 128
    NGOs demand, 123
Egypt
    antidumping (AD) practices, 294t
    coarse grains trade, 198, 201
    rice trade, 148
    wheat trade, 180t, 181, 184
Employment, modern agricultural
    sector, 61-62, 62t
Environment
    developing countries, 106, 108
    EU policy, 65, 126-128

Environment *(continued)*
    Fourth Ministerial Conference
        (Qatar, 2001), 37, 38
    and multifunctionality, 63
    NGO coalition, 113, 119-124, 130
    sugar industry, 173
    and trade liberalization, 2, 3,
        113-117, 357, 359
Environmental impact assessments,
    NGOs demand, 122
Environmental Protection Agency
    (EPA), environmental
    demands, 119-124
European Commission. *See* European
    Union (EU)
"European model of agriculture," 62-63
European Union (EU)
    Agenda 2000, 19, 60-61, 67, 332,
        336-337, 357
    agricultural export policy, 3, 4
    AMS notifications, 205-206
    animal welfare policy, 64, 65-66,
        67, 358
    antidumping (AD) practices, 289,
        290t, 292, 293t, 300t, 302
    beef hormone treatment, 48, 65, 307
    Blair House agreement, 39, 46, 47,
        57
    CAP negotiations, 4, 20, 46, 357
    coarse grains trade, 197-198, 201,
        204, 206, 210, 211
    domestic supports, 17-18, 18i, 132
    early agricultural policy, 55-57
    and environment, 65, 126-128
    expansion pressures, 59-61
    export subsidies, 228, 228t, 229t,
        230t
    food safety, 20, 21, 64, 65, 66, 67,
        358
    Fourth Ministerial Conference
        (Qatar, 2001), 36, 38, 41-42,
        45, 48, 49, 50
    GMO policy, 64, 65, 66
    green box policies, 16, 17, 58
    modern agricultural sector, 61-62,
        62t
    multifunctionality policy, 18-19,
        63-65, 67, 126, 358
    PTAs, 273
    rice trade, 142t, 148, 152

European Union (EU) *(continued)*
    as RTA, 2
    RTAs/FTAA analysis, 262, 262t,
        267, 267t, 268, 269
    SPS, 127, 314
    sugar industry, 156, 157t, 166, 166t,
        167t, 171, 172t
    supply management, 28
    Third Ministerial Conference
        (Seattle, 1999), 90
    and URAA, 19-21, 57-59
    wheat trade, 10i, 17i, 177, 178, 179,
        180t, 181-182, 184, 187
    WTO negotiations, 1, 75
Export competition
    Fourth Ministerial Conference
        (Qatar, 2001), 43
    Ministerial conferences, 30
    rice trade negotiations, 149, 151
    United States policy, 82t, 83-84
    WTO, 20
Export credits
    coarse grains trade, 198, 208
    as export subsidy, 43
    rice trade negotiations, 151
    wheat trade, 178, 184
    URAA, 43, 200
Export Enhancement Program (EEP)
    coarse grain trade, 204
    rice trade, 147
    United States program, 57, 82t, 83
    wheat trade, 183
Export subsidies
    CGE projections, 246, 251-252,
        251i
    coarse grain, 197, 203-205
    cotton, 217
    developing countries, 91-92, 93, 101
    EU, 56-57, 59, 61
    Fourth Ministerial Conference
        (Qatar, 2001), 37, 38, 42-44
    GATS, 100
    Japanese policy, 128
    NGOs demand, 123-124
    reduction of, 227-228
    rice, 151
    sugar, 166, 166t
    United States, 57, 77, 78, 124-125,
        358
    URAA, 37, 41, 245

Export subsidies *(continued)*
volume/value analysis, 229-243,
236i, 237i, 238i, 239i
volume/value limits, 227-228
wheat, 177, 178, 182-184, 186
Export taxes
coarse grains, 198, 211
Japanese policy, 128
rice, 151
United States, 77-78

Federal Agriculture Improvement and
Reform (FAIR) Act (1996)
coarse grains program, 206-207
cotton program, 217
price volatility, 343, 348-350
U.S. agricultural policy, 9, 46, 85,
147, 174, 341, 342
U.S. sugar programs, 159-161
WATSIM assumptions, 332
Fertilizers, agricultural use, 119
Fifth Ministerial Conference (2003),
proposed WTO trade
negotiations, 29
Fiji, Cairns Group, 129
Finland
antidumping (AD) practices, 290t,
300t
rice trade, 148
Fischler, F., and multifunctionality, 63,
64, 65
Flo-Sun, international trade, 26, 27
Food, Agriculture, Conservation, and
Trade Act (FACTA) (1990)
price volatility, 349
U.S. policy, 341
U.S. sugar programs, 159-160
Food Alliance Labels, 123
Food and Agricultural Act (1981), U.S.
sugar programs, 159
Food and Agricultural Organization
(FAO), World Food Model
(WFM), 323
Food and Agricultural Organization
Statistical Databases
(FAOSTAT), CGE
simulation, 248-249

Food and Agricultural Policy Research
Institute (FAPRI)
GMS (GMS) model, 322-323
WATSIM assumptions, 331
Food and Drug Administration,
genetically modified foods,
132
Food for Peace (FFP), United States, 83
Food safety
CUSTA, 282
EU, 20, 21, 64, 65, 66, 67, 358
Fourth Ministerial Conference
(Qatar, 2001), 37, 38, 44, 48
Ministerial conferences, 30
Food security
Japanese policy, 128
Norwegian policy, 129
rice trade, 141
United States, 77-78
Food Security Act (1985)
U.S. policy, 57, 341
U.S. sugar programs, 159-160
Food Security Declaration, URAA, 44
Foreign direct investment (FDI)
developing countries, 101, 108
IPRs, 103
trade barriers, 97-98
Former Soviet Union (FSU)
coarse grains trade, 199
sugar trade, 157t
wheat trade, 24, 25i, 177, 180t, 181,
182, 184-185
Fourth Ministerial Conference (Qatar,
2001)
preparations for, 35-51
WTO trade negotiations, 29, 302
France, wheat trade, 184
"Free access" commitments, GATS, 98
Free Trade Area of the Americas
(FTAA)
formation of, 259, 260
as RTA, 2
RTAs/FTAA analysis, 263-269,
264t, 266t, 267t
WTO negotiations, 1
Freedom to Farm Act (1996), U.S.
agricultural policy, 31
Friends of the Earth, environmental
demands, 119-124

GARCH
price volatility measurement, 342,
344, 345i, 346i, 347-348, 350i
price volatility results, 350-352,
350t, 351i, 352t
General Agreement on Tariffs and
Trade (GATT)
Article XI, 44
dumping rules, 288-289
negotiating asymmetries, 102, 108
PTAs, 273
SPS agreement, 305, 306-313, 316
TRIPS agreement, 105
WTO negotiations, 1-2, 3, 57, 89
General Agreement on Trade in
Services (GATS)
developing countries, 101, 108
negotiation issues, 98-100
Genetically modified organisms (GMOs)
coarse grains trade, 198, 210
EU policy, 64, 65, 66
Fourth Ministerial Conference
(Qatar, 2001), 48-49
Ministerial conferences, 29, 30, 38
NGOs demand, 121
rice trade, 151-152
SPS Agreement, 314-315
U.S. sugar policy, 173
wheat industry, 187
Germany, antidumping (AD) practices,
294t
Gilbert, John, 5
Global Modeling System (GMS),
FAPRI, 322-323
Global Trade Analysis Project
(GTAP-4), CGE simulations,
245-246, 247-248, 248t,
249-250, 261
Global Wheat Policy Simulation
Model, 188-194, 191t, 192t
Globalism, opposition to, 73-74
Grain
market changes, 24
price volatility, 342, 343, 344t, 345i,
349-351, 350t, 352t, 353i
supply management, 27, 28. *See*
*also* Coarse grains
Green box policy
developing countries, 101
disaster relief, 14

Green box policy *(continued)*
and EU, 58
Fourth Ministerial Conference
(Qatar, 2001), 46-47
Japanese policy, 128
United States, 77
WTO, 15-17, 22, 132
"Green" goods, 118
Greenpeace USA
environmental demands, 119-124
environmental harmonization, 114
Group of Three, RTA, 260

"Hard law," 307
Hayes, Rita Derrick, 76-77
Hazardous waste, regulation of, 120
Hertel, Thomas W., 5
High-fructose corn syrup (HFCS)
U.S. sugar importation, 159
U.S.-Mexico trade, 12, 12t,
162-163, 165, 168, 170-171
Hoekman, Bernard M., 5
Hoffman, Linwood, 5
Hudson, Darren, 5
Huff, Karen M., 6
Hungary
coarse grains trade, 204
export value commitments, 229t

Iceland, coarse grains trade, 206
India
antidumping (AD) practices, 291t,
292-293, 293t, 294t, 299, 301t
CGE projection, 95, 96i
coarse grains trade, 198, 202-203
cotton trade, 217
and MFAs, 218
rice trade, 141, 147
sugar industry, 156, 157t, 166t, 167,
171
Indonesia
antidumping (AD) practices, 291t,
301t
FDI restrictions, 97-98
rice trade, 141, 142t, 148, 150
wheat trade, 177
"Industrial flight," 120

Injury determination, antidumping
(AD) practices, 288, 296, 302
Institute for Agricultural and Trade
Policy, environmental
demands, 119-124
Intellectual property rights (IPRs), TRIPS
agreement, 103-104, 105
Internal supports. *See* Domestic
supports
International Barley and Malt Coalition
for Free Trade, 210
International Food Policy Research
(IFPRI), IMPACT model, 323
International Model for Policy Analysis
of Agricultural Commodities
and Trade (IMPACT), IFPRI,
323
International Monetary Fund,
WATSIM assumptions, 331
Intra-industry trade (IIT), textiles,
220-221
Iowa Beef Packers Incorporated (IBPI),
international trade, 26
Iraq, wheat trade, 184
Israel
antidumping (AD) practices, 291t,
301t
coarse grains trade, 206
PTAs, 273

Japan
agricultural trade policy, 76
agriculture reform, 255
AMS notifications, 205-206
antidumping (AD) practices, 290t,
294t, 300t, 302
CGE projection, 95, 96i
coarse grains trade, 197-198, 199,
202, 206, 208
domestic supports, 132
environmental trade policy, 128-129
Fourth Ministerial Conference
(Qatar, 2001), 36, 38
nontrade issues, 75
rice trade, 144-146, 146i, 148,
149-150, 151, 152
sugar industry, 156, 157t
Varietal Testing case, 307, 308
wheat trade, 180t, 181, 182, 187

Jordan, wheat trade, 184
Josling, Tim, 5

Kennedy, P. Lynn, 5, 6
Koo, Won W., 5, 6
Krugman, Paul, 113

Labeling, biotechnology products, 121,
131, 210, 314-315
Labor, CGE simulation, 246-247, 248t
Labor standards, developing countries,
105-106, 108
Lamb, United States policy, 81
Lebanon, IPRs, 103
Less developed countries (LDC). *See
also* Developing countries
"civil society" criticism, 113
cotton trade, 218
trade liberalization, 114-117
Less favored areas (LFA), CAP, 19
Livestock
and coarse grains trade, 198
price volatility, 342, 343, 344t, 345i,
349-350, 352, 354i
Lome Convention, EU-ACP sugar
agreement, 155, 156, 166
Long-term agreements (LTA), wheat
trade, 184-185

MacSharry, Ray, 57
"Mad cow" disease, 30. *See also*
Bovine Spongiform
Encephalopathy (BSE)
Malaysia
antidumping (AD) practices, 291t,
301t
coarse grains trade, 201
rice trade negotiations, 150
Manufacturing, developing countries,
93-94, 94i
Market access
coarse grains trade, 198, 201
EU policy, 58-59
Fourth Ministerial Conference
(Qatar, 2001), 37-39

Market access *(continued)*
   Ministerial conferences, 30
   rice trade negotiations, 149-150
   WTO, 20
Martin, Will, 5
Matching grants, developing countries,
   100-101
Meat
   and coarse grains consumption, 198,
    199
   CUSTA, 282
   export subsidies, 228
MERCOSUR. *See* Southern Cone
   Common Market
   (MERCOSUR)
Methyl tertiary-butyl ether (MTBE),
   NGOs demand, 120
Mexico
   agricultural trade liberalization,
    118-119
   antidumping (AD) practices, 289,
    290t, 292, 293t, 300t, 302
   coarse grains trade, 198, 199, 201
   NAFTA coarse grain trade, 199
   NAFTA sugar trade, 155, 162-165,
    164t, 174
   NAFTA wheat trade, 185
   as PTA, 273
   PTA break period analysis, 275,
    277t, 278, 278t, 279-281,
    280t, 282
   rice trade, 143t, 144, 149, 150
   RTAs/FTAA analysis, 262, 262t,
    267t, 268
   sugar industry, 11-12, 12t, 155,
    157t, 162-165,164t, 166t,
    167, 168-170, 169i, 170i,
    173, 174
   wheat trade, 180t, 181, 184
Miljkovic, Dragan, 6
Miner, William H., 21
Minimum access (MA) commitments,
   rice trade, 144-145, 146
Ministry of Agriculture, Forestry, and
   Fisheries (MAFF), Japanese
   rice trade, 145
Minton, Tara M., 5
Moore, Mike, on WTO decision-
   making procedure, 29
Morocco, wheat trade, 180t, 181

Multi-fiber arrangement (MFA)
   composition effect, 116
   cotton industry, 218, 220, 221, 222
   quotas, 95, 102
   trade barriers, 263
Multifunctionality
   Brazilian policy, 132
   EU policy, 18-19, 63-64, 67, 126,
    358
   Fourth Ministerial Conference
    (Qatar, 2001), 46-47
   Japanese policy, 128
   Ministerial conferences, 30
   NGOs position, 123-124, 130
   Norwegian policy, 129
   rice trade negotiations, 152
   United States policy, 132
Multilateral environmental agreements
   (MEAs)
   EU policy, 127-128
   NGOs demand, 120, 130
Multilateral trade agreements
   CGE analysis, 259-260, 262, 262t,
    263-269
   prospects for, 261
Multinational corporation, agricultural
   products, 26-27
Myanmar, rice trade, 150


Natcher, William C., 6
National Enquiry Points, SPS
   Agreement, 309
National Environmental Policy Act,
   environmental assessment,
   122
National Farmers Union, dumping
   complaints, 14
National Notification Authorities, SPS
   Agreement, 309
National security, and U.S. trade
   policy, 72, 74-75
National Wildlife Federation,
   environmental demands,
   119-124
Natural Resources Defense Council,
   environmental demands,
   119-124

New Zealand. *See also* Oceania
  antidumping (AD) practices, 290t,
    293t, 299, 300t
  biotech labels, 210
  CGE projection, 95, 96i
  meat exports, 81
  Pacific-5 member, 250, 254
Nimon, Wesley, 5
NISA program, WTO policies, 16
Nongovernmental organizations
  (NGOs)
  environmental demands, 119-124, 130
  market access, 38
  trade liberalization criticism, 113
  and United States policy, 126
  and WTO, 73
Nontariff barriers (NTBs)
  EU policy, 65-66, 126
  Ministerial conferences, 30
  rice trade, 149, 151-152
  URAA, 20-21, 36-37, 245
North American Free Trade Agreement
  (NAFTA)
  agricultural trade liberalization, 119
  break period analysis, 275, 278,
    279-281, 282
  coarse grain trade, 199
  creation of, 274
  policy harmonization, 9, 11-12
  PTAs, 273, 282-283
  regionalism, 282
  rice trade, 144
  as RTA, 2, 260
  RTAs/FTAA analysis, 262, 264t,
    265, 266, 266t, 269
  sugar trade, 155, 162-165, 164t,
    168-171, 169i, 170i, 173, 174
Norway
  agricultural trade policy, 76
  coarse grains trade, 206
  environmental trade policy, 129
  export value commitments, 229t
  export volume commitments, 230t
Notifications, SPS Agreement,
  309-310, 310t

Oats, world market, 198, 204
Oceania, RTAs/FTAA analysis, 262,
  262t, 267, 267t, 268. *See also*
  Australia; New Zealand

Office of the U.S. Trade Representative
  (USTR), environmental
  demands,
  119-124
Open borders, agricultural policy, 80,
  86
Organization for Economic
  Cooperation and
  Development (OECD)
  AGLINK, 322, 323
  agricultural tariffs, 90, 91, 91t, 92i
  environmental criteria, 114
  European farm sector, 62
  export credits, 43, 208-209
  manufactured goods tariffs, 94, 95t
  rice trade negotiations, 151
  service sector, 98, 100


Pacific Environment and Resource
  Center, environmental demands,
  119-124
Pacific-5 CGE projections, 250, 254,
  255i
Pakistan
  cotton trade, 217
  wheat trade, 184
Paul, Rodney, 6
Peace clause
  EU policy, 58
  United States policy, 82
  URAA, 21, 50, 200
Peanuts
  supply management, 27, 28
  United States policy, 81, 358
Peru
  antidumping (AD) practices, 291t,
    301t
  coarse grains trade, 202
Pesticides, agricultural use, 119
Philippines
  antidumping (AD) practices, 291t,
    301t
  Cairns Group, 129
  coarse grains trade, 202
  rice trade, 148, 150
Poland
  antidumping (AD) practices, 290t,
    299, 300t
  export volume commitments, 230t

Pollution, 3, 117
Pork
   price supports, 9
   supply management, 27
Precautionary principle
   EU policy, 127, 128
   NGOs demand, 121, 130, 131
Preferential Trade Areas (PTAs)
   description of, 273
   and NAFTA, 282-283
   trade obstacles, 357
Price discrimination, WTO policy, 13
Price pooling, export subsidy, 43
Price volatility, trade liberalization,
   342-343, 352
Producer Subsidy Equivalents (PSE),
   wheat industry, 186, 187
Production processes and methods (PPM)
   EU policy, 127
   NGOs demand, 120-121, 131
Progress, trade negotiations, 71, 76
Protectionism, agricultural policy, 2,
   14, 32

Queensland Sugar Corporation (QSC),
   167, 171

Railroads, transportation subsidies,
   183-184
Ranchers-Cattlemen Action Legal
   Foundation, dumping
   complaints, 14
Red box policies, 16
Regional trade agreements (RTAs)
   CGE analysis, 259-260, 262, 262t,
    263-269
   CGE projections, 246, 254-256,
    255i
   studies of, 274
   trade issue, 2, 357
Regionalism
   definition of, 282
   and trade liberalization, 6
   WTO negotiations, 1
Regionalization
   definition of, 282
   SPS Agreement, 313

Regmi, Anita, 5
"Regulatory takings," 120
Rent seekers, farm policy, 10, 27
Rice
   commodity policy, 5
   trade characteristics of, 141
   world market, 141, 144i, 144-149
Romania
   antidumping (AD) practices, 294t
   coarse grains trade, 198, 202
Rude, James, 6
Ruiz, Lilian, 5
Russia. *See also* Former Soviet Union
   (FSU)
   coarse grains trade, 202, 208
   non-WTO member, 209
   STE policy, 30
   sugar industry, 156
Rye, world market, 198, 204

Safeguard Clause, URAA, 21
Safeguards, GATS, 100
Sanitary and Phytosanitary Agreement
   (SPS)
   biotechnology, 314-315
   CUSTA, 282
   EU policy, 127, 314
   Fourth Ministerial Conference
    (Qatar, 2001), 38, 48
   SPS committee, 308-310, 316
   standards, 310-313, 315-316
   United States policy, 121, 125
   URAA, 305
   WTO, 20, 131, 200, 305-306, 316
Saudi Arabia, coarse grains trade, 198,
   202, 208
"Savings clause," Cartagena Protocol,
   121
Scale effect, environmental impact,
   115, 116, 117
Schmitz, Andrew, 5, 13
Service sector, trade barriers, 90, 97-98
Sierra Club
   environmental demands, 119-124
   environmental harmonization, 114
Singapore
   antidumping (AD) practices, 290t,
    300t
   Pacific-5 member, 250, 254

Single-desk selling agencies, 43
"Soft law," 307
Soft price discrimination, WTO policy, 13
Software development, developing countries, 98
Sorghum, world market, 198
South Africa
  antidumping (AD) practices, 291t, 292-293, 293t, 301t
  Cairns Group, 129
  coarse grains trade, 202, 204, 208
  export subsidy reduction, 228, 229t, 230t
  export volume commitments, 230t
  sugar policies, 166t, 166-167
South Korea
  agricultural policy, 76, 255
  antidumping (AD) practices, 289, 290t, 292, 294t, 300t, 302
  biotech labels, 210
  coarse grains trade, 197, 198, 199, 201, 202, 206
  FDI restrictions, 97-98
  rice trade, 144-145, 146, 147, 149-150, 151, 152
  SPS Agreement, 308-309
  sugar industry, 156, 157t
  wheat trade, 177, 180t, 181, 182, 184
Southern Cone Common Market (MERCOSUR)
  coarse grain trade, 202
  rice trade, 144
  RTA, 260
  RTAs/FTAA analysis, 262, 262t, 264t, 265, 266, 266t, 267t, 268
Soviet Union. *See* Former Soviet Union (FSU)
Soybeans
  price volatility, 344t, 345i, 349, 350, 350t, 351, 352, 353i
  world market, 141
Special Safeguard Clause
  EU policy, 66
  URAA, 21, 203
SPS Committee, dispute resolution, 308-310
Standards Code. *See* Agreement on Technical Barriers to Trade (Tokyo, 1979)

State trading, definition of, 2
State trading enterprises (STE)
  coarse grains trade, 198, 203, 211
  Fourth Ministerial Conference (Qatar, 2001), 37, 38, 41, 43-44
  Ministerial conferences, 29, 30
  rice trade negotiations, 149, 150, 151
  sugar policies, 166t, 167, 171-172
  trade issue, 2-3, 13-14
  United States, 13, 77-78, 83-84, 358
  URAA, 200, 207-208
  wheat industry, 186
Stickiness, rice characteristic, 141
Stolper-Samuelson (SS) theory, textiles, 221
Sugar
  commodity policy, 5
  export subsidies, 228
  and multinational corporations, 26
  and NAFTA, 155, 162-165, 164t, 168-171, 169i, 170i, 173, 174
  supply management, 27, 28
  world market, 156, 157t, 158i, 159, 159i
  United States policy, 81, 358
  U.S.-Mexico agreement, 11-12
Sugar Industry Act (1991), Australia, 167
Sumner, Daniel A., 5
Supply Administration of the Republic of Korea (SAROK), rice trade, 147
Sweden
  antidumping (AD) practices, 291t
  rice trade, 148
Switzerland
  coarse grains trade, 206
  export value commitments, 229t
  export volume commitments, 230t


Taiwan
  antidumping (AD) practices, 294t
  coarse grains trade, 198, 199, 201, 208
  non-WTO member, 209
  rice trade, 149, 150
  wheat trade, 177, 180t, 181

Taiwan Agricultural Yearbook, CGE
    simulation, 248-249
Tapioca, coarse grains trade, 198
Tariff escalation
    coarse grain sector, 197, 202
    URAA, 245
Tariff rate quotas (TRQs)
    Chinese wheat, 187, 187t, 191
    coarse grain sector, 197, 198, 201,
        202-203, 211
    developing countries, 93
    EU agriculture, 58-59, 61
    Fourth Ministerial Conference
        (Qatar, 2001), 37, 38, 40-41
    rice trade, 148, 149
    trade obstacles, 357
    URAA, 37, 202, 203
    U.S. agriculture, 77, 78, 81, 83
    U.S. sugar programs, 160, 161, 162,
        173, 174
Tariffication
    rice trade, 145
    URAA, 37, 40, 92, 245
Tariffs
    CGE projection, 95-96, 96i, 97, 246,
        250, 251i, 251-252
    developing countries, 90-91, 91t,
        92i, 94-96, 95t, 96i
    EU agriculture, 58, 126
    Fourth Ministerial Conference
        (Qatar, 2001), 39
    industrial sector, 93-97, 94i, 95t
    rice trade, 142t-143t, 147
    U.S. policy, 81
    U.S. sugar industry, 161
    wheat trade, 177
Technical Barriers to Trade (TBT)
        agreement
    EU policy, 127
    NGOs demand, 121, 123, 131
    SPS Agreement, 305, 311, 316
    WTO, 20
Technique effect, environmental
        impact, 115-116, 117
Technology, agricultural prices, 23, 24,
        24i, 25i
Textiles. *See also* Multi-fiber
        arrangement (MFA)
    and cotton production, 220-221
    GATT, 89
    tariffs on, 94

Thailand
    antidumping (AD) practices, 291t,
        294t, 301t
    FDI restrictions, 97-98
    rice trade, 147
    sugar industry, 156, 157t
    wheat trade, 177
Third Ministerial Conference (Seattle,
        1999)
    environmental issues, 113-114
    WTO trade negotiations, 29, 35, 50,
        89-90, 104, 260-261
Thornsbury, Suzanne D., 5
Tobacco, supply management, 27, 28
Trade barriers, agricultural policy, 2,
        9-14
Trade diversion, RTAs/FTAA analysis,
        264, 264t, 268-269
Trade liberalization
    agricultural prices, 23, 24
    and agriculture, 118-119, 342
    antidumping practices (AD),
        287-288, 292-293
    and environment, 113-117
    Chinese wheat simulation, 188-194,
        191t, 192t
    developing countries, 96-97
    and environmental impact, 113, 114,
        115
    Fourth Ministerial Conference
        (Qatar, 2001), 37-38
    impact of, 27-28
    Japan, 128-129
    LDCs, 114-117
    Ministerial conferences, 29-30, 89
    national environmental views,
        124-130
    NGOs demands, 119-124
    Norway, 129
    obstacles to, 357-360
    rice trade, 141, 144-152
    RTAs/FTAA analysis, 259-260,
        262, 262t
    URAA, 37, 58
Trade Related Investment Measures
        (TRIMs), developing
        countries, 101
Trade-related aspects on Intellectual
        Property Rights (TRIPS),
        WTO, 20, 102

Transatlantic Free Trade Area, RTA, 2
Transparency, NGOs demand, 122,
    126, 130
Transportation effect, environmental
    impact, 115, 116-117
Treaty of Rome, agricultural policy, 56
Tunisia, wheat trade, 180t, 181, 184
Turkey
    antidumping (AD) practices, 291t,
        301t
    coarse grains trade, 202, 208
    cotton industry, 26-27, 217
    export value commitments, 229t
    export volume commitments, 230t
Tweeten, Luther, 17

Ukraine
    antidumping (AD) practices, 294t
    sugar industry, 156
United Kingdom, antidumping (AD)
    practices, 294t
United States
    agricultural policy, 1, 3-4, 9, 10-11,
        31, 46, 358
    agricultural sector, 61-62, 62t
    agricultural trade, 119, 341-342
    AMS notifications, 205-206
    antidumping (AD) practices, 289,
        290t, 292, 293t, 293-294,
        294t, 299, 300t, 302
    beef-hormone treatment, 48, 65, 307
    Blair House agreement, 39, 46, 47, 57
    Bush agricultural policy, 72, 74,
        75-80, 84-85
    CGE strategy projections, 250-256,
        251i, 253i, 255i
    and Chinese wheat, 191, 191t,
        193-194, 194t
    coarse grains trade, 197, 198-199,
        204-205, 209, 210
    cotton trade, 215, 216i, 216-217,
        218-220, 219i, 220i
    environmental trade policy, 124-126
    export subsidies, 228, 228t, 229t,
        230t
    Food Security Act, 57
    Fourth Ministerial Conference
        (Qatar, 2001), 36, 38, 39,
        41-42, 43, 45, 48, 49, 50

United States *(continued)*
    green box policies, 16-17, 47
    Mexican sugar trade, 11-12, 12t
    Ministerial conferences, 29
    multifunctionality policy, 132
    NAFTA coarse grain trade, 199
    NAFTA sugar trade, 155, 162-165,
        168-171, 173, 174
    NAFTA wheat trade, 185
    Pacific-5, 250, 254
    protectionism, 14-15, 32
    PTA break period analysis, 275,
        277t, 278t, 278-281
    rice trade, 143t, 146, 147, 150
    RTAs/FTAA analysis, 262, 262t,
        267t, 269
    STE policy, 13, 77-78, 83-84, 358
    sugar industry, 155, 156, 157t,
        159-173, 169i, 170i, 172t, 173
    supply management, 27-28
    Third Ministerial Conference
        (Seattle, 1999), 90
    Varietal testing case, 307, 308
    wheat trade, 10-11, 10i, 17i, 24,
        177, 178, 180t, 181-182, 183,
        184, 186-187, 187t
    zero option policy, 76, 78-80
"Upward harmonization," 114-115
Uruguay Round (UR)
    EU policy, 57
    nontrade issues, 75
    "one size fits all" policy, 102
    United States policy, 76-80
Uruguay Round Agreement on
        Agriculture (URAA)
    accomplishments of, 5, 36-38,
        199-200, 201, 202, 203-204,
        205, 211, 245, 321, 359, 360
    antidumping rules, 287
    coarse grain trade, 197, 198,
        199-200
    colored boxes policy, 15-17, 22, 85
    domestic supports, 16, 18, 205,
        249-250
    and EU, 19-21, 57-59
    export credits, 209
    export subsidies, 203-204
    export subsidies reduction, 227-228,
        249
    market access, 40, 201

Uruguay Round Agreement on
    Agriculture (URAA)
    *(continued)*
    peace clause, 21, 50, 82
    rice trade, 144-149
    SPS Agreement, 305, 306-311, 316
    sugar industry, 155, 161
    tariffs, 210
    TRIPS, 103-104
    TRQs, 202, 203
    WATSIM assumptions, 332
    wheat trade, 177-178, 183, 185
    WTO negotiations, 1-2, 35, 36, 245
Uzbekistan, cotton trade, 217


Valuation, antidumping (AD) practices,
    288, 295-296, 302
Vasavada, Utpal, 5
Venezuela
    antidumping (AD) practices, 291t,
        301t
    coarse grains trade, 201
    wheat trade, 180t, 181
Vertical markets, multinational
    corporations, 26-27
Vietnam
    coarse grains trade, 208
    non-WTO member, 209
    rice trade negotiations, 150
Voluntary export restraints (VERs)
    agreements, GATT, 102
von Lampe, Martin, 6


Wahl, Thomas, 5
Wailes, Eric J., 5
Weaver, Robert D., 6
Welfare effect
    Chinese accession, 252-254, 253i
    domestic supports elimination,
        251-252, 251i
    RTAs, 254-256, 255i
    RTAs/FTAA analysis, 267-268,
        267t

Western Grain Transportation Act
    (WGTA), Canadian wheat
    industry, 183
Wheat
    and Chinese impact, 186, 188-191
    export subsidies, 228
    grain classes, 178-179
    trade policies, 177
    world market, 9, 10i, 17i, 24, 141,
        177-182
World Agricultural Trade Simulation
    (WATSIM) model
    assumptions, 331-332
    creation of, 322
    criteria, 323-324
    modeling system, 324-331, 325i,
        329t, 330i, 337
    projections, 332-333, 334t, 335-337
World Bank, WATSIM assumptions,
    331
World Food Model (WFM), FAO, 323
World Food Summit, 321
World Trade Organization (WTO)
    accessions to, 187-194, 209,
        252-254, 253i
    antidumping (AD) practices, 287
    coarse grain trade, 197-198
    color boxes policies, 15-16
    compliance costs, 102-103
    cotton industry, 215, 218
    current negotiations, 1-2
    description of, 72-73
    environmental criticism, 113
    environmental standards, 106, 108
    export subsidies reduction, 227-228
    Fifth Ministerial Conference, 29
    Fourth Ministerial Conference
        (Qatar, 2001), 29, 35-51
    future negotiations, 1, 5, 19, 75,
        359-360
    labor standards, 105-106, 108
    member domestic policy, 14-15
    member participation, 106-107,
        315t, 315-316
    members policy, 4
    mission statement, 71
    NGOs demand, 120
    rice tariffs, 142t-143t

World Trade Organization (WTO)
(*continued*)
STE policy, 13-14
sugar industry, 171, 174
Third Ministerial Conference
(Seattle, 1999), 29, 35, 50,
89-90, 104, 113-114
transparency demand, 122, 126,
130
wheat industry, 186-187

World Wildlife Fund, environmental
demands, 119-124

Zero option policy, United States, 76,
78-80
Zero-for-zero strategy, barley trade,
210-211

## SPECIAL 25%-OFF DISCOUNT!

### Order a copy of this book with this form or online at:

*http://www.haworthpressinc.com/store/product.asp?sku=4616*

# AGRICULTURAL TRADE POLICIES
# IN THE NEW MILLENNIUM

_____in hardbound at $67.46 (regularly $89.95) (ISBN: 1-56022-932-2)

_____in softbound at $37.46 (regularly $49.95) (ISBN: 1-56022-933-0)

Or order online and use Code HEC25 in the shopping cart.

COST OF BOOKS_____

OUTSIDE US/CANADA/
MEXICO: ADD 20%_____

POSTAGE & HANDLING_____
*(US: $5.00 for first book & $2.00
for each additional book)
Outside US: $6.00 for first book
& $2.00 for each additional book)*

SUBTOTAL_____

IN CANADA: ADD 7% GST_____

STATE TAX_____
*(NY, OH & MN residents, please
add appropriate local sales tax)*

**FINAL TOTAL_____**
*(If paying in Canadian funds,
convert using the current
exchange rate, UNESCO
coupons welcome)*

☐ **BILL ME LATER:** ($5 service charge will be added)
(Bill-me option is good on US/Canada/Mexico orders only;
not good to jobbers, wholesalers, or subscription agencies.)

☐ Check here if billing address is different from
shipping address and attach purchase order and
billing address information.

Signature_____

☐ **PAYMENT ENCLOSED: $_____**

☐ **PLEASE CHARGE TO MY CREDIT CARD.**

☐ Visa ☐ MasterCard ☐ AmEx ☐ Discover
☐ Diner's Club ☐ Eurocard ☐ JCB

Account # _____

Exp. Date_____

Signature_____

Prices in US dollars and subject to change without notice.

NAME_____

INSTITUTION_____

ADDRESS_____

CITY_____

STATE/ZIP_____

COUNTRY_____ COUNTY (NY residents only)_____

TEL_____ FAX_____

E-MAIL_____

May we use your e-mail address for confirmations and other types of information? ☐ Yes ☐ No
We appreciate receiving your e-mail address and fax number. Haworth would like to e-mail or fax special
discount offers to you, as a preferred customer. **We will never share, rent, or exchange your e-mail address
or fax number.** We regard such actions as an invasion of your privacy.

### Order From Your Local Bookstore or Directly From
**The Haworth Press, Inc.**
10 Alice Street, Binghamton, New York 13904-1580 • USA
TELEPHONE: 1-800-HAWORTH (1-800-429-6784) / Outside US/Canada: (607) 722-5857
FAX: 1-800-895-0582 / Outside US/Canada: (607) 722-6362
E-mailto: getinfo@haworthpressinc.com
PLEASE PHOTOCOPY THIS FORM FOR YOUR PERSONAL USE.

http://www.HaworthPress.com                                    BOF02